AF588260

MS&A

Modeling, Simulation and Applications

Volume 22

MS&A publishes advanced textbooks and research-level monographs that will illustrate the scientific foundations of the modeling and simulation process as well as concrete instances of its role in addressing complex and relevant problems in everyday life. Mathematical modeling aims to describe through mathematics the different aspects of the real world, their dynamics and their interaction. Numerical simulation provides accurate and certified solutions to complex mathematical models by means of scientific computing. Modeling and numerical simulation have become the road-map for mathematics to develop and analyze novel techniques to solve problems in basic sciences (such as physics, chemistry, biology) and engineering, environmental, life and social sciences. The purpose of this series is to host high level contributions describing the interplay among mathematical analysis, numerical analysis and scientific computing, advanced programming techniques, control and optimization, validation, verification and testing. This interplay makes the modeling and numerical simulation process as a whole a unique and effective tool for applied sciences as well as for enhancing technological innovation. The series has already published some successful books (relevant also from a print/online sales perspective), and has planned to publish in the next 2 years about 10 new works focusing on the most significant emerging areas. The Series in indexed in SCOPUS. Volumes of the series are indexed in Web of Science - Thomson Reuters.

Pablo Braz e Silva • Enrique Fernandez-Cara •
Marko Antonio Rojas-Medar

Analysis and Control of the Variable Density Incompressible Navier-Stokes Equations

Pablo Braz e Silva
Department of Mathematics
Universidade Federal de Pernambuco
Recife, Pernambuco, Brazil

Enrique Fernandez-Cara
Facultad de Matematicas
University of Sevilla, Department EDAN
Sevilla, Spain

Marko Antonio Rojas-Medar
Departamento de Matemática
Universidad de Tarapacá
Arica, I - Iquique, Chile

ISSN 2037-5255 ISSN 2037-5263 (electronic)
MS&A
ISBN 978-3-032-14509-3 ISBN 978-3-032-14510-9 (eBook)
https://doi.org/10.1007/978-3-032-14510-9

This work was supported by MICIU/AEI (Spain) (PID2020-114976GB-I00, PID2024-158206NB-I00).

This Springer imprint is published by the registered company Springer Nature Switzerland AG
The registered company address is: Gewerbestrasse 11, 6330 Cham, Switzerland

The first author dedicates this book to Paulo and Miguel, for inspiring his courage and teching him love.

The second author dedicates this book to Rosa Echevarria, for one million reasons he has in mind.

The third author dedicates this book to his mother Drina Medar, his grandson Maxi and his professor and friend Jose Luiz Boldrini for his unconditional support.

Preface

Objectives, Audience, and Motivation

The main objective of this book is to provide an introductory study to the variable density incompressible Navier-Stokes equations. We will deal with their motivation, the main related mathematical problems, the main techniques used to solve and/or control the systems, and the main open questions arising in the subject. Additionally, the detailed description of some (more or less) elementary numerical techniques is given.

The book is intended to be useful to master's and PhD students and young researchers interested by the analysis and control of nonlinear partial differential equations (PDEs), in particular those concerned with applications to mechanics, engineering, biology, economics, etc. We have tried to adapt the explanations, the proofs of the results and the proposed exercises to this level.

We consider questions concerning fluid mechanics that are

- In part, purely theoretical (existence, uniqueness, regularity, stability)
- In part, connected with applications (some particular solutions, some particular geometrical situations, the behavior with respect to some parameters, the limit as space dependent initial densities tend to a constant, the limit as viscosity tends to zero)
- Motivated by control theory (optimal control of variable density Navier-Stokes fluids, controllability problems)

There are several reasons to be interested by these topics:

1. Many incompressible fluids are nonhomogeneous (i.e., they possess a nonconstant mass density) and Newtonian. This is what we find when we deal with oceans, rivers, large containers, etc. In fact, in our discussions we will try to clarify the choice of the PDEs for many flow problems. In particular, oceans can be modeled by systems of PDEs similar to the 2D variable density Navier-Stokes equations. Also, in rivers, it is usual and interesting to incorporate coupling with other variables.

2. The techniques that we present extend those for the classical Navier-Stokes equations and can be applied to many situations associated with other nonlinear phenomena arising in science and engineering. For instance, the existence results rely on standard approximation schemes for PDEs (Galerkin, semi-Galerkin, ...), energy estimates, compactness results in spaces that take the form $L^p(0, T; B)$, regularity theory for elliptic and parabolic PDEs, etc.
3. Some arguments that can be used in the context of constant density Navier-Stokes analysis do not work well here. We believe that it is interesting to enhance the differences and address possible remedies (if they exist).
4. A collection of open problems (not all of them of the same difficulty) can be found in the field. Some of them are recalled in the following chapters. Hopefully, this will arouse interest to young researchers.

Prerequisites

The book is intended to be appropriate for graduate students in mathematics, especially if their interests concern the theoretical and/or numerical analysis of nonlinear PDEs and even more if they want to understand how PDEs can be used to describe fluids. We hope it will be readable, understandable, and useful for someone familiar with

- Basic linear functional analysis: Banach and Hilbert spaces; orthogonal projectors and the Lax-Milgram Theorem; strong and weak convergence; bounded and compact linear operators; basic spectral theory
- Basic results in measure theory, distributions, and Sobolev spaces: Lebesgue L^p spaces and properties; distributions and their derivatives; Sobolev spaces and their properties
- Basic existence and regularity theories for linear elliptic and parabolic PDEs
- (For the results in Chap. 4) Basic numerical analysis and fundamentals of computer science
- (For the results in Chap. 6) Basic results of the Calculus of Variations: minimization of continuous convex functions, associated Euler and Euler-Lagrange characterizations, etc.

Some of these results will be recalled in the text when needed. Well-known references that can give support are [1–8].

Some Comments on the Contents

In Chap. 1, we present the physical motivations and formulations of the main problems.

The main objectives are to justify the utility of the PDEs, indicate how and why the particular PDEs corresponding to a precise situation are chosen, and, accordingly, address the physical interpretation.

We recall the fundamental problem in continuum mechanics and present its formulation in terms of PDEs. Thus, we indicate that the main objective is to determine the mechanical state of a medium at all times (in the future), using the known description of the present status and the intrinsic physical properties.

We also give justifications of the existence of the main variables, the stress tensor and the external force fields, and other sources from the viewpoint of continuum mechanics. Additionally, we explain briefly how laminar and turbulent flows behave and the way this can be described.

This chapter contains a discussion on the relevance of the theoretical and numerical analysis of the equations of fluid mechanics. We have also mentioned a collection of real-life applications. Finally, we have tried to summarize the most important achievements obtained up to date and the impact they have had and probably will have in the future.

Chapter 2 is devoted to recall the mathematical background needed in this book. The main aim is to review the most basic and visited results in the field with as much as possible self-contained proofs.

We begin with a discussion on Lebesgue spaces and their main properties. Then, real-valued and Banach space-valued distributions are considered. We also recall the definitions and properties of Sobolev spaces, the usual compact embeddings, *De Rham's Lemma*, and other fundamental tools.

We revisit the basic existence, uniqueness, and regularity theory of linear elliptic and parabolic PDEs. This is completed with a detailed analysis of the Stokes operator and the stationary Stokes equations.

Chapter 3 deals with global weak solutions for the variable density Navier-Stokes PDEs.

We present the most general results, under low regularity of the data. They are proved following well-known arguments.

Thus, we introduce semi-Galerkin approximations, deduce *a priori* estimates, apply compact embedding results, extract convergent sequences, and, finally, pass to the limit to achieve the proof.

We also include similar results and/or comments on problems in unbounded domains, for variable viscosity fluids, for coupled systems of the Boussinesq kind, etc. In all the related existence theorems, the scheme of the proof is more or less the same: a first step concerns the existence of approximations, then appropriate uniform estimates are established, and finally the desired claim is deduced after taking limits.

Chapter 4 is devoted to show how we can get accurate numerical approximations of the variable density Navier-Stokes equations.

We have given priority to numerical techniques relying on time discretization through characteristics and spatial approximation with mixed finite elements. The main reason is that, after some work related to the backwards in time computation of characteristics, the systems that remain to be solved are linear.

Also, we give details on the numerical simulation of the method, and we briefly recall other possible approximation strategies.

In Chap. 5, strong solutions appear.

It is shown that, as expected, if we increase the regularity of the data we can deduce the existence (and uniqueness) of a more regular solution, at least locally in time.

We have tried to explain that the key point is to establish "good" or "strong" estimates of the semi-Galerkin approximations, stronger than those in Chap. 3. For three-dimensional fluids this is only possible during a short time interval, unless the data are regular and small (this is a major open question even for constant density fluids).

We also analyze the two-dimensional case, where strong global in time solutions can be found.

At the end of the chapter, we review (among others) some recent results where the initial data of the velocity are chosen in Besov spaces. They lead to local in time existence and uniqueness and may be useful to motivate future research.

In Chap. 6, we deal with several control problems for the nonhomogeneous Navier-Stokes PDEs.

We describe the two main strategies (optimal control and controllability) and give details on what can be done at present. We also mention several practical situations that motivate the formulation and analysis of these problems.

For the considered optimal control problems, we recall the main questions of interest (existence and uniqueness, characterization, and computation), and we apply appropriate techniques to give satisfactory answers.

More precisely, for the existence of optimal control-states, we use suitable variants of the classical principle of the Calculus of Variations. The characterization of optimal couples is achieved by introducing adjoint states; this is the key to reformulate the necessary optimality conditions as a coupled system of PDEs whose solution can be obtained in various ways. On the other hand, some iterative algorithms for the computation of optimal controls and states are also mentioned.

In this chapter, we also state and solve some relevant time optimal control problems. They are a little more complicated than the (more or less standard) problems considered before. However, techniques of the same kind can be adapted to produce satisfactory results.

Concerning controllability, we first provide an overview of the known results at present and we then present some novelties. It should be said that, in this context, the results are more involved and can be reasonably omitted in a first lecture. Furthermore, they are still few in number and considerable work has to be carried out to understand in depth what to do.

We end the chapter with a brief summary of other recent results.

At the end of each chapter, we have proposed a collection of exercises whose resolution can help to understand the arguments and techniques. Some of them are relatively easy; they are marked with one asterisk $(*)$. Other exercises are more involved, may need nontrivial results, and may even lead to new contributions to the field; for these, two warning asterisks appear.

Very frequently, help hints are provided.

Let us terminate this description of the contents with some comments:

- Master's and doctorate students might be interested at first stage in reading carefully Chaps. 1 and 2, which are written in a "textbook" style. There, we have tried to give a presentation of the main tools needed to understand in depth the fundamentals of the analysis and control of equations in fluid mechanics.
- More experienced readers may go directly to Chaps. 3 to 6. They correspond to a more advanced style and contain the most relevant part of the book: existence, uniqueness, regularity, numerical approximation, and control results. It will be noted that the style is different there.
- For the equations in fluid mechanics, together with control problems, it is also very relevant to consider *parameter identification* issues. There, the goal is to recover some unknown or uncertain data (a right-hand side of the PDE, boundary data, etc.) from the information furnished by "observing" the solutions. These questions will not be considered in this book for reasons of space.
- The numerical analysis of the equations is a very important area. It has been the objective of a lot of work since more than 50 years. Again, for reasons of space, it has been mentioned in this book only briefly (see Chap. 4) but, without any doubt, it should have deserved much more attention.

Mathematics and Fluid Mechanics Nowadays

At this point, we will present some considerations on (a part of) the research performed recently in mathematical fluid mechanics. We will also give a few ideas concerning future possibilities in the short and medium terms.

Theoretical Problems

There are several important open problems in fluid mechanics that have motivated a lot of work in the last decades. Let us mention some of them:

- **The Millennium Problem on the Regularity of the Navier-Stokes Equations.** For classical (i.e., constant density) Navier-Stokes equations, the standard initial-boundary value problem possesses at least one weak solution that is unique in the two-dimensional case.

This is true for any spatial domain, including the whole Euclidean space. This is known since the work of Leray, see [9, 10].[1] Furthermore, if the data are regular

[1] Leray gave a name to the weak solutions: he called them *turbulent solutions*.But this was not a good choice, as our description of Turbulence shows, see Sect. 1.6.5.

(for instance C^{∞}), the weak solution is also regular at least for a short time (for all times in the two-dimensional case).

The Sixth Millennium Problem, formulated by C. Fefferman [11], concerns this question. Specifically, it states that one of the following assertions must be proved:

(A) For any smooth initial data with compact support, there exists a smooth solution to the three-dimensional Navier-Stokes system in the whole Euclidean space for all positive time.
(B) There exist smooth initial data and right-hand side functions (external forces) such that the corresponding initial-value problems possess no smooth solution existing for all positive time.

There have been several important attempts to solve this problem. A complete summary prior to 2016 can be found in [12]. We will just mention two important contributions that have inspired and continue to inspire work:

- The paper [13] by Caffarelli, Kohn and Nirenberg, where the authors prove the existence of a class of weak solutions (the so called *suitable* weak solutions) possessing partial regularity properties. Specifically, for these solutions, the set of *singular points*, i.e., points around which regularity is lost, is small in an appropriate satisfactory sense.
- The paper [14] by T. Tao, where he presents a variant of the Navier-Stokes system whose solutions satisfy similar properties for which regularity is lost at finite time.

Of course, for variable density Navier-Stokes systems, the regularity problem is completely open. Actually, to find an "analog" of suitable weak solutions can be the goal of some research activity in the next years.

- **The Uniqueness Problem.**

This is also a problem specific to the three-dimensional case. Indeed, if the dimension is two, the weak solution indicated in the previous paragraph is unique. However, this has been a major open question for a longtime.

Many partial uniqueness results have been proved since several years. Most of them state that, under some regularity assumptions, the solution is unique in the associated class. In the following chapters, we will mention several theorems of this kind for variable density systems.

In the opposite direction, in a recent result, Buckmaster and Vicol [15] have proved that the nonuniqueness of a kind of "weak" solution to the classical Navier-Stokes equations can be asserted under appropriate circumstances. It is however unknown whether there exists at most one solution satisfying the energy inequality, a requirement absolutely exigible from the physical viewpoint.

An interesting direction of research is to try to find minimal regularity assumptions on the solutions to variable density systems that ensure uniqueness, in particular in dimension two. On the other hand, a multiple existence result would also be intriguing.

- **Possible Extensions of Known Results to Other Systems.**

In the following chapters, we will recall a collection of existence, uniqueness, and regularity results that extend to the framework of variable density Navier-Stokes equations' well-known "classical" achievements. There are however many other PDE systems that model fluids with different properties for which some of these achievements are at present unknown.

Let us mention some of them:

- Visco-elastic Oldroyd-like systems. Here, the motion equation is coupled to a nonlinear hyperbolic non-scalar system whose solution quantifies the elastic interaction between particles. In general, the existence of a solution is an open question.
- Quasi-Newtonian PDEs of the dilatant or visco-plastic kind, second grade fluids, etc. Additional open questions are found in the analysis of the existence and uniqueness of a solution to standard initial-boundary value problems, the regularity analysis, the continuous dependence with respect to initial and boundary data, the stability, and the asymptotic behavior as time goes to infinity.
- The Navier-Stokes-Fourier system. We mean here a system where we find a motion equation coupled to a transport-diffusion PDE for the temperature (an internal energy equation) with a heat source term in the right-hand side generated by viscous forces. Although some recent results have been obtained (see [16–18]), there are still things to do in this context.

- **The Analysis of Turbulent Flows.**

The flow of a fluid can be laminar, transitional, or turbulent. In the first case, we are dealing with more or less ordered trajectories along which the particles move at relatively small velocities.

Contrarily, if a fluid is in transitional or turbulent regime, the velocity field and the other physical variables have a tendency to exhibit rapid variations in space and time at a scale so small that direct numerical solution is impossible in practice.

Accordingly, we must introduce averages in the model and settle for computing numerical approximations of averaged quantities. This leads to turbulence modeling.

The theoretical analysis of turbulent models is a complex subject. Indeed, the typical structure of such a model is a motion equation for the averaged velocity and pressure strongly coupled to a set of nonlinear transport-diffusion equations for additional (turbulent) variables, furthermore frequently completed with nonlocal boundary conditions on the solid boundary. This is a source of open questions whose answers could help to understand the underlying physics.

Numerical Problems

From the numerical viewpoint, many relevant problems in fluid mechanics remain open as well. The good news is that, at present, many laminar flow problems can be solved numerically.

- **Recent Advances in Numerical Methods.**

Today, *computational fluid dynamics* (CFD) techniques are successfully applied for car, ship and aerospace design, in aircraft manufacturing, etc. Additionally, they are also used to solve problems coming among others from astrophysics, meteorology, and biomedicine.

Fortunately, the recent advancements in computer technology allow the numerical simulation in many physically and geometrically complex situations. However, there are still important open questions related to heat transfer, combustion modeling, turbulence, and efficient solution methods, and the coupling of CFD and other disciplines requires further research.

To illustrate the topic, here are several motivating directions of research:

- The analysis of blood flow under the suspension of nanoparticles and microorganisms. The main objective is to obtain accurate numerical results of the pressure. This is especially difficult in the case of the flow in an anisotropic artery of curved profile.
- The magnetohydrodynamics of three-dimensional flows through rotating conduits under the effect of viscous dissipation.
- Flows with moving and deforming boundaries. This is the situation found in many important industrial applications, such as wind turbines. The numerical techniques for these flows must include very delicate and complete treatment of meshes. To reduce computational costs is also a challenge.

- **Numerical Simulation of Turbulent Flows.**

A lot of numerical work has been carried out in this context. However, the existing methods (many of them using spectral techniques and/or finite differences and/or finite element methods) are still unable to furnish acceptable results in a large family of industrial problems and many questions are still waiting for satisfactory answers.

Optimization and Control Problems

A very important area in science concerns control. We consider *state systems,* whose solutions are assumed to describe the behavior of phenomena that can have origin in physics, chemistry, biology, etc. The basic idea is then to choose (part of) the data and thus act on the system in order to get a good behavior. Of course, this has to be made precise (and it will be later, in Chap. 6).

The data we fix in order to make the system produce good solutions are usually called *controls*. The associated solutions are called *states*. Thus, the goals are to find, characterize, and compute control-state pairs as "good" as possible.

Let us indicate several ways to accomplish this task and some research lines that generate activity at present:

- **Optimal Control Problems.**

They can be formulated by fixing a family of admissible controls and a *cost function* that in general depends on the data and the associated solutions to the system. In optimal control problems, the goal is thus to find the controls that minimize the cost function in the admissible set.

For realistic models, the system can be a complicated nonlinear collection of PDEs and the problem can become far from trivial. Among other possibilities, let us mention some particular problems of interest:

- Geometric control problems, where one of the data is, for instance, the spatial domain where the state system is assumed to be satisfied. A canonical example corresponds to determining optimal aerodynamical profiles around which a fluid flows. Another example is found when we try to compute the optimal location and shape of a cardiopulmonary bypass.
- Time optimal control problems, where the goal is to find the best (shortest or largest) time to carry out a mission. For instance, we may be interested in computing the best strategy to transport mass from a place to a distant point as fast as possible. Also, a very relevant problem of this kind concerns the determination of therapy strategies for cancer growth that lead to largest survival times.

- **Controllability.**

Here, the goal is to drive or steer the solution to the state system from an initial value to a second *desired* status. These problems are in general much harder and their solution (if it exists) deals with much more "expensive" controls. In other words, to achieve the goal we must generally inject much more energy in the system.

Let us give two interesting (and difficult to handle) examples:

- The autonomous car problem. In a simplified model, we have to find the best direction and speed strategies that drive a car from a starting point to a target.
- Molecular manufacturing problems. This is related to the production of new materials with improved performance from existing materials; it is also related to nanotechnology. The objective is to position a large amount of atoms and molecules adequately. It admits a formulation where a complex graph governed by a system of PDEs is, again, driven by appropriate manipulators (or controllers) from an initial to a desired state.

- **Other Control Strategies.**

In many cases, the previous techniques used individually do not suffice to produce satisfactory results. Then, something more elaborate must be done. Let us mention some related ideas.

Thus, it may be convenient to try to achieve not just one but several control goals. For instance, we may desire to solve *multi-objective* optimal control problems, that is, to get small values for more than one cost function simultaneously. Accordingly, we must define the *equilibria*, i.e., the data that lead to a "commitement" and our goals can be to prove their existence, deduce an optimality characterization, and compute them.

On the other hand, it is relatively frequent to apply optimal control and controllability techniques together. An example is found with the so-called *hierarchical* strategies: some leading controls are in charge of driving the solution to a prescribed value and other following data (the followers) minimize a cost.

Thus, we may be interested in steering the temperature of a room to zero at final time by acting with two controllers with a minimal energy waste; coming back to the autonomous car example, we may desire to travel from a point exactly to another following the shortest path in a minimal time, etc.

- **The Numerical Analysis and Resolution of Control Problems.**

Together with the theoretical analysis of control problems, major relevance must be paid to the related numerical study and simulation. The subject is not easy and demands a lot of effort.

The numerical solution of optimal control problems can be focused from at least two viewpoints: either by determining and computing a minimizing sequence for the cost function (and consequently by invoking optimization tools), or by trying to solve a necessary optimality system. For many realistic problems, both directions lead to nontrivial difficulties. For example, the geometric control of aerodynamical obstacles to three-dimensional variable density Navier-Stokes systems is a hard problem.

The situation is still worse if we speak about numerical controllability. Considerable work has been done. Nevertheless, many questions remain with no satisfactory answer. For example, serious numerical problems appear in connection with nonlinear hyperbolic equations and systems.

- **Incorporating Machine Learning Techniques.**

Finally, let us make reference to a promising viewpoint that is being incorporated to Control Theory at present.

In general terms, machine learning techniques can be described as those designed to solve problems with the help of computers making them previously "learn." This learning relies on experience. That is, in some preliminary steps of the algorithm the computations provide "knowledge" and this is later used in practice as part of the model. Frequently, the result (or benefit) is an essential improvement of the method and the outcomes.

In view of their structure and objectives, optimal control problems are without any doubt natural candidates to incorporating of these ideas.

Recife, Brazil
Sevilla, Spain
Arica, Chile
October 2025

Pablo Braz e Silva
Enrique Fernandez-Cara
Marko A. Rojas-Medar

References

1. Adams, R.A.: Sobolev Spaces. Pure and Applied Mathematics, vol. 65. Academic Press [A subsidiary of Harcourt Brace Jovanovich, Publishers], New York (1975)
2. Boyer, F., Fabrie, P.: Mathematical Tools for the Study of the Incompressible Navier-Stokes Equations and Related Models, volume 183 of Applied Mathematical Sciences. Springer, New York (2013)
3. Brézis, H.: Functional Analysis, Sobolev Spaces and Partial Differential Equations. Universitext. Springer, New York (2011)
4. Ekeland, I., Témam, R.: Convex Analysis and Variational Problems, volume 28 of Classics in Applied Mathematics. Society for Industrial and Applied Mathematics (SIAM), Philadelphia, PA (1999)
5. Evans, L.C.: Partial Differential Equations, volume 19 of Graduate Studies in Mathematics. American Mathematical Society, Providence, RI (1998)
6. Lang, S.: Real and Functional Analysis, volume 142 of Graduate Texts in Mathematics, third edn. Springer, New York (1993)
7. Panton, R.L.: Incompressible Flow. A Wiley-Interscience Publication. Wiley, New York (1984)
8. Zeidler, E.: Nonlinear Functional Analysis and Its Applications. IV. Springer, New York (1988). Applications to mathematical physics, Translated from the German and with a preface by Juergen Quandt
9. Leray, J.: Sur le mouvement d'un liquide visqueux emplissant l'espace. Acta Math. **63**(1), 193–248 (1934)
10. Leray, J.: Selected papers/Oeuvres scientifiques. II. Springer Collected Works in Mathematics. Springer, Heidelberg; Société Mathématique de France, Paris (2014)
11. Fefferman, C.L.: Existence and smoothness of the Navier-Stokes equation. In: The Millennium Prize Problems, pp. 57–67. Clay Math. Inst., Cambridge, MA (2006)
12. Lemarié-Rieusset, P.G.: The Navier-Stokes Problem in the 21st Century. CRC Press, Boca Raton, FL (2016)
13. Caffarelli, L., Kohn, R., Nirenberg, L.: Partial regularity of suitable weak solutions of the Navier-Stokes equations. Comm. Pure Appl. Math. **35**(6), 771–831 (1982)

14. Tao, T.: Finite time blowup for an averaged three-dimensional Navier-Stokes equation. J. Amer. Math. Soc. **29**(3), 601–674 (2016)
15. Buckmaster, T., Vicol, V.: Nonuniqueness of weak solutions to the Navier-Stokes equation. Ann. Math. (2) **189**(1), 101–144 (2019)
16. Chaudhuri, N., Feireisl, E.: Navier-Stokes-Fourier system with Dirichlet boundary conditions. Appl. Anal. **101**(12), 4076–4094 (2022)
17. Chiodaroli, E., Feireisl, E.: On the long-time behavior of solutions to the Navier–Stokes–Fourier system on unbounded domains. J. Lond. Math. Soc. (2) **111**(1), Paper No. e70067 (2025)
18. Feireisl, E.: On strict positivity of the temperature in the Navier-Stokes-Fourier system. Pure Appl. Funct. Anal. **10**(1), 45–57 (2025)

Acknowledgements The authors are indebted to the anonymous referees for their very valuable comments and suggestions that served to improve several versions of this book.

Declarations

Competing Interests The second and third authors were funded, respectively, by MICIN-Spain (Grants PID2020-114976GB-I00, PID2024-158206NB-I00) and Fondecyt-Chile (Grant 1240152). The authors have no conflicts of interest to declare.

Contents

Notation and Terms

- ODE: Ordinary differential equation.
- PDE: Partial differential equation.
- $\mathbb{R}$: The field of real numbers.
- $\mathbb{R}_+$: The set of positive real numbers.
- $\mathbb{R}^N_+$: The set $\mathbb{R}^{N-1} \times \mathbb{R}_+$.
- Ω, ω, U, O, G : Open sets in $\mathbb{R}^m$ for some integer $m \geq 1$.
- $\partial\Omega$: The boundary of Ω.
- $\mathbf{n}(\mathbf{x})$: The unit normal vector at $\mathbf{x} \in \partial\Omega$, outwards directed.
- $\mathbf{a}$, $\mathbf{b}$, ... : N-dimensional vectors with components $a_i, b_j, \ldots$.
- $\mathbf{a} \otimes \mathbf{b}$: The $N \times N$ tensor with components $(\mathbf{a} \otimes \mathbf{b})_{ij} = a_i b_j$.
- $\mathbb{1}_G$: The characteristic function of G, equal to 1 in G and equal to 0 outside G.
- δ_{ij} : *Kronecker's symbol*, which is $\delta_{ij} = 1$ if $i = j$, and $\delta_{ij} = 0$ otherwise.
- $C^0(U)$ and $C^0(\overline{U})$: The spaces of continuous functions $\varphi : U \mapsto \mathbb{R}$ and $\varphi : \overline{U} \mapsto \mathbb{R}$, respectively.
- $C^{0,\alpha}(U)$ and $C^{0,\alpha}(\overline{U})$ for $0 < \alpha \leq 1$: The spaces of Hölder-continuous functions respectively in U and $\overline{U}$. For instance, a function $\varphi : U \mapsto \mathbb{R}$ belongs to $C^{0,\alpha}(U)$ if and only if

$$\sup_{\mathbf{x},\mathbf{x}' \in U} \frac{|\varphi(\mathbf{x}) - \varphi(\mathbf{x}')|}{|\mathbf{x} - \mathbf{x}'|^\alpha} < +\infty.$$

- $C^k(U)$ for $k \geq 1$: The space of functions with continuous partial derivatives of orders up to k in U.
- $C^k(\overline{U})$ for $k \geq 1$: The spaces of restrictions to U ro functions in $C^k(\mathbb{R}^m)$.
- $C^\infty(U)$ (resp. $C^\infty(\overline{U})$): The intersections of all $C^k(U)$ (resp. $C^\infty(\overline{U})$) with $k \geq 1$.
- $C^{k,\alpha}(U)$ and $C^{k,\alpha}(\overline{U})$ for $0 < \alpha \leq 1$ and $k \geq 1$: The spaces of functions, respectively, in $C^k(U)$ and $C^k(\overline{U})$ whose derivatives of order k are Hölder-continuous.
- $\operatorname{Supp} \varphi$: The support of φ, that is, the closure of the set where $\varphi \neq 0$.

- $C_c^k(\Omega)$, $C_c^k(\overline{\Omega})$: The subspace of $C^k(\Omega)$ (resp. $C^k(\overline{\Omega})$) formed by the functions with compact support in Ω (resp. $\overline{\Omega}$).
- $\mathcal{D}(\Omega)$ (or $C_c^\infty(\Omega)$), $\mathcal{D}(\overline{\Omega})$ (or $C_c^\infty(\overline{\Omega})$): The subspace of $C^\infty(\Omega)$ (resp. $C^\infty(\overline{\Omega})$) formed by the functions with compact support in Ω (resp. $\overline{\Omega}$).
- a.e.: "Almost everywhere" (it indicates that a property is satisfied at any point except possibly in a set of zero Lebesgue measure).
- $D(A)$, $N(A)$, and $R(A)$, where A is a linear mapping: the domain, the kernel, and the image of A, respectively.

We will permanently use the well-known *Lebesgue spaces* $L^p(G)$, $L^p(G; L^q(\Omega))$, etc. for $1 \le p, q \le +\infty$ (they are rigorously defined in Chap. 2).

In particular, we will need the Hilbert spaces $L^2(G)$; the corresponding norm and scalar product will be, respectively, denoted by $\|\cdot\|$ and $(\cdot\,,\cdot)$.

In general, the notation will be abridged. Thus, if there is no danger of confusion, $\|\cdot\|_{L^r}$ will stand for $\|\cdot\|_{L^r(G)}$ and even $\|\cdot\|_{L^a(L^b)}$ will stand for $\|\cdot\|_{L^r(G;L^b(\Omega))}$.

The symbols C, K, and R will denote generic positive constants. Sometimes, we will try to emphasize the dependence of (say) C with respect a parameter δ by writing C_δ or $C(\delta)$.

List of Figures

Chapter 1
Physical Motivation

This chapter is devoted to present the main physical ideas leading to the derivation and formulation of the equations that model variable density incompressible Navier-Stokes fluids. We have mainly followed the approach of [5]. We have also incorporated arguments from [25] and [37], among others.

The techniques rely on introducing significative variables, identifying their variations with respect to time by applying the laws of physics and, finally, deducing appropriate systems of PDEs.

1.1 Introduction. The Fundamental Problem in Continuum Mechanics

Mechanics can be defined as the area of physics dealing with body motion and its causes (forces). In particular, *continuum mechanics* is mechanics applied to continuum media (to be specified below) and *fluid mechanics* is continuum mechanics restricted to *fluids,* that is to say, to continuum media not subject to rigid motion.

From a historical point of view, the first ideas concerning fluid mechanics seem to go back at least to the days of ancient Greece, with the seminal work of Archimedes and, in particular, the celebrated *Archimedes' Principle.*[1]

[1] Archimedes (287–212 B.C.) was a physicist, mathematician and engineer. He is considered one of the most important scientists of classical antiquity. Among many other achievements, he postulated the laws of buoyancy and applied them to floating and submerged objects. According to Archimedes' principle, on a body immersed in a fluid the exerted buoyant force is equal to the weight of the fluid that the body displaces; it is postulated to act in the upward direction at the center of mass of the body. A well known legend says that he used this law to detect whether a crown was made of impure gold. After achieving satisfactorily the experiment, he exclaimed "Eureka"; see [17].

P. Braz e Silva et al., *Analysis and Control of the Variable Density Incompressible Navier-Stokes Equations*, MS&A 22, https://doi.org/10.1007/978-3-032-14510-9_1

The first contributions to what can be called *modern fluid mechanics* are due to Leonardo da Vinci, who performed important experiments and got very relevant observations[2] and Torricelli, who invented the barometer.[3]

Later, we find the work of Newton, who introduced and analyzed the concept of viscosity,[4] Pascal, who investigated hydrostatics[5] and Daniel Bernoulli, the first actually involved with mathematical fluid dynamics.[6]

In subsequent steps, there are the contributions of Euler[7] and also D'Alembert, Lagrange, Laplace, Poisson, etc. in the context of inviscid fluids, while viscous flows were explored by Poiseuille, Hagen and, from a mathematical viewpoint, by Navier and Stokes.[8]

---

[2] Leonardo da Vinci (1452–1519) is viewed as the prototypical scientist of Renaissance (engineer, physician, mathematician) and also as a fabulous artist. He was able to formulate the law of conservation of mass for some particular flows. He also performed a lot of related experiments involving waves, jets, etc.; see [36] and [1] for a biography.

[3] Evangelista Torricelli (1608–1647) was an Italian physicist and mathematician. He invented the barometer around 1643. Torricelli filled with mercury a tube of glass, inverted the tube into a basin and observed the vacuum inside. A detailed description of this experiment can be found in [9].

[4] Sir Isaac Newton (1642–1727) was an English physicist, mathematician, theologian, inventor, alchemist, among many other things. A famous sentence on Newton, due to Joseph Louis Lagrange, is the following: "Newton was the greatest genius that ever existed and also the most fortunate, since a system that governs the world can be found only once". In the framework of fluid mechanics, Newton was particularly interested by the law that governs the force of resistance experimented by a body immersed in a fluid. He also postulated that, in a real-world fluid, the internal efforts among particles are proportional to the spatial variations of the velocity. Newton's contributions to fluid mechanics are explained in [8, 33].

[5] Blaise Pascal (1623–1662) was a French mathematician, physicist, theologian, philosopher and writer. He made a lot of contributions to science. Among others, he deduced the *principle of the transmission of the pressure,* nowadays known as *Pascal's Principle.* This law states that the pressure change at any point in a confined incompressible fluid is instantaneously transmitted throughout the fluid in such a way that the same change is observed everywhere. More details are given in [31].

[6] Daniel Bernoulli (1700–1782) was the most prominent member of the Bernoulli family of mathematicians. He made contributions not only in mathematics but also in many other fields: medicine, biology, mechanics, physics, astronomy, etc. I general terms, Bernoulli's Theorem in fluid dynamics states that, if the speed of a fluid increases, we simultaneously observe a decrease of the pressure or a decrease of the potential energy; see [14].

[7] Leonhard Euler (1707–1783) was a Swiss mathematician, physicist and engineer. He made a lot of crucial contributions in many areas of mathematics: differential calculus, graph theory, topology, analytic number theory, etc. The relevance of Euler in mathematics becomes clear in the following statement, due to Pierre-Simon de Laplace: "Read Euler, read Euler, he is the master of us all." The Euler equation for an ideal fluid is a relation between the velocity and the pressure based on Newton's Second Law and seems to be the second PDE found in History; more details are given below. For biographical notes, see [24].

[8] Claude-Louis Navier (1785–1836) was a French engineer and physicist who specialized in mechanics; Sir George Gabriel Stokes (1819–1903) was an Anglo-Irish physicist and mathematician. They independently deduced the Navier-Stokes equations, that are assumed to govern the behavior of incompressible viscous homogeneous fluids.

We must also mention the first analysis of boundary layers by Prandtl and the formulation and study of turbulent flows by Reynolds, Kolmogorov and Geoffrey Taylor.

It is well known that fluid mechanics analysis is fundamental nowadays to describe a lot of important phenomena in the real world. We can mention applications to

- Acoustics, i.e., the science that deals with the study of mechanical waves in gases, liquids and solids.
- Aerodynamics, that is, the study of the motion of air, particularly when it interacts with a solid object such as (for instance) an aircraft. Obviously, the relevance of this field in the present world is enormous. Since several decades, the study of aerodynamics concerns problems with great theoretical and numerical complexity.
- Oceanography, i.e., the science that studies the ocean and includes, e.g., ecosystem dynamics, ocean currents, sea waves, geophysical fluid dynamics, fluxes of pollutant substances, etc.
- Meteorology, i.e., the study of the atmosphere, whose main objective in practice is weather forecasting. Meteorological phenomena can be described by solving appropriate PDEs that involve a few variables in the atmosphere and the ocean: air and water temperature, pressure, density, velocity, salinity, etc.

It is commonly admitted that the main problems in fluid mechanics can be understood only through their formulations in terms of partial differential equations (PDEs). We will try to justify this assertion and get benefits from it.

In order to clarify the situation and fix ideas, let us (first) adopt the viewpoint of a physicist.

To this end, let $\Omega \subset \mathbb{R}^N$ be a non-empty bounded connected open set with $N = 2$ or $N = 3$ and let $T > 0$ be given. It is assumed that a *continuum* occupies the set Ω during the time interval $[0, T]$.

Thus, we accept that the medium under study is composed of *particles*[9] and that there exist functions ρ, $\mathbf{u}$ and w, with[10]

$$\begin{cases} \rho \in C^1(\overline{\Omega} \times [0, T]), \quad \rho > 0, \\ \mathbf{u} \in C^1(\overline{\Omega} \times [0, T]; \mathbb{R}^N), \\ w \in C^1(\overline{\Omega} \times [0, T]), \quad w > 0, \end{cases}$$

such that the following holds:

[9] In this context, a particle must be viewed as a conglomerate of many individual molecules that "live", interact and travel together.

[10] These regularity properties are postulated in order to justify the computations that follow. It will be seen later that they are not essential.

1. The total *mass* of the particles of the medium whose positions at time t are points of the open set $W \subset \Omega$ is given by
$$m(W,t) := \int_W \rho(\mathbf{x},t)\,d\mathbf{x}. \tag{1.1}$$
2. The total *linear momentum* associated to the particles in $W \subset \Omega$ at time t is
$$\mathbf{P}(W,t) := \int_W (\rho\mathbf{u})(\mathbf{x},t)\,d\mathbf{x}. \tag{1.2}$$
3. Finally, the *total energy* associated to the particles in $W \subset \Omega$ at time t is given by
$$E(W,t) := \int_W \left(\frac{1}{2}\rho|\mathbf{u}|^2 + \rho w\right)(\mathbf{x},t)\,d\mathbf{x}. \tag{1.3}$$
Hence, the energy of this set of particles can be written as the sum of two terms: the associated *kinetic energy*
$$E_{\text{kin}}(W,t) := \int_W \left(\frac{1}{2}\rho|\mathbf{u}|^2\right)(\mathbf{x},t)\,d\mathbf{x}$$
and the *internal energy*
$$E_{\text{int}}(W,t) := \int_W (\rho w)(\mathbf{x},t)\,d\mathbf{x}.$$

The functions ρ, $\mathbf{u}$ and w are respectively called the *mass density*, the *velocity field* and the *internal energy distribution per unit mass*. To assume that they exist is justified from a practical point of view by the fact that it is preferable to carry out a macroscopic description of the medium instead of a microscopic study (see more details in the Additional Notes of this chapter).

It is commonly accepted that for each t the functions $\rho(\cdot\,,t)$ and $\mathbf{u}(\cdot\,,t)$ furnish a complete description of the mechanical state of the medium at time t. This explains that the following is known as the "fundamental task in continuum mechanics":

Assume that, for a given medium, the mechanical state at $t = 0$ and the physical properties at any t are known. Then, determine the mechanical state of the medium for all t.

The physical properties of the medium can be frequently expressed as a partial differential system where we find ρ, $\mathbf{u}$, the internal energy w and, eventually, other additional variables, together with appropriate initial and boundary conditions on $\partial\Omega \times (0,T)$. This will be shown in the following sections.

Remark 1.1 Up to now, we have implicitly assumed that the main unknowns of the problem are ρ, $\mathbf{u}$ and w. But are there other possibilities? Note that, in fact, in view of the physical meaning assigned to ρ, $\mathbf{u}$ and w, it is rather tempting to assert that the main variables are ρ, $\mathbf{p} := \rho\mathbf{u}$ and $e := \frac{1}{2}\rho|\mathbf{u}|^2 + \rho w$. □

1.2 Lagrangian Coordinates and the Transport Lemma

In order to be precise in the formulation of the fundamental problem of continuum mechanics, we have to recall some ideas and apply some results from mathematical analysis.

In particular, we have to define the concept of *trajectory*.

For any $\mathbf{x} \in \Omega$, we can consider the corresponding Cauchy problem

$$\begin{cases} \mathbf{y}_t = \mathbf{u}(\mathbf{y}, t) \\ \mathbf{y}(0) = \mathbf{x}, \end{cases} \tag{1.4}$$

where the independent variable t must again be interpreted as time, $\mathbf{y}_t$ denotes derivative with respect to t and, as before, it is assumed that (for instance) $\mathbf{u} \in C^1(\overline{\Omega} \times [0, T]; \mathbb{R}^N)$.

There exists exactly one solution $\mathbf{y} = \mathbf{y}(t)$ for this problem that is maximal to the right, defined in an interval of the form $[0, T_*(\mathbf{x}))$, where $0 < T_*(\mathbf{x}) \le T$. By definition, we say that $t \mapsto \mathbf{y}(t)$ is the trajectory of the fluid particle located at $\mathbf{x}$ at time $t = 0$.

The individual trajectories can be gathered together by introducing *Lagrangian coordinates*. More precisely, let us set

$$O := \{(\mathbf{x}, t) \in \Omega \times [0, T) : 0 \le t < T_*(\mathbf{x})\} \tag{1.5}$$

and let the function $\mathbf{Y} : O \mapsto \mathbb{R}^N$ be defined as follows: for each $(\mathbf{x}, t) \in O$, we set $\mathbf{Y}(\mathbf{x}, t) := \mathbf{y}(t)$, where $\mathbf{y}$ is the solution to the associated problem (1.4).

We then say that $\mathbf{Y}$ is the *flux function* of the medium and the components of $\mathbf{Y}(\mathbf{x}, t)$ are the Lagrangian coordinates at time t of the particle that was initially at $\mathbf{x}$.

Thus, for any open set W with $W \times \{t\} \subset O$,

$$W_{[t]} := \{\mathbf{Y}(\mathbf{x}, t) : \mathbf{x} \in W\}$$

can be viewed as the set of positions at time t of the particles of the medium that were at a point of W at time 0.

Under the regularity conditions imposed to $\mathbf{u}$, it is clear that $\mathbf{Y} \in C^2(O; \mathbb{R}^N)$. In particular, if we denote by $\mathbf{W}^i$ the partial derivative of $\mathbf{Y}$ with respect to x_i, one has $\mathbf{W}^i(\mathbf{x}, t) = \mathbf{w}^i(t)$ for all $(\mathbf{x}, t)$, where the $\mathbf{w}^j$ solve the N-dimensional Cauchy problems

$$\begin{cases} \mathbf{w}^i_t = \sum_{j=1}^{N} \partial_j \mathbf{u}(\mathbf{Y}(\mathbf{x},t),t)\, \mathbf{w}^j, \quad 1 \le i \le N, \\ \mathbf{w}^i(0) = \mathbf{e}^i, \quad 1 \le i \le N. \end{cases}$$

Remark 1.2 Sometimes, it is also useful to consider the so called *inverse Lagrangian coordinates* $\mathbf{X} = \mathbf{X}(\mathbf{x}, t; s)$. They are defined as follows: let $\mathcal{B}$ be given by

$$\mathcal{B} := \{(\mathbf{x}, t, s) \in \Omega \times [0, T) \times [0, T) : 0 \le t, s < T_*(\mathbf{x})\};$$

then, for each $(\mathbf{x}, t, s) \in \mathcal{B}$, we set $\mathbf{X}(\mathbf{x}, t; s) := \mathbf{z}(s)$, where $\mathbf{z}$ is the maximal solution to the right to the Cauchy problem

$$\begin{cases} \mathbf{z}_s = \mathbf{u}(\mathbf{z}, s), \\ \mathbf{z}(t) = \mathbf{x}. \end{cases}$$

Thus, the components of $\mathbf{X}(\mathbf{x}, t; s)$ are the coordinates of the point at which we find at time s the particle that "was" or "will be" at $\mathbf{x}$ at time t. It is obvious that $\mathbf{Y}(\mathbf{x}, t) = \mathbf{X}(\mathbf{x}, 0; t)$. Also,

$$(\mathbf{X}_t + (\mathbf{u} \cdot \nabla \mathbf{X}))(\mathbf{x}, t; s) = 0 \quad \forall (\mathbf{x}, t, s) \in \mathcal{B}, \tag{1.6}$$

as a consequence of the identity $\mathbf{X}(\mathbf{X}(\mathbf{x}, s; t), t; s) \equiv \mathbf{x}$. The inverse Lagrange coordinates will be used in Chap. 4 for numerical purposes. □

The following result is known as the *Transport Lemma:*

Lemma 1.1 *Assume that* $f = f(\mathbf{x}, t)$ *is given, with* $f \in C^1(\Omega \times (0, T))$. *Let* $W \subset \Omega$ *be a non-empty set, let* $T_0 > 0$ *be such that*

$$W \times [0, T_0) \subset \mathcal{O}$$

and set

$$F(t) := \int_{W_{[t]}} f(\mathbf{y}, t)\, d\mathbf{y} \quad \forall t \in [0, T_0). \tag{1.7}$$

Then $F : [0, T_0) \mapsto \mathbb{R}$ *is a well defined* C^1 *function. Moreover,*

$$\frac{dF}{dt}(t) = \int_{W_{[t]}} (f_t + \nabla \cdot (f\mathbf{u}))\, (\mathbf{y}, t)\, d\mathbf{y} \quad \forall t \in [0, T_0). \tag{1.8}$$

Proof It is straightforward to check that F is well defined and C^1, due to the regularity of $\mathbf{u}$ and f.

Let us prove that the identity (1.8) holds.

Let us perform the change of variable $\mathbf{y} = \mathbf{Y}(\mathbf{x}, t)$ inside the integral in (1.7). We find that

$$F(t) = \int_W f(\mathbf{Y}(\mathbf{x}, t), t)\, J(\mathbf{x}, t)\, d\mathbf{x}, \tag{1.9}$$

where

$$J(\mathbf{x}, t) := \left| \det \frac{\partial \mathbf{Y}}{\partial \mathbf{x}}(\mathbf{x}, t) \right|.$$

□

From the definition of the flux function $\mathbf{Y}$, it is not difficult to see that

$$J_t(\mathbf{x}, t) = (\nabla \cdot \mathbf{u})(\mathbf{Y}(\mathbf{x}, t), t)\, J(\mathbf{x}, t) \quad \forall (\mathbf{x}, t) \in O \tag{1.10}$$

(this is called *Liouville's formula)* and

$$J(\mathbf{x}, 0) = 1 \quad \forall \mathbf{x} \in \Omega.$$

Consequently, it follows from (1.9) that

$$\begin{aligned}
\frac{dF}{dt}(t) &= \int_W \left(f_t + \sum_{\ell=1}^{N} \frac{\partial f}{\partial x_\ell} u_\ell + f \sum_{\ell=1}^{N} \frac{\partial u_\ell}{\partial x_\ell} \right) (\mathbf{Y}(\mathbf{x}, t), t)\, J(\mathbf{x}, t)\, d\mathbf{x} \\
&= \int_W (f_t + \nabla \cdot (f\mathbf{u}))\, (\mathbf{Y}(\mathbf{x}, t), t)\, J(\mathbf{x}, t)\, d\mathbf{x} \\
&= \int_{W[t]} (f_t + \nabla \cdot (f\mathbf{u}))\, (\mathbf{y}, t)\, d\mathbf{y}.
\end{aligned}$$

Thus, (1.8) holds. □

It will be seen later that (1.8) is a very useful formula. It means that the value of F may change along time because of two reasons: the (local) change in time of f and the motion of the particles, that are transported by $\mathbf{u}$.

Lemma 1.1 will be used in the following sections to furnish useful expressions of the change in time of several physically relevant quantities.

1.3 The Universal Laws

Let us go back to the fundamental task in continuum mechanics.

In order to describe with precision the physical properties of a medium, we will deduce a set of PDEs that must be satisfied by the corresponding ρ, $\mathbf{u}$ and w.

The physical principles that lead to these equations are of two kinds:

- Universal *conservation laws,* that are common to all media,

- Particular *constitutive laws,* only satisfied by some of them, that can be different at each case.

In this section, we recall the usual conservation laws and we present associated integral formulations. Then, we will show how they lead to a first set of PDEs for the variables of interest.

For simplicity, we will assume that $T_*(\mathbf{x}) = T$ for all $\mathbf{x} \in \Omega$, that is, the set O in (1.5) coincides with $\Omega \times [0, T)$. The first conservation law concerns mass.[11] It asserts that the mass of a set of particles is invariant in time. Since we can quantify the mass as in (1.1), an appropriate way to state this law is the following:

Conservation of Mass *Let $W \subset \Omega$ be an open ball. Then*

$$\frac{d}{dt}\left(\int_{W_{[t]}} \rho(\mathbf{x}, t)\, d\mathbf{x}\right) = 0 \tag{1.11}$$

for allt $\in [0, T)$.

From Lemma 1.1, we can rewrite Eq. (1.11) in the form

$$\int_{W_{[t]}} (\rho_t + \nabla \cdot (\rho \mathbf{u}))\, d\mathbf{x} = 0.$$

Then, since the ball W is arbitrary, the same must hold for any open set $U \subset \Omega$ and any $t \in [0, T)$:

$$\left(\int_U (\rho_t + \nabla \cdot (\rho \mathbf{u}))\, d\mathbf{x}\right)(t) = 0 \quad \forall U \subset \Omega \text{ (open)}, \quad \forall t \in [0, T).$$

Consequently,

$$\rho_t + \nabla \cdot (\rho \mathbf{u}) = 0 \text{ in } \Omega \times (0, T). \tag{1.12}$$

This is a first PDE for ρ and $\mathbf{u}$. It is frequently called the *continuity equation.*

---

[11] It is accepted that a "primitive" or seminal principle of conservation of mass was already known in ancient Greece. It was stated by Empedocles (fourth century BC) and then Epicurus (third century BC) under the following formulation:

"The totality of things has always been and will always continue to be as it is at present".

This principle has played a fundamental role in chemistry since the eighteenth century. It was discovered by Lomonosov. More precisely, in a letter to Euler on July 5th 1748, Lomonosov said that, in his opinion, matter and motion were both preserved. He stated rigorously for the first time this law in 1756, although it was only popularized some time later, in the 1780s, by Antoine Lavoisier. Two crucial consequences of the mass conservation law were the "death" of alchemy and, more important, the birth of the quantitative description of chemical transformations; see [12] for more details.

The second universal principle concerns conservation of linear momentum.[12] It is the following:

Conservation of Linear Momentum *Let $W \subset \Omega$ be an open ball and let us denote by $\mathbf{F}(W_{[t]}, t)$ the resultant of the forces acting on the particles whose positions are in $W_{[t]}$ at time t. Then*

$$\frac{d}{dt}\left(\int_{W_{[t]}} (\rho\mathbf{u})(\mathbf{x}, t)\, d\mathbf{x}\right) = \mathbf{F}(W_{[t]}, t) \tag{1.13}$$

for all $t \in [0, T)$.

This principle is in fact the continuum mechanics version of the celebrated *Newton's second law.*

In our framework, the resultant $\mathbf{F}(W_{[t]}, t)$ must be split as the sum of two vectors:

$$\mathbf{F} = \mathbf{F}_{\rm ten} + \mathbf{F}_{\rm ext},$$

where $\mathbf{F}_{\rm ten}$ is the resultant of the *tension forces* (i.e., those exerted by the particles located outside $W_{[t]}$ on the particles located inside) and $\mathbf{F}_{\rm ext}$ is the resultant of all other external forces.

It is usual and reasonable to assume that $\mathbf{F}_{\rm ten}(W_{[t]}, t)$ and $\mathbf{F}_{\rm ext|}(W_{[t]}, t)$ are respectively given by integrals on $\partial W_{[t]}$ and $W_{[t]}$. More precisely, it is accepted that

$$\mathbf{F}_{\rm ten}(W_{[t]}, t) = \int_{\partial W_{[t]}} \mathbf{T}(W_{[t]}; \mathbf{x}, t)\, d\Gamma \tag{1.14}$$

for some $\mathbf{T} = \mathbf{T}(W_{[t]}; \mathbf{x}, t)$ and

$$\mathbf{F}_{\rm ext}(W_{[t]}, t) = \int_{W_{[t]}} (\rho\mathbf{f})(\mathbf{x}, t)\, d\mathbf{x} \tag{1.15}$$

for some $\mathbf{f} = \mathbf{f}(\mathbf{x}, t)$.[13]

---

[12] It seems that the seminal ideas had been discovered by Aristotle, that asserted that a body can only be held in motion if an external force acts continuously. Around 1350, the French philosopher Jean Buridan introduced the concept of *impetus,* in an attempt to explain motion. His work was later revisited and expanded by Galileo, Leonardo da Vinci, René Descartes and finally Newton. Newton established the famous *second law of motion:*

> "The change of motion of a body is proportional to the applied external force and takes place along the same direction."

More details on the history of this principle can be found in [3, 27].

[13] These assumptions are relatively easy to justify if we accept the hypotheses of continuum mechanics. For more details, see the Additional Notes of this chapter.

It is also usual to assume the following:

(a) On $\partial W_{[t]}$, the field $\mathbf{T}$ is of the form

$$\mathbf{T}(W_{[t]}; \mathbf{x}, t) = \sigma \cdot \mathbf{n}(\mathbf{x}), \tag{1.16}$$

where $\sigma = \sigma(\mathbf{x}, t)$ is a new unknown, a C^1 symmetric matrix-valued function, usually called the *stress tensor* and $\mathbf{n}(\mathbf{x})$ is the unit normal vector at $\mathbf{x} \in \partial W_{[t]}$, exterior to $W_{[t]}$.

Roughly speaking, the i-th column of $\sigma(\mathbf{x}, t)$ provides the resultant of the tension forces acting near a point $\mathbf{x}$ (from outside to inside) on a virtual wall with outer normal vector $\mathbf{e}^i$.

The assumption (1.16)—and the fact that σ must be symmetric—will be justified below; see Sect. 1.6.

(b) The function $\mathbf{f}$ is known and satisfies $\mathbf{f} \in C^0(\overline{\Omega} \times [0, T); \mathbb{R}^N)$ (at least). Indeed, it determines the mechanical action of the external world on the medium and, in principle, must be viewed as something independent of its mechanical state.[14]

Taking into account the conservation of linear momentum (1.13), the previous assumptions and *Gauss Formula,* one gets

$$\begin{aligned}\frac{d}{dt}\left(\int_{W_{[t]}} (\rho\mathbf{u})(\mathbf{x}, t)\, d\mathbf{x}\right) &= \int_{\partial W_{[t]}} \sigma(\mathbf{x}, t) \cdot \mathbf{n}(\mathbf{x})\, d\Gamma + \int_{W_{[t]}} (\rho\mathbf{f})(\mathbf{x}, t)\, d\mathbf{x} \\ &= \int_{W_{[t]}} \left(\nabla \cdot \sigma + \rho\mathbf{f}\right)(\mathbf{x}, t)\, d\mathbf{x},\end{aligned}$$

for all W and t. Thus, in view of Lemma 1.1,

$$\int_{W_{[t]}} ((\rho\mathbf{u})_t + \nabla \cdot (\rho\mathbf{u} \otimes \mathbf{u}))\, d\mathbf{x} = \int_{W_{[t]}} (\nabla \cdot \sigma + \rho\mathbf{f})\, d\mathbf{x}.$$

As before, since the ball W is arbitrary, we obtain that

$$\int_U ((\rho\mathbf{u})_t + \nabla \cdot (\rho\mathbf{u} \otimes \mathbf{u}))\, d\mathbf{x} = \int_U (\nabla \cdot \sigma + \rho\mathbf{f})\, d\mathbf{x}$$

for any non-empty open set $U \subset \Omega$ and any $t \in [0, T)$. This leads to the so called *equation of motion:*

$$(\rho\mathbf{u})_t + \nabla \cdot (\rho\mathbf{u} \otimes \mathbf{u}) = \nabla \cdot \sigma + \rho\mathbf{f} \;\text{ in }\; \Omega \times (0, T). \tag{1.17}$$

[14] However, if we were concerned with physical problems involving more general or complex phenomena, it could be reasonable to assume that $\mathbf{f}$ depends on $\mathbf{u}$, ρ and/or on their spatial derivatives. For instance, this happens if the medium is sensible to thermodynamic or electromagnetic effects.

Conservation of Energy *Let*[15] $W \subset \Omega$ *be an open ball and let us denote by* $P(W_{[t]}, t)$ *the instantaneous power of the work performed by the forces acting on the particles located at the points of* $W_{[t]}$ *at time* t*. Then*

$$\frac{d}{dt}\left(\int_{W_{[t]}} \left(\frac{1}{2}\rho|\mathbf{u}|^2 + \rho w\right) d\mathbf{x}\right) = P(W_{[t]}, t) \tag{1.18}$$

for all $t \in [0, T)$.

The work mentioned in Eq. (1.18) can have origin either in mechanical forces (those considered in $\mathbf{F}_{\text{ten}}(W, t)$ and $\mathbf{F}_{\text{ext}}(W, t)$) or other mechanisms.

Accordingly, it is usual to write that

$$P(W_{[t]}, t) = \int_{\partial W_{[t]}} (\mathbf{T} \cdot \mathbf{u})\, d\Gamma + \int_{W_{[t]}} (\rho \mathbf{f} \cdot \mathbf{u})\, d\mathbf{x} + \int_{\partial W_{[t]}} (-\mathbf{q} \cdot \mathbf{n})\, d\mathbf{x},$$

where $\mathbf{q} = \mathbf{q}(\mathbf{x}, t)$ is a new unknown, called the *heat flux* (or internal energy flux) of the medium. An interpretation of $\mathbf{q}$ is given in Sect. 1.6.

Recalling (1.16), we find that

$$\begin{aligned} P(W_{[t]}, t) &= \int_{\partial W_{[t]}} ((\sigma \cdot \mathbf{u} - \mathbf{q}) \cdot \mathbf{n})\, d\Gamma + \int_{W_{[t]}} (\rho \mathbf{f} \cdot \mathbf{u})\, d\mathbf{x} \\ &= \int_{W_{[t]}} (\nabla \cdot (\sigma \cdot \mathbf{u} - \mathbf{q}) + \rho \mathbf{f} \cdot \mathbf{u})\; d\mathbf{x}, \end{aligned} \tag{1.19}$$

Putting (1.18) and (1.19) together, we obtain the following integral identity:

$$\begin{aligned} &\frac{d}{dt}\left(\int_{W_{[t]}} \left(\frac{1}{2}\rho|\mathbf{u}|^2 + \rho w\right) d\mathbf{x}\right) \\ &\quad = \int_{W_{[t]}} (\nabla \cdot (\sigma \cdot \mathbf{u} - \mathbf{q}) + \rho \mathbf{f} \cdot \mathbf{u})\; d\mathbf{x}. \end{aligned}$$

This must be satisfied for any open ball $W \subset \Omega$ and any $t \in [0, T)$. Now, taking into account Lemma 1.1 and the previous Eqs. (1.12) and (1.17) and arguing as before, we find the so called *energy equation:*

$$(\rho w)_t + \nabla \cdot (\rho w \mathbf{u}) = -\nabla \cdot \mathbf{q} + \sigma : \nabla \mathbf{u} \;\text{ in }\; \Omega \times (0, T). \tag{1.20}$$

---

15 Apparently, the first evidence of conservation of energy was found by Galileo around 1630, in experiments concerning the pendulum, where potential energy is converted into kinetic energy and conversely. This was later confirmed by Christiaan Huygens, in the analysis of collision phenomena. The first mathematical formulation of the energy conservation law is due to Leibniz and concerns mechanical systems of many particles that are assumed not to interact. Since then, a lot of people contributed to clarify the energy concepts and their behavior and transformations: Newton, Johann and Daniel Bernoulli, D'Alembert, Lagrange, etc.; more details can be found for instance in [20, 32].

In this way, we obtain a system of $N + 2$ equations—Eq. (1.12), the N equations (1.17) and the scalar equation (1.20)—involving $\frac{1}{2}(N^2 + 5N + 4)$ unknowns: the scalar unknowns ρ and w, the N components of $\mathbf{u}$, the N components of $\mathbf{q}$ and the $\frac{1}{2}N(N + 1)$ components of σ.

It is not surprising that we get less equations than unknowns. Indeed, so far, we have only used universal laws that must be in principle valid for *all* continua. It is therefore reasonable to expect that they do not suffice by themselves to provide a complete description of the behavior of the media.[16]

In order to get a precise picture, we have to particularize and introduce additional laws.

1.4 Some Particular (Constitutive) Laws

Now, we will consider some particular cases, determined by specific sets of laws. They correspond to classical choices in fluid mechanics and will be presented with increasing complexity.

Homogeneous Incompressible Ideal Fluids *We assume that the medium is a fluid, which means that no rigidity assumption is imposed to the motion of the particles. We will also assume that this fluid is ideal. In other words, we will suppose that there exists a scalar function* $p \in C^1(\overline{\Omega} \times [0, T])$ *such that*

$$\sigma = -p \, \mathrm{Id}. \tag{1.21}$$

Furthermore, we will assume that the fluid is incompressible, i.e., that

$$\frac{d}{dt}\left(\int_{W_{[t]}} d\mathbf{x}\right) = 0 \tag{1.22}$$

for any open set W *and any* $t \in [0, T)$*. Finally, the fluid is assumed to be homogeneous, that is,*

$$\nabla \rho \equiv 0. \tag{1.23}$$

The three assumptions above can be interpreted as follows.

First, (1.21) indicates that the tensor-valued density of tension forces in Eq. (1.14) is normally-directed. In other words, the resultant of the tension forces acting near a point $\mathbf{x}$ on a virtual wall with outer normal vector $\mathbf{e}^i$ is parallel to $\mathbf{e}^i$.

[16] Note that (1.12), (1.17) and (1.20) hold for all kinds of solids (plastic, elastic, hyperelastic or others) and fluids (liquids, gases, etc.).

This is the simplest possible structure of σ; it means that the interaction between particles is mainly frontal and, accordingly, oblique effects, i.e. *friction efforts,* are negligible. The scalar function p is, by definition, the *pressure* of the fluid.

Secondly, (1.22) means that the volume occupied by a set of particles does not change with time. This explains quite well the term *incompressible.*

Finally, assumption (1.23) means that the mass distribution of the medium is constant in space.

The previous constitutive laws are acceptable and realistic for many flow problems. For example, this is the case for the flow of the air around an airfoil, "far" from the rigid body.

Taking into account (1.21) in the motion equation (1.17), we obtain:

$$(\rho \mathbf{u})_t + \nabla \cdot (\rho \mathbf{u} \otimes \mathbf{u}) = -\nabla p + \rho \mathbf{f} \ \text{ in } \ \Omega \times (0, T). \tag{1.24}$$

On the other hand, in view of Lemma 1.1, we see from (1.22) that

$$\int_{W[t]} (\nabla \cdot \mathbf{u})(\mathbf{x}, t) \, d\mathbf{x} = 0$$

for any W and $t \in [0, T)$, whence we get the so called incompressibility equation

$$\nabla \cdot \mathbf{u} = 0 \ \text{ in } \ \Omega \times (0, T). \tag{1.25}$$

Also, we see from (1.17), (1.25), and (1.23) that $\rho_t \equiv 0$. Hence, ρ is a positive constant, that is,

$$\rho \equiv \rho_0. \tag{1.26}$$

In this case, we assume ρ_0 to be known.[17]

It follows from (1.24), (1.25), and (1.26) that the components of the velocity $\mathbf{u}$ and the pressure p have to satisfy the following equations in $\Omega \times (0, T)$:

$$\begin{cases} \mathbf{u}_t + \nabla \cdot (\mathbf{u} \otimes \mathbf{u}) + \dfrac{1}{\rho_0} \nabla p = \mathbf{f}, \\ \nabla \cdot \mathbf{u} = 0. \end{cases} \tag{1.27}$$

These are the (homogeneous) *incompressible Euler equations* of a fluid. They form a system of $N + 1$ nonlinear partial differential equations for $N + 1$ unknowns: the N components of $\mathbf{u}$ and p.[18]

[17] This is reasonable since, at time $t = 0$, ρ must be known.

[18] These equations first appeared in the paper "Principes généraux du mouvement des fluides", by Leonhard Euler, published in *Mémoires de l'Académie des Sciences de Berlin* in 1757. They form the second system of partial differential equations in the literature. Only 5 years before,

An appropriate strategy to provide an answer to the fundamental problem of continuum mechanics in this case is to solve (1.27) completed with appropriate boundary and initial conditions. In a subsequent step, we can try to compute the internal energy distribution w by solving (1.20). Since $\sigma : \nabla \mathbf{u} = -p(\nabla \cdot \mathbf{u}) = 0$, this equation takes now the form

$$w_t + \nabla \cdot (w\mathbf{u}) = -\frac{1}{\rho_0} \nabla \cdot \mathbf{q}. \tag{1.28}$$

Note however that we still have too many unknowns in this equality and consequently additional information has to be used to achieve. We will come back to this point later.

Homogeneous Incompressible Newtonian Fluids *We assume again that the medium is a fluid, but now we suppose that the velocity field* $\mathbf{u}$ *is of class* C^2 *and moreover*

$$\sigma = -p\,\mathrm{Id.} + \mu(\nabla \mathbf{u} + \nabla \mathbf{u}^t) - \frac{2}{3}\mu(\nabla \cdot \mathbf{u})\,\mathrm{Id.}, \tag{1.29}$$

where p *and* μ *are two scalar functions of class* C^1 *(they are respectively called the pressure and the dynamic viscosity). Once more, we assume that the medium is incompressible and homogeneous, that is to say, conditions* (1.25) *and* (1.23) *are satisfied.*

The assumption (1.29) is frequently known as the *Newtonian law* in fluid mechanics. It indicates that the stress tensor σ can be written as the sum of a normal stress tensor $-p\,\mathrm{Id.}$ (just the same we found in the case of an ideal fluid) and an *oblique stress tensor* that only depends on the spatial gradient of $\mathbf{u}$. The latter is due to friction forces.

Note that, in this case, we are assuming that particles interact not only normally but also tangentially; it seems reasonable to suppose that the tangential efforts are generated by friction and consequently depend on the spatial derivatives of the components of the velocities.[19]

The particular forms of the coefficients of the second and third terms in (1.29) will be explained later, in Sect. 1.6.3.

D'Alembert had introduced the one-dimensional wave equation (also called the vibrating string equation); let us complete this historical information by recalling that the two-dimensional and the three-dimensional wave equations were formulated respectively by Euler and Daniel Bernoulli short time later. More details can be found in [15].

[19] In fact, it was Newton the first to postulate this behavior. He moreover assumed that the dependence is linear, as we are doing here.

There are many real situations where the Newtonian law (1.29) is (much) more accurate and convenient than the ideal fluid assumption (1.21). For example, water and most gases behave as Newtonian fluids; see [10, 25] for more details.

Moreover, (1.21) can be viewed as the limit of (1.29) as μ goes to zero. This limit is in some sense singular. For example, in the case of a fluid flow around a solid body, it is known that, for small μ, (1.21) and (1.29) lead to fluids that are very similar outside a well chosen neighborhood of the body. But this is not at all true near the solid walls where, in order to provide good approximations, the velocity profiles corresponding to (1.29) have to be modified significantly.

There are fluids for which (1.29) does not hold. It is usual to say that they are non-Newtonian.

For instance, this happens to most fluids with long molecular chains. As a matter of fact, stirring a non-Newtonian fluid may make "holes" appear or, alternatively, may decrease the viscosity; see [18, 19] for these and other interesting phenomena.

We will mainly deal with the assumption (1.29) in the case where μ is a positive constant.[20] Taking into account (1.29) in Eq. (1.17), we obtain:

$$(\rho\mathbf{u})_t + \nabla\cdot(\rho\mathbf{u}\otimes\mathbf{u}) = -\nabla p + \mu\Delta\mathbf{u} + \rho\mathbf{f} \ \text{ in } \ \Omega\times(0,T).$$

As before, we also have (1.26), that is, the density is a constant $\rho_0 > 0$. Consequently, the components of $\mathbf{u}$ and p must satisfy the following equations in $\Omega\times(0,T)$, where we have set $\nu := \mu/\rho_0$:

$$\begin{cases} \mathbf{u}_t + \nabla\cdot(\mathbf{u}\otimes\mathbf{u}) - \nu\Delta\mathbf{u} + \dfrac{1}{\rho_0}\nabla p = \mathbf{f}, \\ \nabla\cdot\mathbf{u} = 0. \end{cases} \tag{1.30}$$

These are the (homogeneous) *incompressible Navier-Stokes equations*. Usually, the constant ν is called the *kinematic viscosity*.

Again, we have ended up with a system of $N+1$ PDEs for $N+1$ unknowns and an adequate strategy to achieve the fundamental task in continuum mechanics is to first find $\mathbf{u}$ and p from (1.30) and then, if desired, find w from (1.20).

Since $\sigma : \nabla\mathbf{u} = -p(\nabla\cdot\mathbf{u}) + 2\mu D\mathbf{u} : \nabla\mathbf{u} = 2\nu\rho_0 D\mathbf{u} : \nabla\mathbf{u}$ where $D\mathbf{u} := \frac{1}{2}(\nabla\mathbf{u}+\nabla\mathbf{u}^T)$ is the symmetrized gradient, the particular form of the energy equation is the following:

$$w_t + \nabla\cdot(w\mathbf{u}) = -\frac{1}{\rho_0}\nabla\cdot\mathbf{q} + 2\nu D\mathbf{u} : \nabla\mathbf{u}. \tag{1.31}$$

Nonhomogeneous (or Variable Density) Incompressible Newtonian Fluids *Now, we assume that the medium is a Newtonian incompressible fluid but we do not impose* (1.23).

[20] However, in Sect. 3.2, we will briefly consider dynamic viscosities of the form $\mu = \mu(\rho)$.

This will be the main situation considered in the following chapters.

Nonhomogeneous fluids (also called variable density fluids) arise in many flow processes of interest.

This is the kind of fluids we find when we deal with oceans, rivers, large containers, etc. Also, examples can be found related to general circulation models, dealing with the coupled water-atmosphere dynamics.

There, the phenomena can also be influenced by many other physical variables: temperature, salinity, moisture etc.; see for instance [7].

Other variable density problems arise in the modeling of the interaction of several phases, for example droplet impact onto a solid or liquid surface, accurate tracking of interface surfaces between fluids of different densities in multiphase flow problems, etc.; see for example [35].

Arguing as before, we find the system

$$\begin{cases} \rho_t + \nabla \cdot (\rho \mathbf{u}) = 0, \\ (\rho \mathbf{u})_t + \nabla \cdot (\rho \mathbf{u} \otimes \mathbf{u}) - \mu \Delta \mathbf{u} + \nabla p = \rho \mathbf{f}, \\ \nabla \cdot \mathbf{u} = 0, \end{cases} \tag{1.32}$$

which contains $N+2$ PDEs for $N+2$ unknowns: ρ, p and the N components of $\mathbf{u}$. In this case, the resulting energy equation is

$$(\rho w)_t + \nabla \cdot (\rho w \mathbf{u}) = -\nabla \cdot \mathbf{q} + 2\mu D\mathbf{u} : \nabla \mathbf{u}. \tag{1.33}$$

Remark 1.3 A lot of phenomena with origin in physics, chemistry, biology, etc. can be modeled by systems of PDEs by imposing postulates and carrying out computations just as we have done in this and the previous sections. For example, we can describe the growth of a cancer cell population during the *avascular* (o preliminar) phase with similar tools and principles. Thus, let us assume that the status of the tumor at time t is identified by a *normalized* cell density function $c(\cdot\,, t)$ defined and (for instance) of class C^1 in a set of the form $\overline{\Omega} \times [0, T]$, where $\Omega \subset \mathbb{R}^3$ is a bounded open set. Then, it is natural to impose $0 \le c \le 1$ and require c to satisfy the conservation identity

$$\frac{d}{dt} \int_W c(\mathbf{x}, t)\, d\mathbf{x} = - \int_{\partial W} \mathbf{b}(\mathbf{x}, t) \cdot \mathbf{n}\, d\Gamma + \int_W R(\mathbf{x}, t)\, d\mathbf{x} \tag{1.34}$$

for every ball $W \subset \Omega$ and every $t \in [0, T)$, where $\mathbf{b}$ and R respectively stand for the flux of cells and the proliferation source (reproduction minus death). In a simplified situation, it is reasonable to assume that $\mathbf{b}$ and R are given by the following constitutive laws:

- $\mathbf{b} = -D\nabla c$ for some $D > 0$ (Fick's law).
- $R = \rho c(1 - c)$ for some $\rho > 0$ (Logistic law).

Accordingly, we find from (1.34) the following PDE for c:

$$c_t - D\Delta c = \rho c(1 - c).$$

This is called the *Fisher-Kolmogorov* equation. It has been used successfully to describe and compute the evolution of many tumors, in particular, in the case of low-grade gliomas; see [26]. □

1.5 Some Initial-Boundary Value Problems

We will deal in this section with the variable density Navier-Stokes equations, see (1.32). As for any other time-dependent PDEs, this system must be completed with appropriate *boundary* and *initial* conditions.

The boundary conditions describe the behavior of the fluid particles at the points $\mathbf{x} \in \partial\Omega$ for all $t \in (0, T)$. They can be of several kinds and their nature depends on the particular situation we are dealing with. We will now recall some of them.

Dirichlet Boundary Conditions on the Velocity and the Density

$$\mathbf{u} = \mathbf{a} \text{ on } \partial\Omega \times (0, T), \tag{1.35}$$

$$\rho = \overline{\rho} \text{ on } \Sigma_{\text{in}}, \tag{1.36}$$

where $\mathbf{a}$ is a sufficiently smooth vector-valued prescribed function and

$$\Sigma_{\text{in}} := \{(\mathbf{x}, t) \in \partial\Omega \times (0, T) : \mathbf{a}(\mathbf{x}, t) \cdot \mathbf{n}(\mathbf{x}) < 0\},$$

that is, Σ_{in} is the part of $\partial\Omega \times (0, T)$ through which the fluid particles enter the domain.

A particular case, corresponding to $\mathbf{a} = 0$, is the so called *no-slip condition:*[21]

$$\mathbf{u} = 0 \text{ on } \partial\Omega \times (0, T). \tag{1.37}$$

Now, $\Sigma_{\text{in}} = \emptyset$, i.e., no particle enters the domain during $(0, T)$ and (1.36) has no sense. It will be shown in the following chapters that, in this case, no boundary condition on ρ is needed.

[21] The no-slip condition says that the fluid particles closest to a solid boundary adhere to the solid. It indicates that the adhesion forces are greater that the attraction forces between fluid particles and has been checked experimentally for viscous fluids in many cases.

Natural (Free Normal Stress) Boundary Conditions

$$\sigma \cdot \mathbf{n} = 0 \text{ on } \partial\Omega \times (0, T). \tag{1.38}$$

At the points on $\partial\Omega \times (0, T)$ where the fluid is assumed to be leaving the domain, it is natural to assume that the normal component of the stress tensor vanishes. This means that these particles are free, in the sense that they do not receive tension forces from other adjacent particles.

Fourier Boundary Conditions and Incorporation of Rugosity Effects

$$\sigma \cdot \mathbf{n} + K(\mathbf{u} - \mathbf{a}) = 0 \text{ on } \partial\Omega \times (0, T). \tag{1.39}$$

Here, $\mathbf{a}$ is again a prescribed vector field (it may be viewed as the wall velocity) and the coefficient K is positive and may depend on $\mathbf{x}$.

This condition indicates that, on the wall, normal stresses are parallel to the relative velocity of the fluid. In particular, the coefficient K provides quantitative information on the wall *rugosity*.

In fact, the Dirichlet condition (1.35) can be regarded as the limit of (1.39) when rugosity effects are important, i.e., when K is very large. On the other hand, condition (1.38) is the formal limit of (1.39) as $K \to 0^+$, and consequently corresponds to an (ideal) smooth wall.

Dirichlet Boundary Conditions on the Normal Velocity

$$\mathbf{u} \cdot \mathbf{n} = \mathbf{a} \cdot \mathbf{n} \text{ on } \partial\Omega \times (0, T), \tag{1.40}$$

where $\mathbf{a}$ is a sufficiently smooth vector-valued given function.

Sometimes, it is not realistic to have a complete knowledge of $\mathbf{u}$ on the boundary. However, it can be quite natural to assume that the behavior of the normal component of $\mathbf{u}$ is known. Indeed, this is the case when $\partial\Omega$ is a genuine wall, impermeable to the fluid particles. In particular, if we assume that $\partial\Omega$ is at rest, we find the so called *slip condition:*

$$\mathbf{u} \cdot \mathbf{n} = 0 \text{ on } \partial\Omega \times (0, T). \tag{1.41}$$

If (1.41) is imposed, we again have $\Sigma_{\text{in}} = \emptyset$ and, once more, no boundary condition on ρ must be given.

In general, (1.40) is not sufficient to determine the behavior of the fluid near $\partial\Omega$. Consequently, it must be complemented with other boundary conditions (in fact, $N - 1$ additional scalar requirements).

An acceptable way to do this is to impose that

$$\mathbf{u} \cdot \mathbf{n} = \mathbf{a} \cdot \mathbf{n} \text{ and } (\sigma \cdot \mathbf{n})_\tau + K(\mathbf{u} - \mathbf{a})_\tau = 0 \text{ on } \partial\Omega \times (0, T),$$

where, for any vector $\mathbf{g}$, we set $\mathbf{g}_\tau := \mathbf{g} - (\mathbf{g} \cdot \mathbf{n})\mathbf{n}$. It is usual to say that $\mathbf{g}_\tau$ is the *tangential component* of $\mathbf{g}$.

Note that this can be viewed as a modified form of (1.39). Obviously, these conditions indicate that $\partial\Omega$ moves with velocity $\mathbf{a}$, the fluid slips on $\partial\Omega$ and the tangential tension efforts are parallel to the tangential relative velocity. Usually, when $\mathbf{a} = \mathbf{0}$, they are called the *Navier-slip conditions.*[22]

Sometimes, it can be more appropriate to prescribe on (a part of) the boundary the behavior of the pressure p. For a discussion on this topic and some indications on "good" and "bad" boundary conditions for p, see for instance [28, 30]; see also [13] for more details. Let us also mention that, in [2], the authors present a class of boundary conditions for the incompressible Navier-Stokes equations that preserve energy.

In addition to boundary conditions, we must use initial conditions to complement (1.32). It will be seen later that it will suffice to provide values of $\mathbf{u}$ and ρ at the initial time $t = 0$. This is part of our information (recall the statement of the fundamental problem of continuum mechanics in Sect. 1.1). Consequently, we have to impose that

$$\rho(\mathbf{x}, 0) = \rho^0(\mathbf{x}) \ \text{ in } \ \Omega \tag{1.42}$$

and

$$\mathbf{u}(\mathbf{x}, 0) = \mathbf{u}^0(\mathbf{x}) \ \text{ in } \ \Omega \tag{1.43}$$

for some prescribed ρ^0 and $\mathbf{u}^0$.

In the following paragraphs, we will indicate some "typical" situations, where it is easy to understand the role of initial and boundary conditions.

For illustration, we have computed related numerical approximations using the `Freefem` package (see [16]):

- Fluid in a channel after a step, see Fig. 1.1.
- Fluid around a ball, see Figs. 1.2, 1.3, and 1.4.
- Fluid around another non-aerodynamic profile, see Figs. 1.5, 1.6, and 1.7.

Observe that, in all cases, we find *recirculating regions* where the flow is considerably more complex and vortices are created. This gives an idea of the difficulties we can find in the analysis and computation of the solutions.

Detailed information on the numerical techniques used to find these results is given below, in Chap. 4.

---

[22] They appeared for the first time in Navier's paper "Mémoire sur les lois du mouvement des fluides", published in "Mémoires de l'Académie Royale des Sciences de l'Institut de France" in 1823 (no. 6.1823, pp. 389–440).

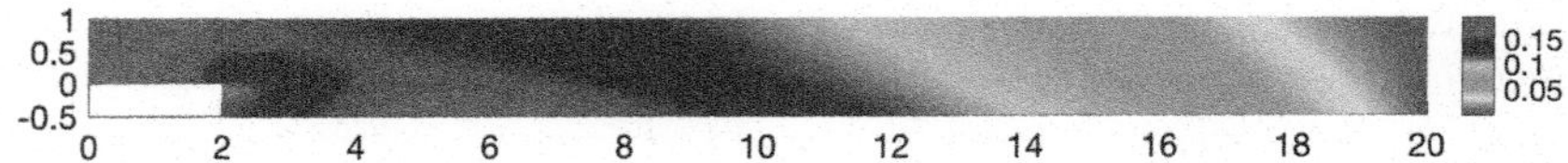

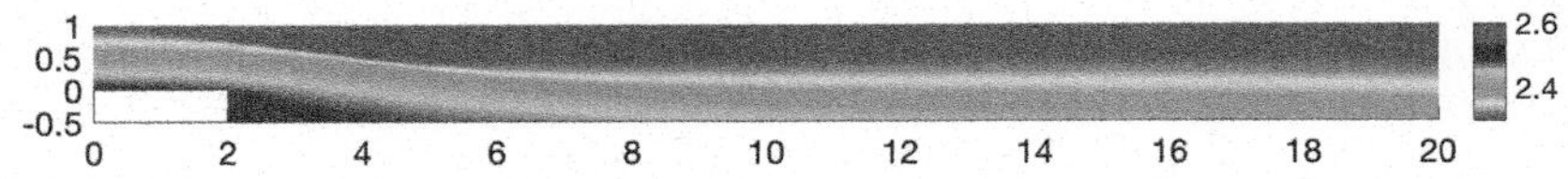

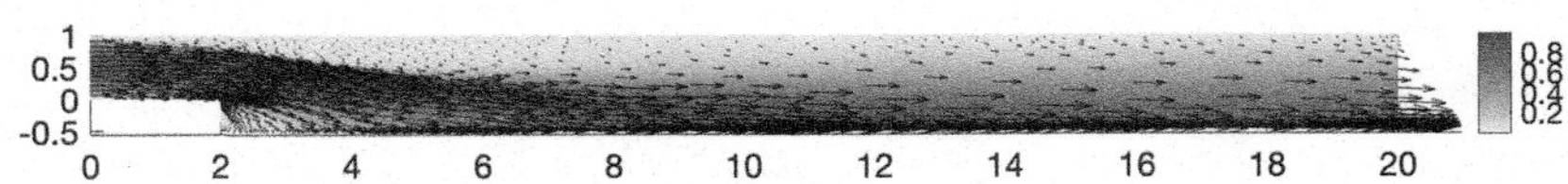

Fig. 1.1 Fluid in a channel after a step. From top to bottom, pressure, density and velocity field

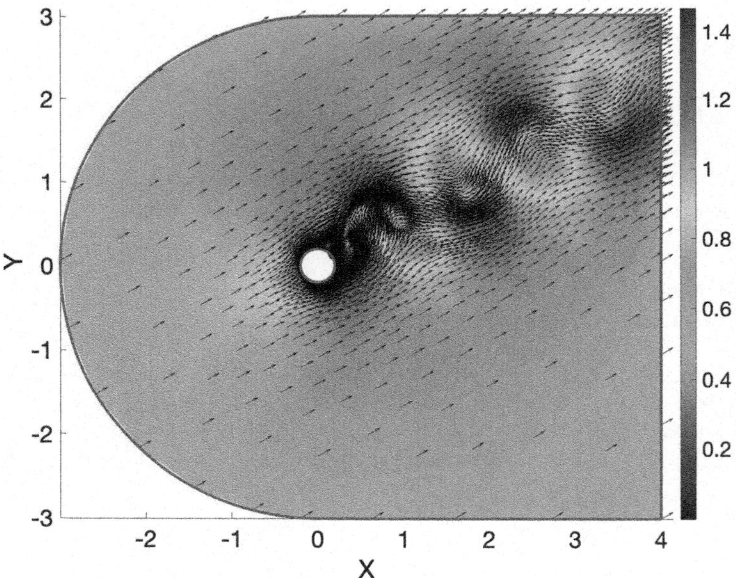

Fig. 1.2 Fluid around a ball: the velocity field

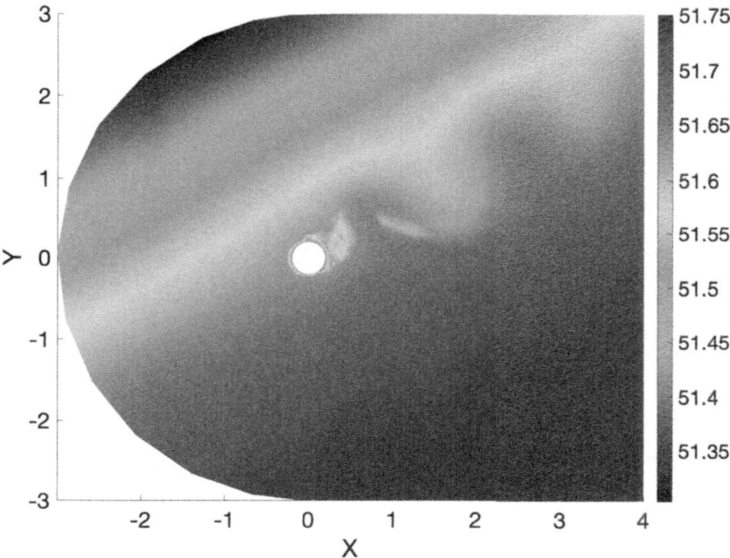

Fig. 1.3 Fluid around a ball: the density

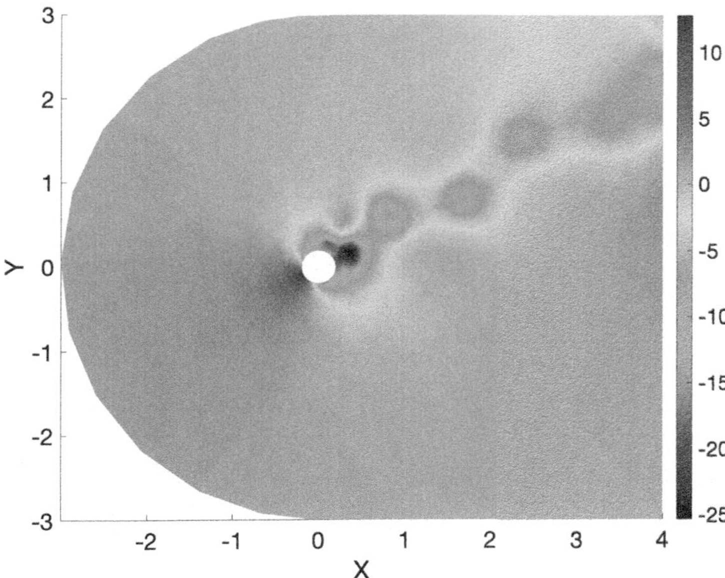

Fig. 1.4 Fluid around a ball: the pressure

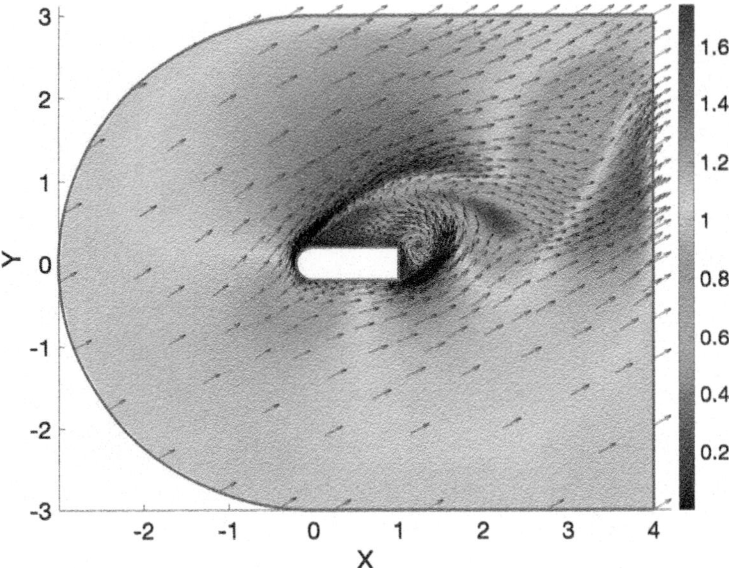

Fig. 1.5 Fluid around a non-aerodynamic profile: the velocity

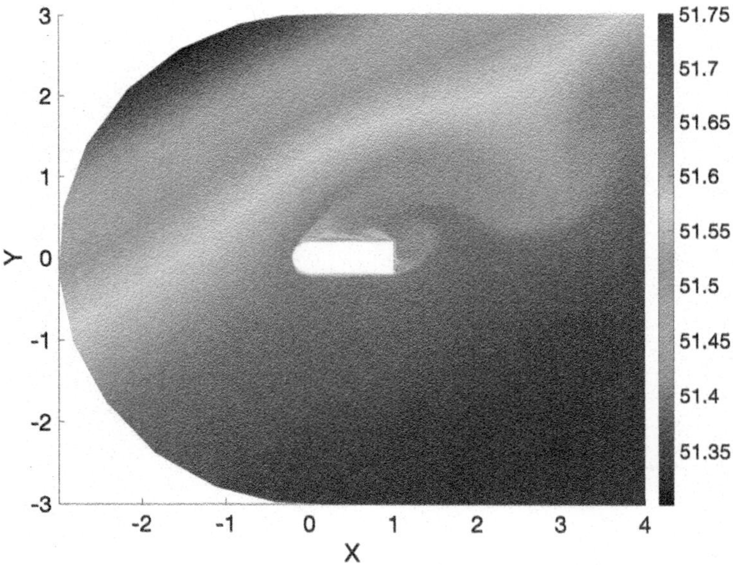

Fig. 1.6 Fluid around a non-aerodynamic profile: the density

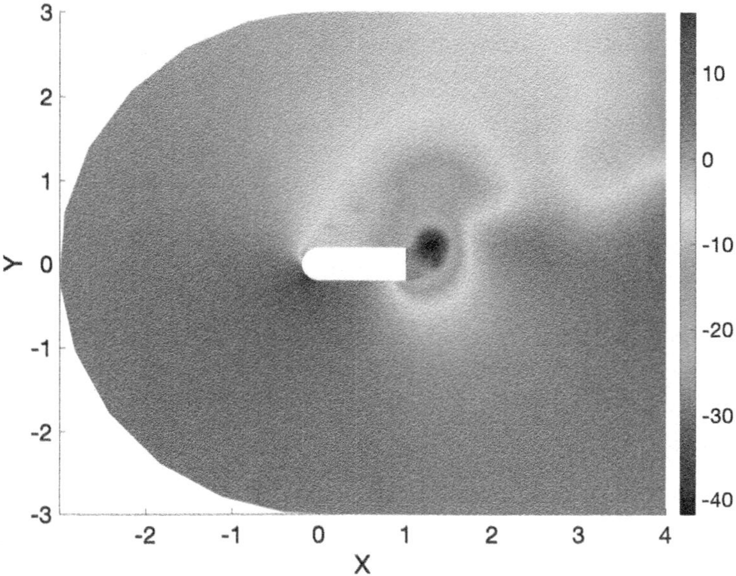

Fig. 1.7 Fluid around a non-aerodynamic profile: the pressure

Remark 1.4 In principle, it makes sense to consider and search for solutions to (1.32) independent of time, that is, corresponding to the permanent flow of a fluid. We are thus making reference to solutions to the stationary system

$$\begin{cases} \nabla \cdot (\rho \mathbf{u}) = 0, \quad \nabla \cdot \mathbf{u} = 0, \\ \nabla \cdot (\rho \mathbf{u} \otimes \mathbf{u}) - \mu \Delta \mathbf{u} + \nabla p = \rho \mathbf{f}, \end{cases} \tag{1.44}$$

that must be completed with boundary conditions on $\mathbf{u}$ and eventually ρ. These equations will be considered at the end of Chap. 5. In particular, it will be seen there that it is not known how to get a "good" formulation of the problem. In other words, to determine boundary conditions and additional requirements that can be added to (1.44) to get a well posed system is at present an open question. □

In the following chapters, we will be concerned with the variable density Navier-Stokes equations (1.32), most cases completed with the no-slip boundary conditions (1.37) and the initial conditions (1.42) and (1.43).

We will end this section with a short justification of the need and convenience of the theoretical analysis of problems like those mentioned before.

As already said, we have adopted the viewpoint of a physicist so far:

- We have accepted the validity of continuum mechanics and have postulated the existence of (main and secondary) variables.
- We have assumed that several sets of balance (universal) and constitutive (particular) laws are satisfied and we have obtained some systems of PDEs.

- The knowledge of the mechanical state at $t = 0$ has been taken into account through appropriate initial conditions.
- Other properties of the flow have led to complementary boundary conditions.

In this way, we have got a mathematical model. We accept that the variables exist and satisfy these equations and additional conditions. In order to get quantitative information, what we have to do is to solve, in practice numerically, the resulting problem.

Let us now adopt the point of view of a mathematician.

Thus, let us deal with the system (1.32), completed with conditions (1.37), (1.42), and (1.43).

In principle, we do not necessarily know whether there exist functions ρ, $\mathbf{u}$ and p satisfying these PDEs and complementary conditions. Our first task is consequently to prove that this is true. If this is the case, we will have a confirmation of the fact that we were not wrong before; that the problem was not over-determined; that we did not introduce contradictory hypotheses, etc. This justifies the need of an *existence* result.

Even in the case that at least one solution exists, we do not know in principle how many there are. Hence, a second task is to prove that the solution is unique. The existence of at most one solution shows that the set of assumptions we have used is *complete* from the logical and physical viewpoints. This is the motivation for a *uniqueness* result.

However, there is still a third important point to check. It could happen to our system to have exactly one solution for each set of prescribed data that is *unstable*. In other words, it can be the case that small perturbations of the data do not lead to small perturbations of the solutions (measured in appropriate norms) and, contrarily, produce oscillations or even blow-up.

Note that, in such a situation, the model is useless (indeed, small perturbations of the data are unavoidable in practice). Consequently, a third task for the mathematician is to prove *stability* in this sense; more precisely, continuous dependence of the solution with respect to the data. So, this is also strongly justified.

Chapters 3 and 5 of this book deal with a theoretical analysis of (1.32), (1.37), (1.42), and (1.42), concerning existence, uniqueness, regularity and stability issues (among others). As already said, Chap. 4 will be devoted to indicate how can we produce numerical approximations.

After achieving the previous steps and assuming that we can assign to each set of data one or more solutions, we may try to consider an additional very relevant question: can we choose the data in such a way that at least one of the associated solutions presents a "desired behavior"?

Of course, we must give a precise sense to this and specify without ambiguity what are the data "at our disposal" and what does "desired behavior" mean. But, assuming this is done, it becomes clear that a positive answer to this question (and, if possible, an explicit method to compute the good data) is of major interest in the field.

These and other similar questions are taken into consideration by *control theory* and will be analyzed in Chap. 5.

1.6 Some Additional Notes

This section is devoted to present some arguments that justify from the physical and geometrical viewpoints the assumptions we have invoked to deduce (1.29). More details can be found for instance in [5, 37].

1.6.1 *Justification of the Existence of* ρ, $\mathbf{u}$ *and* w

As we have already mentioned, each point $\mathbf{x} \in \Omega$ must be identified to a *conglomerate* of particles that travel together, with masses m_i and velocities $\mathbf{v}_i$. These particles are supposed to stay in Ω during the time interval $[0, T]$.

Implicitly, we are assuming in hypothesis (1.1) that, at each $(\mathbf{x}, t)$,

$$\exists \rho(\mathbf{x}, t) := \lim_{\varepsilon \to 0^+} \frac{\sum_{i \in I(\varepsilon,t)} m_i}{V(\varepsilon)},$$

where $I(\varepsilon, t)$ stands for the set of indices corresponding to the particles in $B(\mathbf{x}; \varepsilon)$ at time t and $V(\varepsilon)$ is the corresponding volume of $B(\mathbf{x}; \varepsilon)$. In practice, this means that we are working at a small but not too small scale.

On the other hand, the identities (1.2) mean that the following holds for all $(\mathbf{x}, t)$:

$$\exists \mathbf{u}(\mathbf{x}, t) := \lim_{\varepsilon \to 0^+} \frac{\sum_{i \in I(\varepsilon,t)} m_i \mathbf{v}_i}{\sum_{i \in I(\varepsilon,t)} m_i}. \tag{1.45}$$

Finally, the right interpretation of assumption (1.3) is that the following limit exists for every $(\mathbf{x}, t)$:

$$\exists w(\mathbf{x}, t) := \lim_{\varepsilon \to 0^+} \frac{\sum_{i \in I(\varepsilon,t)} \frac{1}{2} m_i |\mathbf{v}_i'|^2}{\sum_{i \in I(\varepsilon,t)} m_i}, \tag{1.46}$$

where the $\mathbf{v}_i'$ stand for the individual molecular perturbations $\mathbf{v}_i' = \mathbf{v}_i - \mathbf{u}$. Observe that the true total energy associated to the particles in $B(\mathbf{x}; \varepsilon)$ at time t is

$$\begin{aligned} \sum_{i \in I(\varepsilon,t)} \frac{1}{2} m_i |\mathbf{v}_i|^2 = {} & \left(\sum_{i \in I(\varepsilon,t)} \frac{1}{2} m_i \right) |\mathbf{u}|^2 + \sum_{i \in I(\varepsilon,t)} \frac{1}{2} m_i |\mathbf{v}_i'|^2 \\ & + \left(\sum_{i \in I(\varepsilon,t)} m_i \mathbf{v}_i' \right) \cdot \mathbf{u}. \end{aligned} \tag{1.47}$$

In view of (1.45) and the definition of the $\mathbf{v}_i'$, the quotient of the last term divided by $\sum_{i\in I(\varepsilon,t)} m_i$ goes to zero as $\varepsilon \to 0^+$. This justifies (1.46).

1.6.2 Justification of the Existence of a Stress Tensor σ and an External Field $\mathbf{f}$

The tension forces corresponding to the transported domain $W_{[t]}$ are due to the action of the particles located outside $W_{[t]}$ on the particles located inside.

It is therefore reasonable to assume that the resultant $\mathbf{F}_{\rm ten}$ can be computed by integrating a field $\mathbf{T} = \mathbf{T}(W_{[t]}; \mathbf{x}, t)$ along the boundary $\partial W_{[t]}$. It is also reasonable to assume that $\mathbf{T}$ is unknown. Indeed, this field gives information on the way that particles interact.

In view of its physical meaning, it is adequate to admit that $\mathbf{T}$ depends on $W_{[t]}$ through the outwards unit normal vector $\mathbf{n}$, that is,

$$\mathbf{T}(W_{[t]}; \mathbf{x}, t) = \mathbf{s}(\mathbf{n}; \mathbf{x}, t) \tag{1.48}$$

for some tensor-valued function $\mathbf{s}$.

Let us consider a continuum where no external force is applied. Then, an elementary geometric analysis shows that

$$\mathbf{s}(-\mathbf{n}; \mathbf{x}, t) \equiv -\mathbf{s}(\mathbf{n}; \mathbf{x}, t) \text{ and } \mathbf{s}(\mathbf{n}; \mathbf{x}, t) \equiv \sum_{\ell=1}^{3} n_\ell\, \mathbf{s}(\mathbf{e}^\ell; \mathbf{x}, t); \tag{1.49}$$

see Exercise 1.4.

Hence, $\mathbf{T}$ is necessarily of the form $\sigma \cdot \mathbf{n}$ for some $\sigma = \sigma(\mathbf{x}, t)$ and the tension forces can be written in the form

$$\mathbf{F}_{\rm ten}(W_{[t]}, t) = \int_{\partial W_{[t]}} \sigma(\mathbf{x}, t) \cdot \mathbf{n}(\mathbf{x})\, d\Gamma.$$

The stress tensor σ must be symmetric. This is a consequence of the *angular momentum* conservation law.

Indeed, this principle states that the time variation of the angular momentum associated to an arbitrary set of particles must coincide with the *torque* of the corresponding applied forces. Thus, for any open set $W \subset \Omega$ with regular boundary and any time $t \in [0, T)$, we must have

$$\frac{d}{dt}\left(\int_{W_{[t]}} ((\rho\mathbf{u})(\mathbf{x}, t) \times \mathbf{x})\, d\mathbf{x}\right) = \int_{\partial W_{[t]}} ((\sigma \cdot \mathbf{n}) \times \mathbf{x})\, d\Gamma + \int_{W_{[t]}} ((\rho\mathbf{f}) \times \mathbf{x})\, d\mathbf{x}. \tag{1.50}$$

We easily find from (1.50), Lemma 1.1 and the motion equation (1.17) that σ must be symmetric at each $(\mathbf{x}, t)$.

Finally, in the framework of fluid mechanics, it is completely natural to require that σ is continuously differentiable.

In the derivation of the equation of motion (1.17), we have also assumed that the external forces are given by (1.15). Here, $\mathbf{f} = \mathbf{f}(\mathbf{x}, t)$ is the so called external field density per unit mass.

As we have already indicated, it is usual to assume that $\mathbf{f}$ is known since, in the present context, this is an information on the medium that we are supposed to have.

1.6.3 The Origin and Justification of the Formula (1.29)

The ideas behind (1.29) come from the work of Newton.

This formula states that the stress tensor σ is the sum of a normal stress tensor $-p\text{Id}.$ and an oblique stress tensor that depends linearly on the spatial gradient of the velocity field.

We will give in this section a short justification:

• First, note that it is reasonable to assume, at least at first glance, that

$$\sigma(\mathbf{x}, t) = \Sigma_0(\rho(\mathbf{x}, t), w(\mathbf{x}, t), \dots, \frac{\partial u_i}{\partial x_j}(\mathbf{x}, t), \dots), \tag{1.51}$$

where $\Sigma_0 = \Sigma_0(\rho, w, \dots, d_{ij}, \dots)$ is given.

Accordingly, the stress tensor σ depends on the local (pointwise and instantaneous) values of ρ, w, etc. and, in particular, memory effects are being neglected.

In our context, it is also natural to assume that the medium is *isotropic*. That is, from the viewpoint of the mechanical action, a variation of u_i in the direction x_j must be as relevant as a variation of u_j in the direction x_i.

This means that

$$\frac{\partial \Sigma_0}{\partial d_{ij}} \equiv \frac{\partial \Sigma_0}{\partial d_{ji}} \quad \forall i, j.$$

Consequently, introducing $D\mathbf{u} := \frac{1}{2}(\nabla \mathbf{u} + \nabla \mathbf{u}^t)$, we can write that

$$\sigma(\mathbf{x}, t) = \Sigma(\rho(\mathbf{x}, t), w(\mathbf{x}, t), D\mathbf{u}(\mathbf{x}, t))$$

for some $\Sigma = \Sigma(\rho, w, D)$, a symmetric tensor-valued function defined on $\mathbb{R}_+ \times \mathbb{R}_+ \times \mathcal{L}_s(\mathbb{R}^3)$.

• Secondly, a rotation of the reference system must not affect the stress tensor values, that is, we must have

$$\Sigma(\rho, w, R \cdot D \cdot R^{-1}) = R \cdot \Sigma(\rho, w, D) \cdot R^{-1}$$

for any orthogonal matrix R and any (ρ, w, D). Under these conditions, as a consequence of the *Rivlin-Ericksen Theorem* (see for instance [37]),[23] there must exist functions a_0, a_1 and a_2 such that

$$\begin{cases} a_i = a_i(\rho, w, d_1, d_2, d_3), \quad 0 \le i \le 2, \\ d_1 = \text{trace}\, D, \quad d_2 = \text{trace}\, D^2, \quad d_3 = \det D, \\ \Sigma = a_0 \text{Id.} + a_1 D + a_2 D^2. \end{cases} \tag{1.52}$$

Note that, so far, the hypotheses are completely general and reasonable from the physical point of view in a purely viscous context.

• Let us simplify (1.52). More precisely, let us assume that the mapping $D \mapsto \Sigma(\rho, w, D)$ is affine. Then,

$$\sigma = [-p_* + \lambda_*(\nabla \cdot \mathbf{u})]\, \text{Id.} + \mu_* \left(\nabla \mathbf{u} + \nabla \mathbf{u}^t\right) \tag{1.53}$$

for some functions p_*, λ_* and μ_* depending only on ρ and w. This is equivalent to assume in (1.52) that the a_i have the form $a_0 \equiv a_{00} + a_{01} d_1$, $a_1 \equiv a_{10}$ (with a_{00}, a_{01} and a_{10} independent of d_i) and $a_2 \equiv 0$.

If we impose (1.53), we say that the fluid is *Newtonian.*

• Finally, we will assume that the so called *Stokes law* is satisfied. This means that

$$3\lambda + 2\mu \equiv 0. \tag{1.54}$$

In order to justify this law, it suffices to consider a fluid in a three-dimensional domain of the form $B(0; R) \setminus \overline{B}(0; r)$ with velocity $\mathbf{u}(\mathbf{x}, t) \equiv a\mathbf{x}$, where a is a positive constant. Suppose also constant density and constant total energy: $\rho(\mathbf{x}, t) \equiv$

[23] This important result can be used not only in the context of fluid mechanics but also in many other branches of physics and science to produce acceptable models. The rigorous statement is as follows:

Theorem 1.1 *Let $S(\mathbb{R}^N)$ and $Q(\mathbb{R}^N)$ respectively denote the sets of symmetric and orthogonal square matrices of order N and let the function $\Sigma : S(\mathbb{R}^N) \mapsto S(\mathbb{R}^N)$ be given. Then the following two assertions are equivalent:*

1. *For every $D \in S(\mathbb{R}^N)$ and every $R \in Q(\mathbb{R}^N)$, one has*

$$\Sigma(R \cdot D \cdot R^{-1}) = R \cdot \Sigma(D) \cdot R^{-1}.$$

2. *There exist functions $\alpha_0, \ldots, \alpha_{N-1} : \mathbb{R}^N \mapsto \mathbb{R}$ such that*

$$\Sigma(D) = \sum_{j=0}^{N-1} \alpha_j(d_1, \ldots, d_N)\, D^j,$$

where the d_i stand for the algebraic invariants of D, that is, the coefficients of the associated characteristic polynomial.

For the proof, see Exercise 1.6.

ρ_0 and $w(\mathbf{x}, t) \equiv w_0$. In such a situation, it is natural to assume that there is no friction and, consequently, viscous efforts are identically zero.

Accordingly, a simple computation shows that (1.54) necessarily holds. This gives:

$$\sigma = -p_* \mathrm{Id.} + \mu_* \left(\nabla \mathbf{u} + \nabla \mathbf{u}^t - \frac{2}{3} (\nabla \cdot \mathbf{u}) \mathrm{Id.} \right), \tag{1.55}$$

where p_* and μ_* are functions of ρ and w.

Therefore, (1.29) possesses a justification and can be used as a constitutive law to deduce (1.32).

In the previous considerations, we have simplified the argument at two main levels:

1. In (1.51), as already noticed, it was assumed that the value of the stress tensor at $(\mathbf{x}, t)$ only depends on the values at $(\mathbf{x}, t)$ of ρ, $\nabla \mathbf{u}$ and w. However, for fluids with a sufficiently complex molecular structure, it has been checked experimentally that memory or elastic effects become important and the functional dependence is not instantaneous.
   This leads to other more complicated constitutive laws and other PDE systems where some new unknowns must be taken into account.
   For example, in the case of constant density fluids, a celebrated example is the so called *Oldroyd system*
   $$\begin{cases} \mathbf{u}_t + (\mathbf{u} \cdot \nabla)\mathbf{u} - \nu \Delta \mathbf{u} + \dfrac{1}{\rho_0} \nabla p = \nabla \cdot \tau + \mathbf{f}, \quad \nabla \cdot \mathbf{u} = 0, \\ \tau_t + (\mathbf{u} \cdot \nabla)\tau + \mathbf{G}(\tau, \nabla \mathbf{u}) + a\tau = bD\mathbf{u}, \end{cases}$$
   where $a, b > 0$,
   $$\begin{cases} \mathbf{G}(\tau, \nabla \mathbf{u}) := \tau \cdot W\mathbf{u} - W\mathbf{u} \cdot \tau - \alpha(D\mathbf{u} \cdot \tau + \tau \cdot D\mathbf{u}), \\ W\mathbf{u} := \dfrac{1}{2}(D\mathbf{u} - D\mathbf{u}^T), \quad \alpha \in [-1, 1] \end{cases} \tag{1.56}$$
   and $\tau = \tau(\mathbf{x}, t)$ is a symmetric matrix-valued variable, called the *elastic extra-stress* tensor.
   For more details, see for instance [19, 29].
2. The nonlinear in D terms in (1.52) have been discarded in (1.53). For instance, if we accept quadratic terms in the expression of Σ, we arrive at a different constitutive law:
   $$\Sigma = a_0(\rho, w, d_1, d_2)\mathrm{Id.} + a_1(\rho, w, d_1)D + a_2(\rho, w)D^2. \tag{1.57}$$
   Fluids for which this is assumed to hold are usually called *second grade fluids*. For homogeneous incompressible media, the resulting PDE system takes the form

$$\begin{cases} \mathbf{u}_t + (\mathbf{u}\cdot\nabla)\mathbf{u} - \nabla\cdot\sigma + \dfrac{1}{\rho_0}\nabla p = \mathbf{f}, \ \ \nabla\cdot\mathbf{u} = 0, \\ \sigma = \nu_1(D\mathbf{u} + D\mathbf{u}^T) + \nu_2(D\mathbf{u} + D\mathbf{u}^T)^2. \end{cases} \tag{1.58}$$

More information can be found for example in [6].

1.6.4 The Energy Equations for Ideal and Newtonian Fluids

Let us indicate the final forms of the energy equation for the incompressible fluids considered in Sect. 1.4.[24]

For ideal incompressible homogeneous fluids, in view of (1.12), (1.20), (1.21), and (1.26), after some manipulations, we found (1.28). For Newtonian incompressible homogeneous fluids, the situation is a little more complex. Taking into account (1.12), (1.20), (1.29), and (1.26), we saw that w must satisfy Eq. (1.31). Finally, for Newtonian incompressible nonhomogeneous fluids, we found Eq. (1.33).

In all cases, we still have less equations than unknowns for the computation of w. In order to complete our information and close the system, it is again necessary to use additional laws, this time coming from *thermodynamics*.

An usual additional assumption is *Fourier's law*. It states that the heat flow $\mathbf{q}$ is proportional to the spatial gradient of the temperature, that is,

$$\mathbf{q} = -\kappa\nabla\theta \tag{1.59}$$

for some positive κ, which is called the *heat diffusion coefficient*. This is in agreement with common experience: heat is expected to go from high to low temperature regions; the identity (1.59) expresses this fact and, also, that this process is achieved as fast as possible.[25]

---

[24] Strictly speaking, conservation of energy can be understood only if we accept that heat and mechanical work are interchangeable forms of energy. This was first discovered by the German surgeon Julius Robert von Mayer in 1842 and, independently, by the English physicist and mathematician James Prescott Joule in 1843 in a series of experiments. In particular, von Mayer deduced this principle in India, where he found that people possessed darker red colored blood and deduced that they used to consume less oxygen (less energy). The explanation was that they did not need so much energy to keep a comfortable body temperature. The caloric theory used to postulate that heat could neither be created nor destroyed; contrarily, conservation of energy asserts that heat and mechanical work are interchangeable. For more information, we refer to [32].

[25] Indeed, the gradient $\nabla\theta$ provides, at first approximation, the direction of maximal change rate of θ. In the context of thermodynamics, this property was first discovered and formulated by Joseph Fourier in 1822, who concluded the following: "The heat flux resulting from thermal conduction is proportional to the magnitude of the temperature gradient and opposite to it in sign." In many other areas of science, similar observations are used to obtain related laws. Thus, we can

On the other hand, the function w is usually assumed to be a linear function of the temperature, that is,

$$w = c_0\theta \tag{1.60}$$

for some positive c_0, usually called the *specific heat coefficient*.[26,27]

Taking into account (1.59) and (1.60), Eqs. (1.28), (1.31), and (1.33) can be respectively rewritten as

$$c_0(\theta_t + \mathbf{u}\cdot\nabla\theta) - k\Delta\theta = 0,$$

with $k := \kappa/\rho_0$,

$$c_0(\theta_t + \mathbf{u}\cdot\nabla\theta) - k\Delta\theta = 2\nu D\mathbf{u} : \nabla\mathbf{u}$$

and

$$c_0\left((\rho\theta)_t + \nabla\cdot(\rho\theta\mathbf{u})\right) - \kappa\Delta\theta = 2\mu D\mathbf{u} : \nabla\mathbf{u}.$$

1.6.5 Laminar and Turbulent Flows. Basic Ideas

We will finish this chapter with a brief presentation of the fundamental ideas in Turbulence.

For simplicity, consider a homogeneous, incompressible, Newtonian fluid governed by the system

$$\begin{cases} \rho_0(\mathbf{u}\cdot\nabla)\mathbf{u} - \mu\Delta\mathbf{u} + \nabla p = 0 & in\ \Omega, \\ \nabla\cdot\mathbf{u} = 0 & in\ \Omega, \\ \mathbf{u} = \mathbf{a} & on\ \partial\Omega, \end{cases} \tag{1.61}$$

where $\Omega \subset \mathbb{R}^3$ is a bounded, regular, connected domain and $\mathbf{a}$ is a constant vector.

Here, we are assuming that the flow of this fluid is *stationary*, that is, the variables $\mathbf{u}$ and p are independent of t.

speak of Fick's law for the diffusion of a chemical product and Darcy's law for fluid flows in porous media; see Exercise 1.9.

[26] In fact, (1.60) can be viewed as a definition of the temperature θ.

[27] In (1.59) and (1.60), we have assumed for simplicity that κ and c_0 are constant, but it is also meaningful and reasonable to suppose that they depend on $\mathbf{x}$ and/or t. Actually, in many realistic models, it must be assumed that κ or c_0 (or both) change with local or global values of θ or $\nabla\theta$.

Such a fluid can flow in two completely different ways:

- For "small" data **a** or "large" kinematic viscosities $\nu = \mu/\rho_0$, the velocity field and pressure are regular and, roughly speaking, the fluid particles follow more or less ordered trajectories. It is then said that the flow is *laminar.*
- Contrarily, for sufficiently large **a** or sufficiently small ν, both the velocity and the pressure exhibit extremely rapid variations or oscillations in space and the particles seem to have a chaotic behavior. In this case, we say that the flow is *turbulent.*

The *transition* from the laminar to the turbulent regime can be explained if we consider a dimensionless reformulation of system (1.61).

More precisely, let us introduce a *characteristic length* L and a *characteristic velocity* U of the problem. For example, we can take

$$L = \text{diameter of } \Omega, \quad U = |\mathbf{a}|.$$

To fix ideas, assume that $0 \in \Omega$ and set

$$\Omega^* = \frac{1}{L}\Omega, \quad \mathbf{a}^* = \frac{1}{U}\mathbf{a}, \quad \mathbf{x}^* = \frac{1}{L}\mathbf{x}, \quad \mathbf{u}^* = \frac{1}{U}\mathbf{u}, \quad p^* = \frac{1}{\frac{1}{2}\rho_0 U^2}\, p.$$

Then the dimensionless variables $\mathbf{u}^*$ and p^* satisfy

$$\begin{cases} (\mathbf{u}^* \cdot \nabla^*)\mathbf{u}^* - \dfrac{1}{\mathrm{Re}}\Delta^*\mathbf{u}^* + \nabla^* p^* = 0 & \text{in } \Omega^*, \\ \nabla^* \cdot \mathbf{u}^* = 0 & \text{in } \Omega^*, \\ \mathbf{u}^* = \mathbf{a}^* & \text{on } \partial\Omega^*, \end{cases} \tag{1.62}$$

where ∇^*, Δ^*, etc. denote partial derivative operators with respect to the new spatial variable and Re is the *Reynolds number,* a dimensionless quantity given by

$$\mathrm{Re} = \frac{UL\,\rho_0}{\mu}.$$

It is well known that system (1.62) possesses at least one solution $\{\mathbf{u}, p\}$ for each $\mathrm{Re} > 0$; see for example [34]. For small Re, this solution is furthermore unique and coincides with the asymptotic limit of the similar time-dependent problem as $t \to +\infty$, independently of the prescribed initial data.

This is the mathematical realization of the laminar regime.

For large Re, a much more complex situation is found. On the basis of what is known in the ODE and PDE contexts, as Re grows it is expected that, in a first step, bifurcation phenomena appear, uniqueness is lost and several stationary solutions exist, with different stability and attractiveness properties.

This can be interpreted as an evidence of *transition to turbulence.*

For even larger Re, fully turbulent behavior is expected.

Turbulence is mainly characterized by *irregularity* (that is, lack of smoothness of the velocity field and the other variables), *rotationality* (i.e., vortex generation) and the occurrence of *energy cascades.* In fact, turbulent flows are always very irregular and this is why turbulence models rely on statistics; the three-dimensional vortex generation mechanism proper of turbulent flows is usually known as *vortex stretching.*

Usually, a hierarchy of eddies over a wide range of length scales can be found. Most of the kinetic energy of the turbulent motion is contained in the large-scale structures. The energy "travels" (and decays) from these large-scale structures to smaller scale structures by an inertial and essentially inviscid mechanism. This process continues, creating smaller and smaller structures; eventually, very small structures are created, molecular diffusion becomes important and viscous dissipation of energy finally takes place (the scale at which this happens is called the *Kolmogorov length scale*).

Some examples of turbulent flows are the following:

- The smoke rising from a cigarette,
- The external flow over a car, an airplane or a ship,
- The motion of matter in stellar atmospheres,
- The motion of blood in heart vessels, etc.

Note that an atmospheric cyclone is rotational but its shape does not allow vortex generation, whence it is not turbulent; on the other hand, oceanic flows are essentially irrotational and, consequently, not turbulent.

For the mathematical analysis of these phenomena, see for instance [11, 34].

It is usual to view a turbulent flow as the superposition of a mean flow (essentially regular) and a field of velocity fluctuations and eddies. Indeed, a crucial property of turbulent flows is that, usually, related *averaged* variables with good behavior can be defined.

Thus, let us assume for instance that the flow is, as in (1.62), stationary and let us omit the super-index $*$. Then, it makes sense to introduce a *mean velocity field* $\overline{\mathbf{u}}$ and a *mean pressure* $\overline{p}$, with

$$
\begin{aligned}
\overline{\mathbf{u}}(\mathbf{x}) = (\mathbb{E}\mathbf{u})(\mathbf{x}) &:= \lim_{\varepsilon\to 0} \frac{1}{V(\varepsilon)} \int_{B(\mathbf{x};\varepsilon)} \mathbf{u}(\mathbf{x}')\,d\mathbf{x}', \\
\overline{p}(\mathbf{x}) := (\mathbb{E}p)(\mathbf{x}) &:= \lim_{\varepsilon\to 0} \frac{1}{V(\varepsilon)} \int_{B(\mathbf{x};\varepsilon)} p(\mathbf{x}')\,d\mathbf{x}'
\end{aligned}
$$

($V(\varepsilon)$ denotes the volume of $B(\mathbf{x};\varepsilon)$) and try to find a PDE system for $\overline{\mathbf{u}}$ and $\overline{p}$.

Unfortunately, this is not easy: the nonlinear term in (1.62) prevents us to obtain equations for $\overline{\mathbf{u}}$ and $\overline{p}$ with the same structure. In fact, if we write

$$
\mathbf{u} = \overline{\mathbf{u}} + \mathbf{u}', \quad p = \overline{p} + p'
$$

and we assume that $\mathbb{E}$ and the partial derivatives can be exchanged, we find that

$$\begin{cases} \underbrace{(\overline{\mathbf{u}} \cdot \nabla)\overline{\mathbf{u}} + \mathbb{E}\left((\mathbf{u}' \cdot \nabla)\mathbf{u}'\right) - 1}_{\mathrm{Re}\Delta\overline{\mathbf{u}} + \nabla\overline{p} = 0} & \text{in } \Omega, \\ \nabla \cdot \overline{\mathbf{u}} = 0 & \text{in } \Omega, \\ \overline{\mathbf{u}} = \mathbb{E}\mathbf{a} & \text{on } \partial\Omega \end{cases}$$

and we need some extra information to express $\mathbb{E}\left((\mathbf{u}' \cdot \nabla)\mathbf{u}'\right)$ in terms of $\overline{\mathbf{u}}$. This is the first main goal of *turbulence modeling.*

It is usual to write that $\mathbb{E}\left((\mathbf{u}' \cdot \nabla)\mathbf{u}'\right) = -\nabla \cdot \mathbb{R}$, where $\mathbb{R} = -\mathbb{E}\left(\mathbf{u}' \otimes \mathbf{u}'\right)$ is the so called *Reynolds tensor;* for a more complete information on this theory, see for instance [22, 23].

In the most simple turbulence models, we just assume that $\mathbb{E}\left((\mathbf{u}' \cdot \nabla)\mathbf{u}'\right)$ can be approximated by a new diffusion term:

$$\mathbb{E}\left((\mathbf{u}' \cdot \nabla)\mathbf{u}'\right) \approx -\nabla \cdot (\nu_T D\overline{\mathbf{u}}), \quad \text{that is } R = \nu_T D\mathbf{u}, \tag{1.63}$$

where $\nu_T = \nu_T(\mathbf{x}, t)$ is given by an empirical law. The assumption (1.63) is known as *Boussinesq closure hypothesis* and the scalar ν_T is called the *turbulent (or eddy) viscosity.* More details and many other models can be found in [4, 21].

1.7 Exercises of Chapter 1

Exercise 1.1* Prove the identity (1.6).

HINT: Use that $\mathbf{X}(\mathbf{X}(\mathbf{x}, s; t), t; s) \equiv \mathbf{x}$ and, consequently, the derivatives with respect to t of both sides coincide.

Exercise 1.2* Prove Liouville's formula (1.10).

HINT: Use that the time derivative of $J(\mathbf{x}, t)$ can be written as a sum of determinants where we only differentiate each time one row. Also, use that the i-th determinant is equal to

$$\partial_i u_i(\mathbf{Y}(\mathbf{x}, t), t)\, J(\mathbf{x}, t) + K(\mathbf{x}, t),$$

where $K(\mathbf{x}, t)$ is the determinant of a singular matrix.

Exercise 1.3* Assume that $N = 2$, let the couple $(\mathbf{u}, p)$ satisfy (1.30) and let the functions $\psi = \psi(\mathbf{x}, t)$ and $\omega = \omega(\mathbf{x}, t)$ be such that

$$\mathbf{u} := \nabla \times \psi = (\partial_2\psi, -\partial_1\psi) \text{ and } \omega := \nabla \times \mathbf{u} = \partial_1 u_2 - \partial_2 u_1.$$

Rewrite (1.30) as a system of PDEs for ψ and ω (ψ is called a stream function and ω is the vorticity of the fluid).

HINT: Apply the curl operator $\nabla\times$ to the equation satisfied by $\mathbf{u}$ and p.

Exercise 1.4** Justify with the analysis of a fluid in a particular geometric situation that the tensor $\mathbf{s}$ in (1.47) satisfies (1.49).

HINT: First, consider the open set $W(a) = W_1(a) \cup W_2(a) \cup S(a)$, where

$$W_1(a) = \{\mathbf{x} \in \mathbb{R}^3 : |x_i| < a, \quad x_1 < 0\}, \quad W_2(a) = \{\mathbf{x} \in \mathbb{R}^3 : |x_i| < a, \quad x_1 > 0\}$$

and

$$S(a) = \{\mathbf{x} \in \mathbb{R}^3 : |x_i| < a, \quad x_1 = 0\},$$

write that

$$\mathbf{F}_{\rm ten}(W(a), t) = \int_{\partial W_1(a)} \mathbf{s}(\mathbf{n}; \mathbf{x}, t)\, d\Gamma + \int_{\partial W_2(a)} \mathbf{s}(\mathbf{n}; \mathbf{x}, t)\, d\Gamma$$

and also

$$\mathbf{F}_{\rm ten}(W(a), t) = \int_{\partial W(a)} \mathbf{s}(\mathbf{n}; \mathbf{x}, t)\, d\Gamma$$

and take limits as $a \to 0$. It is found that $\mathbf{s}(-\mathbf{e}^1; \mathbf{0}, t) = -\mathbf{s}(\mathbf{e}^1; \mathbf{0}, t)$; an adapted argument shows that we also have $\mathbf{s}(-\mathbf{n}; \mathbf{x}, t) = -\mathbf{s}(\mathbf{n}; \mathbf{x}, t)$ for any $\mathbf{n}$ and any $\mathbf{x}$. Then, consider a tetrahedron $Q(a_1, a_2, a_3)$ with vertices $(0, 0, 0)$, $(a_1, 0, 0)$, $(0, a_2, 0)$ and $(0, 0, a_3)$, where the $a_\ell > 0$; write that

$$\mathbf{F}_{\rm ten}(Q(a_1, a_2, a_3), t) = \int_{A_0} \mathbf{s}(\mathbf{n}; \mathbf{x}, t)\, d\Gamma + \sum_{\ell=1}^{3} \int_{A_\ell} \mathbf{s}(\mathbf{n}; \mathbf{x}, t)\, d\Gamma$$

(where A_0 and the A_ℓ are the faces respectively opposed to the origin and the other vertices) and use that $\mathbf{F}_{\rm ten}(Q(a_1, a_2, a_3), t) \to \mathbf{0}$ as $a_1, a_2, a_3 \to 0$. In the limit, one finds that

$$\mathbf{s}(\sum_{\ell=1}^{3} \alpha_\ell \mathbf{e}^\ell; \mathbf{0}, t) = \sum_{\ell=1}^{3} \alpha_\ell\, \mathbf{s}(\mathbf{e}^\ell; \mathbf{0}, t)$$

for some α_ℓ that only depend on the a_i. A similar identity can be found at any other point $\mathbf{x} \neq \mathbf{0}$.

Exercise 1.5* Using (1.50), Lemma 1.1 and (1.17), prove that σ is symmetric.

HINT: One has

$$\begin{aligned}\frac{d}{dt} \int_{W_{[t]}} (\rho\mathbf{u}) \times \mathbf{x} &= \int_{W_{[t]}} ((\rho\mathbf{u})_t \times \mathbf{x} + (\mathbf{u} \cdot \nabla)((\rho\mathbf{u}) \times \mathbf{x})) \\ &= \int_{W_{[t]}} ((\rho\mathbf{f} + \nabla \cdot \sigma - \nabla \cdot (\rho\mathbf{u} \otimes \mathbf{u})) \times \mathbf{x} + (\mathbf{u} \cdot \nabla)((\rho\mathbf{u}) \times \mathbf{x}))\,.\end{aligned}$$

Compare with

$$\int_{W_{[t]}} \rho \mathbf{f} \times \mathbf{x} + \int_{\partial W_{[t]}} (\sigma \cdot \mathbf{n} \times \mathbf{x})\, d\Gamma$$

and deduce that $\sigma = \sigma^T$.

Exercise 1.6** Prove Rivlin-Ericksen's Theorem (Theorem 1.1) for $N = 3$.

HINT: First, prove that, for any function $f : S(\mathbb{R}^3) \mapsto \mathbb{R}$ satisfying

$$f(R \cdot D \cdot R^{-1}) = f(D) \quad \forall D \in S(\mathbb{R}^3),\ \forall R \in Q(\mathbb{R}^3),$$

there exists $g : \mathbb{R}^3 \mapsto \mathbb{R}$ such that $f(D) = g(\operatorname{tr} D, \operatorname{tr} D^2, \det D)$ for all $D \in S(\mathbb{R}^3)$. Then, prove that there exists $R \in Q(\mathbb{R}^3)$ such that $R \cdot \Sigma(D) \cdot R^{-1}$ and $R \cdot D \cdot R^{-1}$ are both diagonal and the components s_i of $R \cdot \Sigma(D) \cdot R^{-1}$ are functions of the components of $R \cdot D \cdot R^{-1}$, that is, the eigenvalues λ_i of D. To conclude, it suffices to prove that, for some functions $f_j = f_j(\lambda_1, \lambda_2, \lambda_3)$ $(0 \leq j \leq 2)$, one has

$$s_i = f_0(\lambda_1, \lambda_2, \lambda_3) + f_1(\lambda_1, \lambda_2, \lambda_3)\lambda_i + f_2(\lambda_1, \lambda_2, \lambda_3)\lambda_i^2 \quad \text{for} \quad 1 \leq i \leq 3$$

since, if this is the case, the f_j will only depend of $\operatorname{tr} D$, $\operatorname{tr} D^2$ and $\det D$. But the existence of the f_j can be established by solving appropriate linear (Vandermonde) systems.

Exercise 1.7* Using (1.52), deduce that the PDEs satisfied by a viscous fluid for which the stress tensor depends on the spatial derivatives of the velocity (linearly and) quadratically is (1.58) (this is called a second-grade fluid).

Exercise 1.8* Deduce the Stokes law (1.54) for a viscous Newtonian fluid.

HINT: Consider a fluid in an annulus governed by the velocity field $\mathbf{u}(\mathbf{x}, t) \equiv a\mathbf{x}$ and impose that the associated tensor of viscous forces vanishes, that is, $\lambda^*(\nabla \cdot \mathbf{u})\mathrm{Id.} + \mu^*(\nabla \mathbf{u} + \nabla \mathbf{u}^T) = 0$.

Exercise 1.9* Prove that, if $\Omega \subset \mathbb{R}^N$ is a non-empty open set, $T > 0$, $(\mathbf{x}, t) \in Q := \Omega \times (0, T)$, $h > 0$ is given and sufficiently small, θ is (for instance) of class C^2 in a neighborhood of $(\mathbf{x}, t)$ and $\nabla\theta(\mathbf{x}, t) \neq 0$, the direction vector of size h that produces the steepest spatial descent of θ at $(\mathbf{x}, t)$ is $\hat{\mathbf{d}} = -(h/|\nabla\theta(\mathbf{x}, t)|)\, \nabla\theta(\mathbf{x}, t)$.

HINT: Use Taylor expansions of θ at points of the form $(\mathbf{x} - h\mathbf{d}, t)$ with $|\mathbf{d}| = 1$ and $(\mathbf{x} - h\hat{\mathbf{d}}, t)$.

References

1. Bambach, C.C.: Leonardo da Vinci Rediscovered. Vol. 1, The Making of an Artist: 1452–1500. Yale University Press, New Haven (2019)
2. Bothe, D., Köhne, M., Prüss, J.: On a class of energy preserving boundary conditions for incompressible newtonian flows. SIAM J. Math. Anal. **45**, 3768–3822 (2013)
3. Brackenridge, J.B.: The Key to Newton's Dynamics. University of California Press, Berkeley, CA (1995). The Kepler problem and the ıt Principia, Containing English translations from the Latin by Mary Ann Rossi of Sections 1, 2 and 3 of book one from the first (1687) edition of Newton's ıt Mathematical principles of natural philosophy
4. Chacón-Rebollo, T., Lewandowski, R.: Mathematical and Numerical Foundations of Turbulence Models and Applications. Modeling and Simulation in Science, Engineering and Technology. Birkhäuser/Springer, New York (2014)
5. Chorin, A.J., Marsden, J.E.: A Mathematical Introduction to Fluid Mechanics, volume 4 of Texts in Applied Mathematics, 3rd edn. Springer, New York (1993)
6. Cioranescu, D., Girault, V., Rajagopal, K.R.: Mechanics and Mathematics of Fluids of the Differential Type, volume 35 of Advances in Mechanics and Mathematics. Springer, Cham (2016)
7. Cullen, M.: The Mathematics of Large-Scale Atmosphere and Ocean. World Scientific Publishing, Hackensack, NJ (2021)
8. Darrigol, O., Frisch, U.: From Newton's mechanics to Euler's equations. Phys. D **237**(14–17), 1855–1869 (2008)
9. Driver, R.D.: Torricelli's law—an ideal example of an elementary ODE. Amer. Math. Mon. **105**(5), 453–455 (1998)
10. Faber, T.E.: Fluid Dynamics for Physicists. Cambridge University Press, Cambridge (1997). Reprint of the 1995 original
11. Foias, C., Manley, O., Rosa, R., Témam, R.: Navier-Stokes Equations and Turbulence, volume 83 of Encyclopedia of Mathematics and its Applications. Cambridge University Press, Cambridge (2001)
12. Fradlin, B.N., Birštein, V.M.: M. V. Lomonosov's letter to L. Euler dated 5 July 1748 and L. Euler's reply dated 24 August 1748. In: History and Methodology of the Natural Sciences, No. XX (Russian), pp. 22–26. Moskov. Gos. Univ., Moscow (1978)
13. Galdi, G.P.: An Introduction to the Mathematical Theory of the Navier-Stokes Equations. Steady State Problems. Springer, New York (2011)
14. Grattan-Guinness, I.: Daniel Bernoulli and the varieties of mechanics in the 18th century. Nieuw Arch. Wiskd. (5) **1**(3), 242–249 (2000)
15. Gray, J.: Change and Variations—A History of Differential Equations to 1900. Springer Undergraduate Mathematics Series. Springer, Cham (2021). ©2021
16. Hecht, F.: New developments in freefem++. J. Numer. Math. **20**(3–4), 251–265 (2012)
17. Henshaw, J.M.: An Equation for Every Occasion. Johns Hopkins University Press, Baltimore, MD (2014). Fifty-two formulas and why they matter, With contributions from Steven Lewis
18. Irgens, F.: Rheology and Non-Newtonian Fluids. Springer, Cham (2014)
19. Joseph, D.D.: Fluid Dynamics of Viscoelastic Liquids, volume 84 of Applied Mathematical Sciences. Springer, New York (1990)
20. Knudsen, O.: Electromagnetic energy and the early history of the energy principle. In: No Truth Except in the Details, volume 167 of Boston Stud. Philos. Sci., pp. 55–78. Kluwer Acad. Publ., Dordrecht (1995)
21. Launder, B.E., Reynolds, W.C., Rodi, W.: Turbulence Models and Their Applications. Vol. 2, volume 56 of Collection de la Direction des Études et Recherches d'Électricité de France ESE [Collection of the Department of Studies and Research of Électricité de France]. Éditions Eyrolles, Paris (1984). With a French summary
22. Lesieur, M.: Turbulence in Fluids, volume 40 of Fluid Mechanics and its Applications, 3rd edn. Kluwer Academic Publishers Group, Dordrecht (1997)

23. Manneville, P.: An introduction to nonlinear dynamics and complex systems. In: Instabilities, Chaos and Turbulence. Imperial College Press, London (2004).
24. Mikhailov, G.K.: Leonhard Euler and his contribution to the development of rational mechanics. Adv. Mech. **8**(1), 3–58 (1985)
25. Panton, R.L.: Incompressible Flow. A Wiley-Interscience Publication. Wiley, New York (1984)
26. Pérez-García, V.M., et al.: Delay effects in the response of low-grade gliomas to radiotherapy: a mathematical model and its therapeutical implications. Math. Med. Biol. **32**(3), 307–329 (2015)
27. Pourciau, B.: Newton's interpretation of Newton's second law. Arch. Hist. Exact Sci. **60**(2), 157–207 (2006)
28. Rempfer, F.: On boundary conditions for incompressible Navier-Stokes problems. Appl. Mech. Rev. **59**, 107–126 (2006)
29. Renardy, M., Hrusa, W.J., Nohel, J.A.: Mathematical Problems in Viscoelasticity, volume 35 of Pitman Monographs and Surveys in Pure and Applied Mathematics. Longman Scientific & Technical, Harlow (1987)
30. Sani, R.L., Shen, J., Pironneau, O., Gresho, P.M.: Pressure boundary condition for the time-dependent incompressible Navier-Stokes equations. Int. J. Numer. Methods Fluids **50**(6), 673–682 (2006)
31. Smith, D.E.: Recent publications: reviews: Pascal, The life of genius. Amer. Math. Mon. **44**(5), 325–327 (1937)
32. Smith, C.: A new chart for British natural philosophy: the development of energy physics in the nineteenth century. Hist. Sci. **16**(34, part 4), 231–279 (1978)
33. Stein, E.: The History of Theoretical, Material and Computational Mechanics-mathematics Meets Mechanics and Engineering, volume 1 of Lecture Notes in Applied Mathematics and Mechanics. Springer, Heidelberg (2014). Including lectures from the Association for Applied Mathematics and Mechanics (GAMM) Special Sessions held in Karlsruhe in 2010, Graz in 2011 and Darmstadt in 2012
34. Témam, R.: Infinite-dimensional Dynamical Systems in Mechanics and Physics, volume 68 of Applied Mathematical Sciences, 2nd edn. Springer, New York (1997)
35. Tryggvason, G., Scardovelli, R., Zaleski, S.: Direct Numerical Simulations of Gas-liquid Multiphase Flows. University Press, Cambridge (2011)
36. Vasari, G.: The Lives of the Artists (Oxford World's Classics) Reissue edition. Oxford University Press, Oxford, UK (2008)
37. Zeidler, E.: Nonlinear Functional Analysis and Its Applications. IV. Springer, New York (1988). Applications to mathematical physics, Translated from the German and with a preface by Juergen Quandt

Chapter 2
Mathematical Background

In this chapter, we recall concepts and results required for the analysis of the PDEs of fluid mechanics and, in particular, the variable density Navier-Stokes equations.

We will review the main properties of integrable functions and distributions. We will also recall the basic properties of the Stokes operator and other tools needed in the following chapters.

2.1 Notation and Preliminary Results

In this section, we recall some fundamental concepts and results used in PDE theory. We will usually provide explanations and/or interpretations. In almost all cases, the proofs will be given either explicitly in the text or suggested as exercises. More details on the results presented here can be found, for instance, in [1, 16, 29, 32, 34].

In the sequel, it will be assumed that $\Omega \subset \mathbb{R}^N$ is a non-empty (bounded or not) connected open set, with $N \leq 3$. We will frequently suppose that the boundary $\partial\Omega$ is regular enough, which usually means Lipschitz-continuous (and maybe something more, in some of the results that follow). This will be specified when needed.

For any continuous function $\phi : \Omega \mapsto \mathbb{R}$, the support of ϕ is by definition the set

$$\text{supp}(\phi) := \overline{\{\mathbf{x} \in \Omega : \phi(\mathbf{x}) \neq 0\}}.$$

We will denote by $C_c^\infty(\Omega)$ or $\mathcal{D}(\Omega)$ the linear space formed by all the C^∞ functions $\phi : \Omega \mapsto \mathbb{R}$ with compact support in Ω.

We will denote by $\mathcal{L}(\Omega)$ the set of Lebesgue-measurable functions on Ω.[1]

[1] Henri Léon Lebesgue (1875–1941) was a French mathematician. His most famous results concern measure theory and integration, although he also made contributions in other fields. In 1902 he defined a new concept of integral (the Lebesgue integral), generalizing the classical notion and

P. Braz e Silva et al., *Analysis and Control of the Variable Density Incompressible Navier-Stokes Equations*, MS&A 22, https://doi.org/10.1007/978-3-032-14510-9_2

Clearly, this is a (real) vector space with respect to the usual laws: the sum of functions in $\mathcal{L}(\Omega)$ and the product of a real number by a function in $\mathcal{L}(\Omega)$ belong to $\mathcal{L}(\Omega)$.

Recall that, if $v : \Omega \mapsto \mathbb{R}$ is measurable and nonnegative, we can always define the *integral* of v with respect to the Lebesgue measure in Ω. It is either a nonnegative real number or $+\infty$. It will be denoted in this book by

$$\int_\Omega v(\mathbf{x})\,d\mathbf{x}, \quad \int_\Omega v\,d\mathbf{x} \quad \text{or simply} \quad \int_\Omega v.$$

By definition, one has

$$\int_\Omega v := \sup\{I(z) : z \in \mathcal{Z}(\Omega),\ 0 \le z \le v\},$$

where:

- $\mathcal{Z}(\Omega)$ denotes the set of measurable *step functions,* that is, functions of the form $z = \sum_{i=1}^{I} z_i \mathbb{1}_{A_i}$ with $I \ge 1$, $z_i \in \mathbb{R}$ and the sets $A_i \subset \Omega$ Lebesgue-measurable and
- For any such z, we have used the notation $I(z) := \sum_{i=1}^{I} z_i |A_i|$, where $|A_i|$ is the measure of A_i.

If $v : \Omega \mapsto \mathbb{R}$ is measurable but not necessarily nonnegative, we can consider the nonnegative measurable functions $v_+ := \max(v, 0)$ and $v_- := -\min(v, 0)$ and their integrals

$$\int_\Omega v_+ \quad \text{and} \quad \int_\Omega v_-.$$

If both are finite, it is said that v is *integrable.* In that case, the integral of v is by definition the quantity

$$\int_\Omega v := \int_\Omega v_+ - \int_\Omega v_-.$$

making it possible to compute "areas" bounded by nonsmooth curves. The basic idea was to identify the sets of points where a given function takes prescribed values and then measure the sizes of these sets. This is completely different from assigning values to prescribed sets, as is done when one considers the Riemann integral. The Lebesgue integral played a key role in the motivation and growth of functional analysis. It is also fundamental in Fourier analysis, probability calculus and other areas. However, it seems that Lebesgue was not aware of the relevance and usefulness of his discovery; he conceived it as a purely theoretical object; see more details in [10].

The integration process has the following basic properties:

- Let $\mathcal{L}^1(\Omega)$ be the set of measurable and integrable functions on Ω. Then, $\mathcal{L}^1(\Omega)$ is a linear subspace of $\mathcal{L}(\Omega)$ and $v \mapsto \int_\Omega v$ is a linear form on $\mathcal{L}^1(\Omega)$.
- If $u, v \in \mathcal{L}(\Omega)$, the integral of one of them makes sense and $u = v$ *almost everywhere* (a.e.) in Ω, then the integral of the other one is also meaningful and both integrals coincide.
- If $u \in \mathcal{L}(\Omega)$, $u \geq 0$ a.e. in Ω and $\int_\Omega u = 0$, then $u = 0$ a.e.
- If $u \in \mathcal{L}^1(\Omega)$, then

$$\left| \int_\Omega u \right| \leq \int_\Omega |u|. \tag{2.1}$$

All the integrals mentioned in the sequel, unless explicitly stated, are carried out with respect to the Lebesgue measure. Moreover, very frequently the integration element $d\mathbf{x}$ will be omitted.

The following are fundamental properties of integrable functions:

- *Lebesgue's Dominated Convergence Theorem:* If $\{v_n\}$ is a sequence in the space $\mathcal{L}^1(\Omega)$ that converges pointwise a.e. in Ω to v and there exists $w \in \mathcal{L}^1(\Omega)$ such that $\sup_n |v_n(\mathbf{x})| \leq w(\mathbf{x})$ a.e., then $v \in \mathcal{L}^1(\Omega)$ and

$$\lim_{n \to +\infty} \int_\Omega |v_n - v| = 0.$$

In particular,

$$\lim_{n \to +\infty} \int_\Omega v_n = \int_\Omega v.$$

- *Fubini's Theorem:* Let $\Omega \subset \mathbb{R}^N$ and $D \subset \mathbb{R}^M$ be non-empty open sets, let $f : \Omega \times D \mapsto \mathbb{R}$ be given and assume that $f \in \mathcal{L}^1(\Omega \times D)$, that is, f is measurable and integrable with respect to the product Lebesgue measure in $\Omega \times D$. Then the following holds:

  1. For $\mathbf{x}$ a.e. in Ω, the function $\mathbf{y} \mapsto f(\mathbf{x}, \mathbf{y})$ is measurable and integrable in D.
  2. The function $\mathbf{x} \mapsto \int_D f(\mathbf{x}, \mathbf{y})\, d\mathbf{y}$, which is well defined a.e. in Ω, is measurable and integrable.
  3. One has

$$\int_\Omega \left(\int_D f(\mathbf{x}, \mathbf{y})\, d\mathbf{y} \right) d\mathbf{x} = \int_{\Omega \times D} f(\mathbf{x}, \mathbf{y})\, d(\mathbf{x}, \mathbf{y}).$$

Obviously, a similar result holds exchanging the roles of $\mathbf{x}$ and $\mathbf{y}$. On the other hand, the assumption $f \in \mathcal{L}^1(\Omega \times D)$ can be replaced by this one: $(\mathbf{x}, \mathbf{y}) \mapsto$

$f(\mathbf{x}, \mathbf{y})$ is measurable and nonnegative. This result is known as *Tonelli's Theorem.*

For the proofs, see for instance [16].

An elementary application of Fubini's Theorem is the following: let us take $n = m = 1$, $\Omega = D = (0, a)$ with $a > 0$ and $f(x, y) = f_0(x, y)\mathbb{1}_{\{y<x\}}$ for some $f_0 \in \mathcal{L}^1((0, a) \times (0, a))$; then

$$\iint_{(0,a)\times(0,a)} f(x, y)\, d(x, y) = \int_0^a \left(\int_0^x f_0(x, y)\, dy \right) dx$$
$$= \int_0^a \left(\int_y^a f_0(x, y)\, dx \right) dy.$$

In this chapter and the following ones, we will work with classes of measurable functions determined by the following equivalence relation in $\mathcal{L}(\Omega)$:

$$u \sim v \Leftrightarrow u(\mathbf{x}) = v(\mathbf{x}) \text{ a.e. in } \Omega.$$

For brevity, we will frequently refer to them as *functions.*

Obviously, the sum of these "functions" and the product of any of them by a real number are well defined operations and provide new functions of the same kind. Accordingly, we can speak of the linear space of (classes of) measurable functions on Ω.

Also, if $v \in \mathcal{L}^1(\Omega)$, we can speak of the integral of the associated class, that is by definition give by $\int_\Omega v$.

For any p satisfying $1 \le p < +\infty$, we will denote by $L^p(\Omega)$ the set of (classes of) measurable functions $v : \Omega \mapsto \mathbb{R}$ such that $|v|^p$ is integrable, i.e., satisfy

$$\|v\|_{L^p} := \left(\int_\Omega |v|^p \right)^{1/p} < +\infty.$$

On the other hand, for any function $v : \Omega \mapsto \mathbb{R}$, the *essential supremum* of v is the quantity

$$\text{ess sup}_\Omega(v) = \inf\{M : v(\mathbf{x}) \le M \text{ a.e. in } \Omega\}.$$

We will denote by $L^\infty(\Omega)$ the set of (classes of) measurable functions essentially bounded in Ω, that is, satisfying

$$\|v\|_{L^\infty} := \text{ess sup}_\Omega |v| < +\infty.$$

It is well known that, for any $p \in [1, +\infty]$, $L^p(\Omega)$ is a linear space for the usual algebraic rules and the real-valued mapping $v \mapsto \|v\|_{L^p}$ is a norm in $L^p(\Omega)$.

For $p = 1$ and $p = +\infty$, this is rather obvious. For $1 < p < +\infty$, the key tool is the so called *Minkowski's inequality*

$$\left(\int_\Omega |u+v|^p\right)^{1/p} \le \left(\int_\Omega |u|^p\right)^{1/p} + \left(\int_\Omega |v|^p\right)^{1/p} \quad \forall u, v \in L^p(\Omega). \tag{2.2}$$

In turn, this is a consequence of the *Hölder's inequality:* if p' is the *conjugate* to p, i.e., the unique exponent satisfying

$$1 < p' < +\infty, \quad \frac{1}{p} + \frac{1}{p'} = 1,$$

then

$$\int_\Omega |u\, v| \le \left(\int_\Omega |u|^p\right)^{1/p} \left(\int_\Omega |v|^{p'}\right)^{1/p'} \quad \forall u \in L^p(\Omega),\ \forall v \in L^{p'}(\Omega). \tag{2.3}$$

For any $p \in [1, +\infty)$, endowed with the norm $\|\cdot\|_{L^p}$, $L^p(\Omega)$ is a *separable Banach space.* This means the following:

- As a metric space, $L^p(\Omega)$ is *complete,* i.e., any Cauchy sequence in $L^p(\Omega)$ converges in this space. This is a consequence of *Lebesgue's Dominated Convergence Theorem;* see, for instance, [16] for a proof and more details.
- As a topological space, it is *separable,* i.e., there exists a set $E \subset L^p(\Omega)$ that is *dense* and *countable.* For example, if Ω is bounded, a set E with these properties is the family of polynomial functions on Ω with rational coefficients.

In particular, when $p = 2$, $L^2(\Omega)$ is a separable Hilbert space, since $\|\cdot\|_{L^2}$ is induced by the scalar product

$$(u, v) := \int_\Omega u\, v$$

(this notation will be kept in the sequel).

On the other hand, endowed with the norm $\|\cdot\|_{L^\infty}$, $L^\infty(\Omega)$ is a (non-separable) Banach space.

Note that, when $p = 1$, one has $p' = +\infty$ and Hölder inequality becomes

$$\int_\Omega |u\, v| \le \left(\int_\Omega |u|\right) \left(\operatorname{ess\,sup}_\Omega |v|\right) \quad \forall u \in L^1(\Omega),\ \forall v \in L^\infty(\Omega).$$

Under appropriate regularity assumptions on $\partial\Omega$, together with the $L^p(\Omega)$, we can introduce the spaces $L^p(\partial\Omega)$, where measurability and integrability holds with respect to the *surface measure* on $\partial\Omega$.

For completeness, the process is described in Sect. 2.6.

Let $v : \Omega \mapsto \mathbb{R}$ be a measurable function. It will be said that v is *locally integrable in* Ω if, for every bounded open set $\omega \subset\subset \Omega$, the restriction of v to ω is integrable. The set of locally integrable functions in Ω will be denoted by $\mathcal{L}^1_{\text{loc}}(\Omega)$ and the corresponding space of classes of functions will be denoted by $L^1_{\text{loc}}(\Omega)$.

In a similar way, for any p with $1 < p \leq +\infty$, we can speak of the spaces $\mathcal{L}^p_{\text{loc}}(\Omega)$ and $L^p_{\text{loc}}(\Omega)$.

Recall that $\mathcal{D}(\Omega)$ (resp. $\mathcal{D}(\overline{\Omega})$) stands for the subspace of $C^\infty(\Omega)$ (resp. $C^\infty(\overline{\Omega})$) formed by the functions with compact support in Ω (resp. $\overline{\Omega}$). Actually, the spaces $\mathcal{D}(\Omega)$ and $L^1_{\text{loc}}(\Omega)$ must be respectively viewed as the smallest and biggest function spaces that have interest in the PDE context. All the interesting spaces of real-valued (classes of) functions defined in Ω satisfy

$$\mathcal{D}(\Omega) \subset X(\Omega) \subset L^1_{\text{loc}}(\Omega). \tag{2.4}$$

For every $p \in [1, +\infty]$, $L^p_{\text{loc}}(\Omega)$ is a *Fréchet space* (i.e., a complete metrizable locally convex space) for the following family of seminorms:

$$\pi_{p,\omega}(v) = \|v\|_{L^p(\omega)} \quad \forall v \in L^p_{\text{loc}}(\Omega), \quad \forall \omega \in \mathcal{K}(\Omega),$$

where $\mathcal{K}(\Omega)$ denotes the family of all bounded open sets $\omega \subset\subset \Omega$. Moreover, if $p < +\infty$, $L^p_{\text{loc}}(\Omega)$ is separable.

Note that, in $L^p_{\text{loc}}(\Omega)$, a sequence $\{v_n\}$ converges to a function v if (and only if) one has

$$\|v_n - v\|_{L^p(\omega)} \to 0 \quad \forall \omega \in \mathcal{K}(\Omega).$$

It is clear that the $L^p_{\text{loc}}(\Omega)$ are ordered algebraically and topologically by p. More precisely, one has

$$L^p_{\text{loc}}(\Omega) \hookrightarrow L^q_{\text{loc}}(\Omega) \quad \forall p, q \in [1, +\infty] \text{ with } p \geq q,$$

where the embeddings are continuous.

2.2 Real-Valued Distributions, Sobolev Spaces and Their Properties

The first goal of this section is to recall the definition of real-valued distributions, to explain the role they play in the context of PDEs and to review their main properties.

The foundations of the theory are mainly due to Sergei L. Sobolev[2] and Laurent Schwartz.[3]

In a second part of the section, we will recall the definition and basic properties of Sobolev spaces.

First of all, let us endow the linear space $\mathcal{D}(\Omega)$ with the following notion of convergence: it will be said that a sequence $\{\phi_n\}$ converges to a function ϕ in $\mathcal{D}(\Omega)$ if

- There exists a compact set $K \subset \Omega$ such that $\text{supp}\, \phi_n \subset K$ for all n.
- For any multi-index $\alpha = (\alpha_1, \ldots, \alpha_N)$, one has $\partial^\alpha \phi_n \to \partial^\alpha \phi$ uniformly in K.

In that case, we write that $\phi_n \to \phi$ in $\mathcal{D}(\Omega)$.

Usually, $\mathcal{D}(\Omega)$ is called the space of *test functions* on Ω.

There exists exactly one topology on $\mathcal{D}(\Omega)$ which is *compatible* with its linear structure such that the associated convergence notion is the previous one; see [21]. A simple and very intuitive description of this topology is given in [29] in terms of a suitable family of seminorms.

Let us recall some properties of $\mathcal{D}(\Omega)$. First of all, let us see that $\mathcal{D}(\Omega)$ is not the trivial space:

Lemma 2.1 *Let* $\mathbf{x}_0 \in \mathbb{R}^N$ *and* $r > 0$ *be given. There exist functions* $\phi \in \mathcal{D}(\mathbb{R}^N)$ *such that* $\phi > 0$ *in* $B(x_0; r)$ *and* $\text{Supp}\,(\phi) = \overline{B}(x_0; r)$.

The proof is immediate: let ψ be given by

$$\psi(s) := e^{1/s} \mathbb{1}_{\{s<0\}} \quad \forall s \in \mathbb{R}.$$

Then, $\psi \in C^\infty(\mathbb{R})$ and, in order to prove the lemma, it suffices to take

$$\phi(\mathbf{x}) = \psi(|\mathbf{x} - \mathbf{x}_0|^2 - r^2) \quad \forall \mathbf{x} \in \mathbb{R}^N.$$

---

[2] Sergei Lvovich Sobolev (1908–1989) was a Russian mathematician. He introduced notions that are now fundamental in several areas of mathematics. In particular, he was the first to speak of and use generalized functions (later known as distributions). His main motivations were the study of Cauchy problems for hyperbolic PDEs and the analysis of discontinuous solutions to the equations in elasticity theory. The theory of distributions must be regarded nowadays as a fundamental tool to study differential problems. The ideas and methods of Sobolev, further developed by Laurent Schwartz, were studied intensively and gained applications in ODEs and PDEs and, more generally, lots of models in mathematical physics and computational mathematics. Sobolev extended the classical notion of differentiation to distributions, expanding the range of applications of the classical techniques introduced by Newton and Leibniz. His embedding and trace theorems have become some of the most important results in this context. This will become clear in the following sections and chapters. See [35] for more details.

[3] Laurent Schwartz (1915–2002) was a French mathematician. His main achievement was the rigorous formulation of the theory of distributions, giving a well-defined meaning for objects such as the *Dirac delta function,* extending the definition and properties of the *Fourier transform* and leading to fundamental results in the theory of PDEs. He was awarded the Fields Medal in 1950. More details can be found in [19].

For any $A \subset \mathbb{R}^N$ and $\delta > 0$, we will set $A_\delta := \{\mathbf{x} \in \mathbb{R}^N : \text{dist}(\mathbf{x}, A) \le \delta\}$. Then A_δ coincides with the union of all closed balls $\overline{B}(\mathbf{x}; \delta)$ with $\mathbf{x} \in A$ and one has the following:

Proposition 2.1 *Let $K \subset \Omega$ be a non-empty compact set. For any sufficiently small $\delta > 0$, there exist functions $\phi \in \mathcal{D}(\Omega)$ such that $0 \le \phi \le 1$ and $\phi = 1$ in K_δ.*

Proof First, let $\zeta \in \mathcal{D}(\mathbb{R}^N)$ be such that $\zeta > 0$ in $B(0; 1)$ and $\zeta = 0$ outside $B(0; 1)$. In view of Lemma 2.1, such a function ζ exists.

After multiplication by a positive constant (if this is needed), it can be assumed that

$$\int_{\mathbb{R}^N} \zeta = 1. \tag{2.5}$$

Let $\varepsilon > 0$ be given and let us set

$$\zeta_\varepsilon(\mathbf{x}) := \varepsilon^{-N} \zeta(\varepsilon^{-1}\mathbf{x}) \quad \forall \mathbf{x} \in \mathbb{R}^N. \tag{2.6}$$

It will be said that $\{\zeta_\varepsilon\}$ is a *regularizing sequence* in $\mathbb{R}^N$.

Note that, if $\delta > 0$ is sufficiently small, K_δ is non-empty. Let us set

$$\varphi_\varepsilon(\mathbf{x}) := \int_{K_{2\varepsilon}} \zeta_\varepsilon(\mathbf{x} - \mathbf{y})\, d\mathbf{y} \quad \forall \mathbf{x} \in \Omega.$$

Then, if ε is sufficiently small, φ_ε fulfills the desired properties. Indeed, φ_ε is well defined and C^∞ in Ω and

$$0 \le \varphi_\varepsilon(\mathbf{x}) \le \int_{\mathbb{R}^N} \zeta_\varepsilon(\mathbf{x} - \mathbf{y})\, d\mathbf{y} = 1 \quad \forall \mathbf{x} \in \Omega.$$

For $\varepsilon > 0$ small enough, $K_{3\varepsilon}$ is a compact subset of Ω. On the other hand, if $\mathbf{x} \in K_\varepsilon$, one has

$$\varphi_\varepsilon(\mathbf{x}) = \int_{\mathbb{R}^N} \zeta_\varepsilon(\mathbf{x} - \mathbf{y}) \mathbb{1}_{K_{2\varepsilon}}(\mathbf{y})\, d\mathbf{y} = \int_{\mathbb{R}^N} \zeta_\varepsilon(\mathbf{y}) \mathbb{1}_{K_{2\varepsilon}}(\mathbf{x} - \mathbf{y})\, d\mathbf{y} = 1,$$

while, if $\mathbf{x} \notin K_{3\varepsilon}$, the ball $B(\mathbf{x}; \varepsilon)$ and $K_{2\varepsilon}$ are disjoint and

$$\varphi_\varepsilon(\mathbf{x}) = \int_{\mathbb{R}^N} \zeta_\varepsilon(\mathbf{y}) \mathbb{1}_{K_{2\varepsilon}}(\mathbf{x} - \mathbf{y})\, d\mathbf{y} = 0.$$

Therefore, in order to achieve the proof, it suffices to take $\varepsilon = \delta/3$ and $\phi := \varphi_\varepsilon$. □

Theorem 2.1 *For any $p \in [1, +\infty)$, the space $\mathcal{D}(\Omega)$ is dense in $L^p(\Omega)$.*

Proof Let $p \in [1, +\infty)$ and $v \in L^p(\Omega)$ be given. We will prove that, for each $\kappa > 0$, there exists a function $\varphi \in \mathcal{D}(\Omega)$ satisfying

$$\|v - \varphi\|_{L^p} \leq \kappa. \tag{2.7}$$

First, for any $\delta > 0$ we denote by $K(\delta)$ the compact set

$$K(\delta) := \{\mathbf{x} \in \Omega : |\mathbf{x}| \leq \frac{1}{\delta}, \quad \text{dist}\ (\mathbf{x}, \partial\Omega) \geq \delta\}.$$

Set $v_\delta := v\, \mathbb{1}_{K(\delta)}$. Then

$$\|v - v_\delta\|_{L^p} = \left(\int_\Omega |v|^p\, \mathbb{1}_{\Omega \setminus K(\delta)}\, d\mathbf{x} \right)^{1/p}$$

converges to zero as $\delta \to 0$, in view of Lebesgue's Dominated Convergence Theorem. Therefore, by choosing δ small enough, we get

$$\|v - v_\delta\|_{L^p} \leq \frac{\kappa}{2}. \tag{2.8}$$

Now, let δ be fixed such that inequality (2.8) holds and let us take $K = K(\delta)$ in Proposition 2.1.

For each $\varepsilon > 0$, we consider the set $K_\varepsilon := \{\mathbf{x} \in \Omega : \text{dist}\ (\mathbf{x}, K) \leq \varepsilon\}$ and the function w_ε given by

$$w_\varepsilon(\mathbf{x}) := \int_{\mathbb{R}^N} \zeta_\varepsilon(\mathbf{x} - \mathbf{y})\, v_\delta(\mathbf{y})\, d\mathbf{y},$$

where $\{\zeta_\varepsilon\}$ is the regularizing sequence introduced in the proof.

Then, if ε is sufficiently small, the function $\varphi = w_\varepsilon$ satisfies the desired properties.

Indeed, the value of w_ε at $\mathbf{x}$ is well defined for all $\mathbf{x} \in \Omega$ and $w_\varepsilon \in C^\infty(\Omega)$. Moreover, K_ε is a compact subset of Ω. If $\mathbf{x} \notin K_\varepsilon$, the ball $B(\mathbf{x}; \varepsilon)$ does not intersect K and

$$w_\varepsilon(\mathbf{x}) = \int_{\mathbb{R}^N} \zeta_\varepsilon(\mathbf{y}) v_\delta(\mathbf{x} - \mathbf{y})\, d\mathbf{y} = 0.$$

Thus, for ε small enough, $w_\varepsilon \in \mathcal{D}(\Omega)$.

Finally,

$$\begin{aligned}\|v_\delta - w_\varepsilon\|^p &= \int_\Omega \left| \int_{B(0;\varepsilon)} \zeta_\varepsilon(\mathbf{y})(v_\delta(\mathbf{x}) - v_\delta(\mathbf{x}-\mathbf{y}))\,d\mathbf{y} \right|^p d\mathbf{x} \\ &\le \int_\Omega \|\zeta_\varepsilon\|^p_{L^{p'}} \left(\int_{B(0;\varepsilon)} |v_\delta(\mathbf{x}) - v_\delta(\mathbf{x}-\mathbf{y})|^p\,d\mathbf{y} \right) d\mathbf{x} \\ &\le C \int_{B(0;\varepsilon)} \left(\int_\Omega |v_\delta(\mathbf{x}) - v_\delta(\mathbf{x}-\mathbf{y})|^p\,d\mathbf{x} \right) d\mathbf{y}\end{aligned}$$

for some C independent of ε, in view of Fubini's Theorem.

The last quantity goes to zero as $\varepsilon \to 0$. This is (once more) a consequence of Lebesgue's Theorem, since it can be written in the form

$$\int_{\mathbb{R}^N} \left(\int_\Omega |v_\delta(\mathbf{x}) - v_\delta(\mathbf{x}-\mathbf{y})|^p\,d\mathbf{x} \right) \mathbb{1}_{B(0;\varepsilon)}(\mathbf{y})\,d\mathbf{y}.$$

Consequently, if ε is sufficiently small, one has

$$\|v_\delta - w_\varepsilon\|_{L^p} \le \frac{\kappa}{2}. \tag{2.9}$$

From the inequalities (2.8) and (2.9), we get (2.7) for $\varphi = w_\varepsilon$ and this ends the proof. □

Corollary 2.1 *Let $u \in L^2(\Omega)$ be such that*

$$\int_\Omega u\,\varphi = 0, \quad \forall \varphi \in \mathcal{D}(\Omega). \tag{2.10}$$

Then $u = 0$ a.e. in Ω.

Proof The proof is immediate: in view of Theorem 2.1, $\mathcal{D}(\Omega)$ is a dense subspace of $L^2(\Omega)$; therefore, its orthogonal complement is the null space and, consequently, u vanishes a.e. □

Actually, this result also holds under the much weaker assumption $u \in L^1_{\text{loc}}(\Omega)$, as the following proposition shows:

Proposition 2.2 (Du Bois-Reymond Lemma) *Assume that $u \in L^1_{\text{loc}}(\Omega)$ and one has* (2.10). *Then $u = 0$ a.e. in Ω.*

Proof First, let us assume that Ω is bounded and $u \in L^1(\Omega)$. Let $\varepsilon > 0$ be given. We know that there exists $\psi_\varepsilon \in \mathcal{D}(\Omega)$ such that

$$\|u - \psi_\varepsilon\|_{L^1} \le \varepsilon.$$

Therefore, for each $\varphi \in \mathcal{D}(\Omega)$, we have

$$\left|\int_{\Omega} \psi_\varepsilon \varphi\right| = \left|\int_{\Omega} (\psi_\varepsilon - u)\varphi\right| \leq \varepsilon \|\varphi\|_{L^\infty}.$$

Let K_+ and K_- be the sets

$$K_+ := \{\mathbf{x} \in \Omega : \psi_\varepsilon(\mathbf{x}) \geq \varepsilon\}, \quad K_- := \{\mathbf{x} \in \Omega : \psi_\varepsilon(x) \leq -\varepsilon\}.$$

Then, K_+ and K_- are disjoint compact subsets of Ω and it is not difficult to prove using Proposition 2.1 that there exists a function $\varphi_\varepsilon \in \mathcal{D}(\Omega)$ satisfying

$$-1 \leq \varphi_\varepsilon \leq 1, \quad \varphi_\varepsilon = 1 \text{ in } K_+, \quad \varphi_\varepsilon = -1 \text{ in } K_-.$$

Set $K := K_+ \cup K_-$. Then

$$\begin{aligned}\int_K |\psi_\varepsilon| = \int_K \psi_\varepsilon \varphi_\varepsilon &= \int_\Omega \psi_\varepsilon \varphi_\varepsilon - \int_{\Omega\setminus K} \psi_\varepsilon \varphi_\varepsilon \\ &\leq \varepsilon + \int_{\Omega\setminus K} |\psi_\varepsilon| \\ &\leq (1 + |\Omega|)\varepsilon,\end{aligned}$$

whence

$$\|u\|_{L^1} \leq \|u - \psi_\varepsilon\|_{L^1} + \|\psi_\varepsilon\|_{L^1} \leq (2 + |\Omega|)\varepsilon.$$

Since ε is arbitrarily small, we get that $u = 0$ a.e.

Now, let $\Omega \subset \mathbb{R}^N$ be an arbitrary connected open set (not necessarily bounded) and let $u \in L^1_{\text{loc}}(\Omega)$ be given. For each $n \in \mathbb{N}$ with $n \geq 1$, we set

$$\Omega_n := \{\mathbf{x} \in \Omega : |\mathbf{x}| < n, \quad \text{dist }(\mathbf{x}, \mathbb{R}^N \setminus \Omega) > \frac{1}{n}\}.$$

The set $\overline{\Omega}_n$ is a compact subset of Ω, for all n. Moreover,

$$\overline{\Omega}_n \subset \Omega_{n+1} \ \forall n \geq 1 \text{ and } \cup_{n\geq 1} \Omega_n = \Omega.$$

We can argue as before for each Ω_n and deduce that u vanishes a.e. in Ω_n. Hence, due to the properties of the sets Ω_n, it follows that $u = 0$ a.e. in Ω. □

These technical results allow to characterize the dual space $(L^p(\Omega))'$. In other words, we have the following:

Theorem 2.2 *For any $p \in [1, +\infty)$, the space $(L^p(\Omega))'$ is isometrically isomorphic to $L^{p'}(\Omega)$. More precisely, for any continuous linear form $\ell \in (L^p(\Omega))'$, there exists exactly one $g \in L^{p'}(\Omega)$ such that*

$$\langle \ell, f \rangle_{(L^p)', L^p} = \int_\Omega g\, f \quad \forall f \in L^p(\Omega). \tag{2.11}$$

The mapping $\ell \mapsto g$ is an isometric isomorphism from $(L^p(\Omega))'$ onto $L^{p'}(\Omega)$. Consequently, for $1 < p < +\infty$, the Banach space $L^p(\Omega)$ is reflexive.

For the proof, we will use an additional technical lemma (see Exercise 2.11):

Lemma 2.2 (Clarkson's Inequalities) *Assume that $p \in (1, +\infty)$ and $f, g \in L^p(\Omega)$. Then:*

- *If $p \geq 2$,*

$$\|\frac{1}{2}(f+g)\|_{L^p}^p + \|\frac{1}{2}(f-g)\|_{L^p}^p \leq \frac{1}{2}(\|f\|_{L^p}^p + \|g\|_{L^p}^p).$$

- *On the other hand, if $1 < p < 2$,*

$$2^{p-1}(\|f\|_{L^p}^p + \|f\|_{L^p}^p) \leq \|f+g\|_{L^p}^p + \|f-g\|_{L^p}^p \leq 2(\|f\|_{L^p}^p + \|f\|_{L^p}^p).$$

Proof of Theorem 2.2 Let us first assume that $p \in [2, +\infty)$. Consider the mapping $\sigma : L^{p'}(\Omega) \mapsto (L^p(\Omega))'$ defined by $\sigma g = \ell$, where ℓ is given by (2.11).

Clearly, σ is well defined, linear continuous and injective on $L^{p'}(\Omega)$.

It is also onto. To show this, let $\ell \in L^{p'}(\Omega)$ be given and consider the following extremal problem:

$$\begin{cases} \text{Minimize } \mathcal{E}(f) := \dfrac{1}{p} \displaystyle\int_\Omega |f|^p - \langle \ell, f \rangle_{(L^p)', L^p} \\ \text{subject to } f \in L^p(\Omega). \end{cases} \tag{2.12}$$

Assume that $\hat{f}$ solves problem (2.12). Then

$$\int_\Omega |\hat{f}|^{p-2} \hat{f}\, f = \langle \ell, f \rangle_{(L^p)', L^p} \quad \forall f \in L^p(\Omega). \tag{2.13}$$

Indeed, for any $f \in L^p(\Omega)$ and any $\varepsilon > 0$, we must have $\frac{1}{\varepsilon}(\mathcal{E}(\hat{f} + \varepsilon f) - \mathcal{E}(\hat{f})) \geq 0$, whence

$$\frac{1}{\varepsilon p} \int_\Omega \left(|\hat{f} + \varepsilon f|^p - |\hat{f}|^p \right) - \langle \ell, f \rangle_{(L^p)', L^p} \geq 0.$$

In view of Lemma 2.2, we can apply Lebesgue's Theorem and deduce that, as $\varepsilon \to 0$, the previous integral converges to the left-hand side of (2.13). Thus, taking limits, one has

$$\int_\Omega |\hat{f}|^{p-2}\hat{f}\cdot f \geq \langle \ell, f\rangle_{(L^p)',L^p}.$$

If we write these inequalities first for f and then for $-f$, we finally get (2.13).

Now, let us define $\hat{g} := |\hat{f}|^{p-2}\hat{f}$. Then, $\hat{g} \in L^{p'}(\Omega)$ and, moreover,

$$\int_\Omega \hat{g}\, f = \langle \ell, f\rangle_{(L^p)',L^p} \quad \forall f \in L^p(\Omega),$$

that is, $\ell = \sigma\hat{g}$, which proves (if $\hat{f}$ exists) that σ is onto.

It remains to show that, for any $\ell \in L^{p'}(\Omega)$, the corresponding problem (2.12) is solvable. To this end, let $\{f_n\}$ be a minimizing sequence for problem (2.12). If $m, n \geq 1$, one has

$$\begin{aligned}
\mathcal{E}(f_m) + \mathcal{E}(f_n) &= \frac{1}{p}\left(\|f_m\|_{L^p}^p + \|f_n\|_{L^p}^p\right) - \langle \ell, f_m + f_n\rangle_{(L^p)',L^p} \\
&\geq 2\left(\frac{1}{p}\|\frac{1}{2}(f_m+f_n)\|_{L^p}^p - \langle \ell, \frac{1}{2}(f_m+f_n)\rangle_{(L^p)',L^p}\right) \\
&\quad + \frac{1}{p}\|\frac{1}{2}(f_m-f_n)\|_{L^p}^p \\
&= 2\mathcal{E}\left(\frac{1}{2}(f_m+f_n)\right) + \frac{1}{p}\|\frac{1}{2}(f_m-f_n)\|_{L^p}^p.
\end{aligned}$$

Consequently,

$$\frac{1}{p\,2^p}\|f_m - f_n\|_{L^p}^p \leq \mathcal{E}(f_m) + \mathcal{E}(f_n) - 2\,\inf \mathcal{E}(f)$$

and this shows that $\{f_n\}$ is a Cauchy sequence in $L^p(\Omega)$.

Therefore, one has $f_n \to \hat{f}$ for some $\hat{f} \in L^p(\Omega)$, whence $\mathcal{E}(f_n) \to \mathcal{E}(\hat{f})$ and, accordingly, $\hat{f}$ solves problem (2.12).

Thus, when $2 \leq p < +\infty$, σ is an isomorphism from $L^p(\Omega)$ onto $L^{p'}(\Omega)$ and the result holds.

Now, let us suppose that $1 < p < 2$.

Let $\tilde{\sigma} : L^p(\Omega) \mapsto (L^{p'}(\Omega))'$ be given by $\tilde{\sigma} f = k$, where k stands for the linear form

$$\langle k, g\rangle_{(L^{p'})',L^{p'}} = \int_\Omega g\, f \quad \forall g \in L^{p'}(\Omega).$$

As before, $\tilde{\sigma}$ is well defined, linear, continuous and injective on $L^p(\Omega)$. Consequently, $L^p(\Omega)$ is isomorphic to $R(\tilde{\sigma})$, which is a subspace of $(L^{p'}(\Omega))'$.

The space $(L^{p'}(\Omega))'$ is reflexive, since it is the dual of the reflexive space $L^{p'}(\Omega)$ (note that $2 < p' < +\infty$). On the other hand, $R(\tilde{\sigma})$ is closed in $(L^{p'}(\Omega))'$. This is a consequence of the fact that

$$\|\tilde{\sigma} f\|_{(L^{p'})'} = \|f\|_{L^p} \quad \forall f \in L^p(\Omega)$$

which, in turn, is a consequence of the following:

$$\left|\langle \tilde{\sigma} f, g\rangle_{(L^{p'})',L^{p'}}\right| = \left|\int_\Omega g\, f\right| \le \|f\|_{L^p}\|g\|_{L^{p'}} \quad \forall f \in L^p(\Omega), \ \forall g \in L^{p'}(\Omega)$$

and

$$\langle \tilde{\sigma} f, |f|^{p-2} f\rangle_{(L^{p'})',L^{p'}} = \int_\Omega |f|^p = \|f\|_{L^p}\||f|^{p-2} f\|_{L^{p'}} \quad \forall f \in L^p(\Omega).$$

Let us prove that $R(\tilde{\sigma})$ is dense in $(L^{p'}(\Omega))'$. This will imply that $R(\tilde{\sigma}) = (L^{p'}(\Omega))'$, and will provide the desired result for $1 < p < 2$.

To this end, it suffices to check that any $g \in L^{p'}(\Omega)$ satisfying

$$\langle \tilde{\sigma} f, g\rangle_{(L^{p'})',L^{p'}} = 0 \quad \forall f \in L^p(\Omega)$$

necessarily vanishes. But this is obvious since, if this is the case, one has

$$0 = \langle \tilde{\sigma}(|g|^{p'-2} g), g\rangle_{(L^{p'})',L^{p'}} = \int_\Omega |g|^{p'}.$$

Finally, let us consider the case $p = 1$. Again, we can define the mapping $\sigma : L^\infty(\Omega) \mapsto (L^1(\Omega))'$ with $\sigma g = \ell$, where ℓ is given by

$$\langle \ell, f\rangle_{(L^1)',L^1} = \int_\Omega g\, f \quad \forall f \in L^1(\Omega).$$

As before, it is easy to check that σ is linear continuous and one-to-one.

Here, we will assume that Ω is bounded. See for instance [5] for a proof in the general case; see also Exercise 2.12. Thus, $L^2(\Omega) \subset L^1(\Omega)$ and the identity mapping from $L^2(\Omega)$ into $L^1(\Omega)$ is a continuous embedding.

Let $\ell \in (L^1(\Omega))'$ be given. Then, $f \mapsto \langle \ell, f\rangle_{(L^1)',L^1}$ is a linear continuous form on $L^2(\Omega)$ and, consequently, there exists $g \in L^2(\Omega)$ such that

$$\langle \ell, f\rangle_{(L^1)',L^1} = \int_\Omega g\, f \quad \forall f \in L^2(\Omega). \tag{2.14}$$

Let us check that $g \in L^\infty(\Omega)$ and, in fact, (2.14) holds for all $f \in L^1(\Omega)$. This will end the proof in this case.

Set $K := \|\ell\|_{(L^1(\Omega))'}$, assume that $\operatorname{ess\,sup}_\Omega |g| > K$ and let $A \subset \Omega$ be a measurable set of positive finite measure such that $|g| \geq K + \varepsilon$ a.e. in A. Then, if we take $f_A := \operatorname{sign}(g)\, \mathbb{1}_A$, we find that $f_A \in L^2(\Omega)$ and

$$\int_\Omega g\, f_A = \int_A |g| \geq (K + \varepsilon)|A|,$$

while

$$\left|\langle \ell, f_A \rangle_{(L^1)', L^1}\right| \leq K \int_\Omega |f_A| = K|A|.$$

Therefore, we cannot have (2.14). This shows that $g \in L^\infty(\Omega)$ and, in fact, $\|g\|_{L^\infty} \leq \|\ell\|_{(L^1(\Omega))'}$.

If $f \in L^1(\Omega)$, in view of Theorem 2.1, there exists a sequence $\{f_n\}$ in $\mathcal{D}(\Omega)$ such that $f_n \to f$ in $L^1(\Omega)$. For each $n \geq 1$, one has

$$\langle \ell, f_n \rangle_{(L^1)', L^1} = \int_\Omega g\, f_n.$$

Since $g \in L^\infty(\Omega)$, we can take limits as $n \to +\infty$ in both sides, which leads to (2.14).

This ends the proof. □

This result does not hold for $p = +\infty$. Actually, the dual of $L^\infty(\Omega)$ "contains" $L^1(\Omega)$ in the sense that we can associate to any $g \in L^1(\Omega)$ a unique continuous linear form $\ell_g \in (L^\infty(\Omega))'$ and get a continuous embedding $g \mapsto \ell_g$. But there exist many linear forms in $(L^\infty(\Omega))'$ that are not of the form ℓ_g. Indeed, if this were not the case, we would have that the dual of $L^\infty(\Omega)$ is separable and this would imply that $L^\infty(\Omega)$ is also separable, which is not true.

We are now ready to recall one of the main concepts needed to analyze and solve problems concerning PDEs and, in particular, problems with origin in fluid mechanics.

Definition 2.1 A distribution on Ω is a linear form $S : \mathcal{D}(\Omega) \mapsto \mathbb{R}$ sequentially continuous with respect to the convergence in $\mathcal{D}(\Omega)$, that is, such that

$$\phi_n \to 0 \ \text{in} \ \mathcal{D}(\Omega) \ \Rightarrow \ S(\phi_n) \to 0.$$

We will denote by $\mathcal{D}'(\Omega)$ the family of all distributions over Ω. Again, $\mathcal{D}'(\Omega)$ is a vector space with respect to the usual operations. The value assigned by $S \in \mathcal{D}'(\Omega)$ to $\phi \in \mathcal{D}(\Omega)$ will be denoted $\langle S, \phi \rangle$.[4]

The space $\mathcal{D}'(\Omega)$ will be endowed with the following notion of convergence: we say that a sequence $\{S_n\}$ converges to S in $\mathcal{D}'(\Omega)$ if

[4] Occasionally, the symbol $\langle \cdot\,, \cdot \rangle$ will also serve to design some duality pairings.

$$\langle S_n, \phi \rangle \to \langle S, \phi \rangle, \quad \forall \phi \in \mathcal{D}(\Omega).$$

In this case, we write $S_n \to S$ in $\mathcal{D}'(\Omega)$. Actually, there exists a vector topology on $\mathcal{D}'(\Omega)$ that induces this definition of convergence.[5]

Given $f \in L^1_{\text{loc}}(\Omega)$, the linear form $S_f : \mathcal{D}(\Omega) \mapsto \mathbb{R}$ defined by

$$\langle S_f, \phi \rangle = \int_\Omega f\phi \quad \forall \phi \in \mathcal{D}(\Omega)$$

is a distribution. Furthermore, the mapping $f \mapsto S_f$ is linear, sequentially continuous and one-to-one, as a consequence of Lemma 2.2. It follows that one can identify $f \in L^1_{\text{loc}}(\Omega)$ and the associated $S_f \in \mathcal{D}'(\Omega)$.

In other words, the space $L^1_{\text{loc}}(\Omega)$ can be viewed as a subspace of $\mathcal{D}'(\Omega)$ and distributions are objects that, in some sense, "generalize" functions. Thus, we can complete (2.4) as follows:

$$\mathcal{D}(\Omega) \subset X(\Omega) \subset L^1_{\text{loc}}(\Omega) \subset \mathcal{D}'(\Omega).$$

A natural question at this point is whether all distributions are of the form S_f for some $f \in L^1_{\text{loc}}(\Omega)$. The answer is no, as the following definition and the subsequent argument show:

Definition 2.2 Let $\mathbf{x}_0 \in \Omega$ be given. The *Dirac distribution at* $\mathbf{x}_0$ is the linear form $\delta_{\mathbf{x}_0}$ defined by

$$\langle \delta_{\mathbf{x}_0}, \phi \rangle = \phi(\mathbf{x}_0) \quad \forall \phi \in \mathcal{D}(\Omega).$$

For every $\mathbf{x}_0 \in \Omega$, $\delta_{\mathbf{x}_0}$ is certainly a distribution. Moreover, there is no $f \in L^1_{\text{loc}}(\Omega)$ such that $S_f = \delta_{\mathbf{x}_0}$. Indeed, assume that there exist $\mathbf{x}_0 \in \Omega$ and $f \in L^1_{\text{loc}}(\Omega)$ such that $S_f = \delta_{\mathbf{x}_0}$. Then we would have

$$\int_\Omega f\phi = \phi(\mathbf{x}_0) \quad \forall \phi \in \mathcal{D}(\Omega).$$

Consider the open set $O = \Omega \setminus \{\mathbf{x}_0\}$. One obviously has

$$\int_\Omega f\phi = 0 \quad \forall \phi \in \mathcal{D}(O).$$

Therefore, $f = 0$ a.e. in O and consequently $S_f = 0$, which is a contradiction.

[5] It suffices to consider the topology induced by the family of seminorms $\{p_\phi\}_{\phi \in \mathcal{D}(\Omega)}$, where $p_\phi(S) := |\langle S, \phi \rangle|$ for any $S \in \mathcal{D}'(\Omega)$ and any $\phi \in \mathcal{D}(\Omega)$. The corresponding seminormed space is sequentially complete; see [29] for more details.

In view of this property of the Dirac distribution, it is usual to say that "$\delta_{\mathbf{x}_0}$ does not belong to $L^1_{\rm loc}(\Omega)$".

In a similar way, we can exhibit distributions supported on m-dimensional manifolds with $m \leq N - 1$ that, again, do not belong to $L^1_{\rm loc}(\Omega)$. For instance, if we take $\Omega = \mathbb{R}^2$ and set

$$\langle S, \phi \rangle := \int_{-\infty}^{+\infty} \phi(0, x_2)\, dx_2 \quad \forall \phi \in \mathcal{D}(\mathbb{R}^2), \tag{2.15}$$

it is again clear that $S \in \mathcal{D}'(\mathbb{R}^2)$ but $S \notin L^1_{\rm loc}(\mathbb{R}^2)$.

Given $f \in L^1_{\rm loc}(\Omega)$, it is usual to denote also by f the distribution S_f. Since we can identify $L^1_{\rm loc}(\Omega)$ to a proper subspace of $\mathcal{D}'(\Omega)$, this notation leads to no ambiguity.

Moreover, the following holds:

Proposition 2.3 *For any p with $1 \leq p \leq +\infty$, one has*

$$\mathcal{D}(\Omega) \hookrightarrow L^p_{\rm loc}(\Omega) \hookrightarrow \mathcal{D}'(\Omega),$$

where the embeddings are sequentially continuous.

We can complete Proposition 2.3 by writing that, for instance,

$$\mathcal{D}(\Omega) \hookrightarrow C^m(\Omega) \hookrightarrow C^0(\Omega) \hookrightarrow L^p_{\rm loc}(\Omega) \hookrightarrow L^1_{\rm loc}(\Omega), \tag{2.16}$$

for all $m \geq 0$ and $p \in [1, +\infty]$, again with sequentially continuous embeddings.[6]

In the sequel, for any multi-index $\alpha = (\alpha_1, \ldots, \alpha_N)$ with nonnegative integers α_i, we will set $|\alpha| := \alpha_1 + \cdots + \alpha_N$.

Definition 2.3 Let $S \in \mathcal{D}'(\Omega)$ be given. For any multi-index α, the (distributional) derivative of S of order α is the distribution $\partial^\alpha S$ defined by

$$\langle \partial^\alpha S, \phi \rangle = (-1)^{|\alpha|} \langle S, \partial^\alpha \phi \rangle \quad \forall \phi \in \mathcal{D}(\Omega).$$

It is easy to check that $\partial^\alpha S$ is a distribution.

The following result shows that we are in fact extending the usual "classical" definition of derivative to $\mathcal{D}'(\Omega)$:

Proposition 2.4 *Let a multi-index α be given and assume that $f \in C^{|\alpha|}(\Omega)$. Then, the distributions $\partial^\alpha S_f$ and $S_{\partial^\alpha f}$ coincide.*

---

[6] Recall that $C^m(\Omega)$ is a Fréchet space for the family of seminorms $\{\mu_K\}$, where

$$\mu_K(\varphi) = \max_{\mathbf{x} \in K,\ |\alpha| \leq m} |\partial^\alpha \varphi(\mathbf{x})| \quad \forall \varphi \in C^m(\Omega)$$

and K runs over the whole family of compact subsets of Ω.

In this way, we see that functions can be regarded as distributions of a particular kind, it is possible to differentiate distributions arbitrarily many times and get new distributions and also that for regular functions distributional and classical derivatives coincide.

However, the distributional derivative of a $L^1_{\rm loc}$ function may not belong to the space $L^1_{\rm loc}(\Omega)$, as the following simple example shows.

Example 2.1 Assume $\Omega = \mathbb{R}^2$ and consider the so called Heaviside function H defined by

$$H(\mathbf{x}) = \begin{cases} 1 & \text{if } x_1 > 0, \\ 0 & \text{if } x_1 < 0. \end{cases} \tag{2.17}$$

Then $H \in L^1_{\rm loc}(\Omega)$ and the (first order) derivative $\partial^{(1,0)}H$ is the distribution S in (2.15). Therefore, $\partial^{(1,0)}H \notin L^1_{\rm loc}(\Omega)$.

In general, given $f \in L^1_{\rm loc}(\Omega)$, we will use the notation $\partial^\alpha f$ to denote the distributional derivative $\partial^\alpha S_f$.

For any multi-index α, the differential operator $\partial^\alpha : \mathcal{D}'(\Omega) \mapsto \mathcal{D}'(\Omega)$ is obviously well defined. It is also linear and sequentially continuous. It will be seen in Chaps. 3 and 5 that this can be crucial for solving a PDE. Indeed, as a consequence of this property, if there are distributions $u_k, f_k \in \mathcal{D}'(\Omega)$ such that, for instance,

$$-\Delta u_k := -\sum_{i=1}^{N} \partial_i^2 u_k = f_k \ \text{ in } \mathcal{D}'(\Omega)$$

for all $k \geq 1$ and $u_k \to u$ and $f_k \to f$ in $\mathcal{D}'(\Omega)$, then

$$-\Delta u = f.$$

The functions in L^p that have derivatives that belong to L^p play a very important role in the modern theory of PDEs. To this respect, let us recall the definition of the *Sobolev space* $W^{m,p}(\Omega)$:

Definition 2.4 Let $m \geq 0$ be an integer and let us assume that $1 \leq p \leq +\infty$. We denote by $W^{m,p}(\Omega)$ the space of functions $v \in L^p(\Omega)$ such that $\partial^\alpha v \in L^p(\Omega)$ for all $|\alpha| \leq m$. The $W^{m,p}(\Omega)$ are linear spaces for the usual laws and Banach spaces for the norms

$$\|v\|_{W^{m,p}} := \left[\sum_{|\alpha| \leq m} \int_\Omega |\partial^\alpha v|^p \right]^{1/p} = \left[\sum_{|\alpha| \leq m} \|\partial^\alpha v\|_{L^p}^p \right]^{1/p}$$

for $1 \leq p < +\infty$ and

$$\|v\|_{W^{m,\infty}} := \sup_{|\alpha|\le m}\left[\operatorname*{ess\,sup}_{x\in\Omega} |\partial^\alpha v(x)|\right] = \sup_{|\alpha|\le m} \|\partial^\alpha v\|_{L^\infty}$$

for $p = +\infty$.

The completeness of the normed spaces $W^{m,p}(\Omega)$ is a direct consequence of the completeness of the $L^p(\Omega)$ and the sequential continuity of the differential operators.

When $m = 0$, one has $W^{0,p}(\Omega) = L^p(\Omega)$ and $\mathcal{D}(\Omega)$ is dense in $W^{0,p}(\Omega)$. But this does not hold in general for $m \ge 1$. In fact, as shown below, the closure of $\mathcal{D}(\Omega)$ in $W^{m,p}(\Omega)$ is used to identify those functions in $W^{m,p}(\Omega)$ that vanish on the boundary $\partial\Omega$, as well as their derivatives of order $\le m-1$.

Thus, let us denote by $W_0^{m,p}(\Omega)$ the closure of $\mathcal{D}(\Omega)$ in $W^{m,p}(\Omega)$. Then, endowed with the norm of $W^{m,p}(\Omega)$, $W_0^{m,p}(\Omega)$ is a new Banach space.

It is usual to write $H^m(\Omega)$ instead of $W^{m,2}(\Omega)$ and $H_0^m(\Omega)$ instead of $W_0^{m,2}(\Omega)$. Then, $H^m(\Omega)$ and $H_0^m(\Omega)$ are Hilbert spaces with respect to the inner product

$$(u, v)_{H^m} := \sum_{|\alpha|\le m} \int_\Omega \partial^\alpha u\, \partial^\alpha v = \sum_{|\alpha|\le m} (\partial^\alpha u, \partial^\alpha v). \tag{2.18}$$

Proposition 2.5 *The Banach space $W^{m,p}(\Omega)$ is separable for all p with $1 \le p < +\infty$ and reflexive for all p with $1 < p < +\infty$.*

For any p with $1 \le p < +\infty$, we will denote by $W^{-m,p'}(\Omega)$ the topological dual of $W_0^{m,p}(\Omega)$, that is, the space of all linear continuous forms on $W_0^{m,p}(\Omega)$ (recall that p' is the conjugate of p).

The space $W^{-m,p'}(\Omega)$ can be identified to a space of distributions as follows:

Proposition 2.6 *For any integer $m \ge 0$ and any $p \in [1, +\infty)$, one has*

$$W^{-m,p'}(\Omega) \cong \{S \in \mathcal{D}'(\Omega) : S = \sum_{|\alpha|\le m} \partial^\alpha v_\alpha,\ v_\alpha \in L^{p'}(\Omega)\},$$

where the identification is carried out through the identity

$$\langle S, u\rangle_{W^{-m,p'},W_0^{m,p}} = \sum_{|\alpha|\le m} (-1)^{|\alpha|} \int_\Omega v_\alpha\, \partial^\alpha u \quad \forall u \in W_0^m(\Omega). \tag{2.19}$$

Proof Let us check that, for any $S \in W^{-m,p'}(\Omega)$, there exist functions $v_\alpha \in L^{p'}(\Omega)$ such that (2.19) holds and, furthermore,

$$\|S\|_{W^{-m,p'}} \le \left(\sum_{|\alpha|\le m} \|v_\alpha\|_{L^{p'}}^{p'}\right)^{1/p'} \quad \text{if } p > 1 \tag{2.20}$$

and

$$\|S\|_{W^{-m,\infty}} \leq \sup_{|\alpha|\leq m} \|v_\alpha\|_{L^\infty} \quad \text{if } p = 1. \tag{2.21}$$

First, note that $W_0^{m,p}(\Omega)$ can be viewed as a subspace of $L^p(\Omega)^\kappa$, where $\kappa = \kappa(m, N)$ is the number of derivatives of order $\leq m$ of a function of N variables.[7] Indeed, the linear mapping

$$h_0 : W_0^{m,p}(\Omega) \mapsto L^p(\Omega)^\kappa, \quad \text{with } h_0(v) = \{\partial^\alpha v\}_{|\alpha|\leq m} \; \forall v \in W_0^{m,p}(\Omega),$$

is an isometric embedding and allows to identify $W_0^{m,p}(\Omega)$ and $R(H_0)$.

Let $S \in W^{-m,p'}(\Omega)$ be given. In view of the Hahn-Banach Theorem, S can be extended to a continuous linear form $\tilde{S}$ on $L^p(\Omega)^\kappa$. From Theorem 2.2, we deduce that there exist functions $\tilde{v}_\alpha \in L^{p'}(\Omega)$ with for $|\alpha| \leq m$ such that

$$\begin{cases} \tilde{S}(\{z_\alpha\}) = \displaystyle\sum_{|\alpha|\leq m} \int_\Omega \tilde{v}_\alpha z_\alpha, \quad \forall \{z_\alpha\} \in L^p(\Omega)^\kappa \\ \text{and } \|S\|_{W^{-m,p'}} = \|\tilde{S}\|_{(L^p(\Omega)^\kappa)'}. \end{cases} \tag{2.22}$$

Then, if we set $w_\alpha := (-1)^{|\alpha|}\tilde{v}_\alpha$ for all α, one has

$$\langle S, v\rangle = \tilde{S}(\{\partial^\alpha v\}) = \sum_{|\alpha|\leq m} (-1)^{|\alpha|} \int_\Omega w_\alpha \partial^\alpha v \quad \forall v \in W_0^{m,p}(\Omega).$$

From (2.22), we also have (2.20) if $p > 1$ and (2.21) if $p = 1$.

On the other hand, it is clear that, if the v_α belong to $L^{p'}(\Omega)$ for $|\alpha| \leq m$, then the linear form given by (2.19) is well defined and continuous and satisfies (2.20) or (2.21) (in fact, with equal signs).

Finally, observe that any linear form in $W^{-m,p'}(\Omega)$ can be viewed as a distribution: first, its restriction to the dense subspace $\mathcal{D}(\Omega)$ is sequentially continuous; secondly, any linear form on $\mathcal{D}(\Omega)$ that is continuous with respect to the topology of $W_0^{m,p}(\Omega)$ can be extended as a unique form in $W^{-m,p'}(\Omega)$.

This ends the proof. □

For $p = 2$ one has $p' = 2$ and instead of $W^{-m,p'}(\Omega)$ we write $H^{-m}(\Omega)$, which is a separable Hilbert space.

In the remainder of this section, it will be assumed that $\Omega \subset \mathbb{R}^N$ is bounded and $\partial\Omega$ is *regular enough*. To our purposes, if $N \geq 2$, it will be sufficient to suppose (at least for the moment) that $\partial\Omega$ is Lipschitz-continuous and leaves Ω locally at one side; see the definition of this regularity notion in Sect. 2.6.

For any function v defined on $\overline{\Omega}$ and any set $G \subset \overline{\Omega}$, we will denote by $v|_G$ "the restriction of v to G". Also, we will denote by $C^1(\overline{\Omega})$ the linear space of the

[7] That is, $\kappa(m, N) = \binom{N+m}{m}$.

restrictions to $\overline{\Omega}$ of the functions in $C^1(\mathbb{R}^N)$. Clearly, $C^1(\overline{\Omega})$ is a Banach space for the norm

$$\|\phi\|_{C^1} := \|\phi\|_{C^0} + \|\nabla\phi\|_{C^0} = \max_{\mathbf{x}\in\overline{\Omega}} |\phi(\mathbf{x})| + \max_{\mathbf{x}\in\overline{\Omega}} |\nabla\phi(\mathbf{x})|.$$

In a similar way, we can introduce the Banach spaces $C^m(\overline{\Omega})$ (where $m \geq 2$ is an integer) and the linear space $C^\infty(\overline{\Omega}) := \bigcap_{m\geq 1} C^m(\overline{\Omega})$.

Lemma 2.3 *The Banach space $C^1(\overline{\Omega})$ is dense in $H^1(\Omega)$.*

Proof Let $v \in H^1(\Omega)$ be given.

First, note that there exists an extension $\tilde{v} \in H^1(\mathbb{R}^N)$ with compact support such that $\|\tilde{v}\|_{L^2(\mathbb{R}^N)} \leq C\|v\|$ and $\|\nabla\tilde{v}\|_{L^2(\mathbb{R}^N)} \leq C\|\nabla v\|$, where C is independent of v; see Exercise 2.22.

Now, let the functions ζ_ε be given again by (2.6), where $\zeta \in \mathcal{D}(\mathbb{R}^N)$, $\zeta > 0$ in $B(0;1)$, $\zeta = 0$ outside $B(0;1)$ and (2.5) holds. For each $\varepsilon > 0$, let us introduce v_ε as follows:

$$v_\varepsilon(\mathbf{x}) := \int_{\mathbb{R}^N} \tilde{v}(\mathbf{y})\,\zeta_\varepsilon(\mathbf{x}-\mathbf{y})\,d\mathbf{y} \quad \forall \mathbf{x} \in \overline{\Omega}. \tag{2.23}$$

It is not difficult to check that the $v_\varepsilon \in C^\infty(\overline{\Omega})$, $v_\varepsilon \to v$ and $\partial_i v_\varepsilon \to \partial_i v$ strongly in $L^2(\Omega)$ for all $i = 1, \ldots, N$ as $\varepsilon \to 0$; see Exercise 2.23 for details.

Consequently, $\|v_\varepsilon - v\|_{H^1} \to 0$ as $\varepsilon \to 0$ and the proof is done. □

Note that a similar density result can be established in $W^{1,p}(\Omega)$ for all p with $1 \leq p < +\infty$. Also, it can be proved that $C^m(\overline{\Omega})$ is dense in $W^{1,p}(\Omega)$ for all such p ad any $m \geq 1$.

Thanks to Lemma 2.3, it is possible to give a sense to the "restriction to the boundary" of any function in $H^1(\Omega)$. This is the motivation of the following important result:

Theorem 2.3 (Traces in $H^1(\Omega)$) *The linear mapping $v \mapsto v|_{\partial\Omega}$, that is well defined in $C^1(\overline{\Omega})$, can be extended to a unique bounded linear operator $\gamma : H^1(\Omega) \mapsto L^2(\partial\Omega)$, called the trace operator on $\partial\Omega$.*

Moreover, $N(\gamma) = H_0^1(\Omega)$ and $R(\gamma)$ is a nontrivial dense subspace of $L^2(\partial\Omega)$, denoted by $H^{1/2}(\partial\Omega)$. This space, endowed with the norm induced by $H^1(\Omega)$ and γ, is Hilbertian.

Furthermore, γ possesses a continuous inverse on $H^{1/2}(\partial\Omega)$.

Sketch of the Proof The proof of this result is rather technical and, for brevity, we will only indicate some fundamental ideas.

First, it can be shown that the restriction mapping $v \mapsto v|_{\partial\Omega}$ is linear and continuous from $C^1(\overline{\Omega})$ endowed with the norm of $H^1(\Omega)$ into $L^2(\partial\Omega)$; see Exercise 2.24. As a consequence, the trace operator γ exists.

Then, it is immediate that $H_0^1(\Omega) \subset N(\gamma)$ (indeed, γv is given for any $v \in H_0^1(\Omega)$ by the limit in $L^2(\partial\Omega)$ of a sequence of restrictions to $\partial\Omega$ of functions in $\mathcal{D}(\Omega)$).

On the other hand, the following holds:

- If $v \in H^1(\Omega)$ and $\gamma v = 0$, there exist functions $v_n \in C^1(\overline{\Omega})$ such that $v_n \to v$ strongly in $H^1(\Omega)$ as $n \to +\infty$ and $v_n|_{\partial\Omega} = 0$ for all $n \geq 1$.
- Any $w \in C^1(\overline{\Omega})$ such that $w|_{\partial\Omega} = 0$ belongs to $H_0^1(\Omega)$. Indeed, there exist functions $w_\varepsilon \in \mathcal{D}(\Omega)$ with $w_\varepsilon \to w$ in $H^1(\Omega)$ as $\varepsilon \to 0$ (to see this, it suffices to perform extensions, translations and regularizations similar to those in the proof of Lemma 2.3).

These arguments show that $N(\gamma) = H_0^1(\Omega)$.

Finally, the proofs that γ is not onto, $R(\gamma)$ is a proper Hilbert subspace of $L^2(\partial\Omega)$ and γ possesses a continuous inverse are left to the reader; see Exercise 2.25. □

Note that the norm in $H^{1/2}(\partial\Omega)$ is given by definition by

$$\|z\|_{H^{1/2}(\partial\Omega)} := \inf_{v \in H^1(\Omega),\, \gamma v = z} \|v\|_{H^1} \quad \forall z \in H^{1/2}(\partial\Omega).$$

For further use, let us recall the following result:

Proposition 2.7 *Let $G \subset \mathbb{R}^m$ be a non-empty bounded open set and assume that $v \in H^1(G)$. Let us set $v_+ := \max(v, 0)$ and $v_- := -\min(v, 0)$. Then v_+ and v_- belong to $H^1(G)$ and, moreover,*

$$\nabla v_+ = \nabla v \, \mathbb{1}_{\{\mathbf{x} \in G : v(\mathbf{x}) > 0\}} \text{ and } \nabla v_- = -\nabla v \, \mathbb{1}_{\{\mathbf{x} \in G : v(\mathbf{x}) < 0\}} \text{ a.e. in } G. \tag{2.24}$$

Proof Let $v \in H^1(G)$ be given. It is clear that $v_+ \in L^2(G)$. Let us also see that $v_+ \in H^1(G)$ and the first identity in (2.24) is satisfied.

It can be assumed (by density) that $v \in C^1(\overline{G})$. Let $\varepsilon > 0$ be given and let us set

$$p_\varepsilon(s) := \frac{1}{2}\left(\sqrt{s^2 + \varepsilon^2} + s\right) \quad \forall s \in \mathbb{R}.$$

Then, it is true that $p_\varepsilon(v) \in H^1(G)$ and

$$\partial_i p_\varepsilon(v) = p_\varepsilon'(v)\, \partial_i v = \frac{1}{2}\left(\frac{v}{\sqrt{v^2 + \varepsilon^2}} + 1\right) \partial_i v \text{ for } i = 1, \ldots, N.$$

Indeed, for any $\varphi \in \mathcal{D}(G)$, one has

$$\langle \partial_i p_\varepsilon(v), \varphi \rangle = -\int_G p_\varepsilon(v)\, \partial_i \varphi \, d\mathbf{x} = \int_G p_\varepsilon'(v)\, \partial_i v \, \varphi \, d\mathbf{x}.$$

This is true for any small $\varepsilon > 0$. In view of *Lebesgue's Dominated Convergence Theorem,* we can take limits as $\varepsilon \to 0^+$ and find that

$$\langle \partial_i v_+, \varphi \rangle = -\int_G v_+ \, \partial_i \varphi \, d\mathbf{x} = \int_G \mathbb{1}_{\{\mathbf{x}\in G: v(\mathbf{x})>0\}} \partial_i v \, \varphi \, d\mathbf{x}.$$

Since φ is arbitrary $\mathcal{D}(G)$, we get that $v_+ \in H^1(G)$ and $\nabla v_+ = \nabla v \mathbb{1}_{\{\mathbf{x}\in G: v(\mathbf{x})>0\}}$ a.e.

A similar argument shows that $v_- \in H^1(G)$ and the second equality of (2.24) is also satisfied. □

For use in the sequel, note that if H and V are two Banach spaces with $V \hookrightarrow H$ (that is, $V \subset H$ and the identity mapping from V into H is continuous) and V is dense in H, we also have $H' \hookrightarrow V'$, where the embedding is again continuous and dense. Moreover, if the embedding $V \hookrightarrow H$ is compact, the same is true for the embedding $H' \hookrightarrow V'$.

If V and H are Hilbert spaces under these circumstances, after a standard identification of H and H', we can write that $V \hookrightarrow H \hookrightarrow V'$.

Thus, the following holds:

Proposition 2.8 *Let us denote by $H^{-1/2}(\partial\Omega)$ the dual of the space $H^{1/2}(\partial\Omega)$. Then, up to isomorphisms, one has*

$$H^{1/2}(\partial\Omega) \hookrightarrow L^2(\partial\Omega) \hookrightarrow H^{-1/2}(\partial\Omega), \tag{2.25}$$

where the embeddings are dense and continuous.

This means that we can view $H^{-1/2}(\partial\Omega)$ as a space that contains $L^2(\partial\Omega)$ as a dense subspace. In other words, the elements of $H^{-1/2}(\partial\Omega)$ can be viewed as *generalized L^2 functions on $\partial\Omega$.*

Although it will be little used in this book, it is important to mention that the embeddings $H^{1/2}(\partial\Omega) \hookrightarrow L^2(\partial\Omega)$ and $L^2(\partial\Omega) \hookrightarrow H^{-1/2}(\partial\Omega)$ are not only continuous but compact, as long as $\partial\Omega$ is bounded. That is, any bounded set in $H^{1/2}(\partial\Omega)$ (resp. $L^2(\partial\Omega)$) is relatively compact in $L^2(\partial\Omega)$ (resp. $H^{-1/2}(\partial\Omega)$).

Let $E(\Omega)$ be given by

$$E(\Omega) = \{\mathbf{v} \in L^2(\Omega)^N : \nabla \cdot \mathbf{v} \in L^2(\Omega)\}.$$

Then, $E(\Omega)$ is a Hilbert space with respect to the scalar product

$$(\mathbf{u}, \mathbf{v})_E := (\mathbf{u}, \mathbf{v}) + (\nabla \cdot \mathbf{u}, \nabla \cdot \mathbf{v}). \tag{2.26}$$

Here and henceforth, we have denoted by $\nabla \cdot \mathbf{v}$ the *divergence* of $\mathbf{v}$, that is

$$\nabla \cdot \mathbf{v} := \sum_{i=1}^{N} \partial_i v_i .$$

The following holds:

Lemma 2.4 *The space $C^1(\overline{\Omega})^N$ is dense in $E(\Omega)$.*

A consequence of this lemma is the following result:

Theorem 2.4 (Normal Traces in $E(\Omega)$) *The linear mapping $\mathbf{v} \mapsto \mathbf{v} \cdot \mathbf{n}|_{\partial\Omega}$, that is well defined in $C^1(\overline{\Omega})^N$, can be extended to a unique bounded linear operator $\gamma_n : E(\Omega) \mapsto H^{-1/2}(\partial\Omega)$, called the normal trace on $\partial\Omega$. Moreover, for any $\mathbf{v} \in E(\Omega)$ and any $w \in H^1(\Omega)$, one has the so called Stokes formula*

$$(\mathbf{v}, \nabla w) + (\nabla \cdot \mathbf{v}, w) = \langle \gamma_n \mathbf{v}, \gamma w \rangle, \tag{2.27}$$

where $\langle \cdot\,, \cdot \rangle$ denotes the duality pairing for $H^{-1/2}(\partial\Omega)$ and $H^{1/2}(\partial\Omega)$.

Proof Let us endow $C^1(\overline{\Omega})^N$ for a moment with the topology of $E(\Omega)$ and let us consider the linear mapping $\mathbf{v} \mapsto \mathbf{v} \cdot \mathbf{n}|_{\partial\Omega}$, that is obviously well defined from $C^1(\overline{\Omega})^N$ into $H^{-1/2}(\partial\Omega)$ in view of (2.25).

This is a linear continuous mapping, as a straightforward consequence of the identity

$$\int_\Omega (\mathbf{v} \cdot \nabla w + (\nabla \cdot \mathbf{v})\, w) = \int_{\partial\Omega} (\mathbf{v} \cdot \mathbf{n})\, \gamma w \, d\Gamma, \tag{2.28}$$

that must hold for all $w \in H^1(\Omega)$; see Exercise 2.29. As before, we have denoted here by $d\Gamma$ the integration element associated to the surface measure on $\partial\Omega$.

Consequently, the normal trace $\gamma_n : E(\Omega) \mapsto H^{-1/2}(\partial\Omega)$ is well defined.

Also, taking into account the identity (2.28) and the fact that $C^1(\overline{\Omega})^N$ is dense in $E(\Omega)$, we deduce at once (2.27) for any $\mathbf{v} \in E(\Omega)$ and any $w \in H^1(\Omega)$. □

The following is a very important result that concerns the space $H_0^1(\Omega)$:

Theorem 2.5 (Poincaré Inequality) *There exists a constant $C(\Omega) > 0$ such that*

$$\|v\| \le C(\Omega)\|\nabla v\| \quad \forall v \in H_0^1(\Omega). \tag{2.29}$$

Proof It suffices to prove (2.29) when $v \in \mathcal{D}(\Omega)$, since $\mathcal{D}(\Omega)$ is dense in $H_0^1(\Omega)$.

Thus, let φ be given in $\mathcal{D}(\Omega)$ and denote by $\tilde{\varphi}$ its extension by zero to $\mathbb{R}^N$. Then, $\tilde{\varphi} \in \mathcal{D}(\mathbb{R}^N)$ and, for any $\mathbf{x} \in \Omega$, one has

$$\varphi(\mathbf{x}) = \int_{-\infty}^{x_1} \partial_1 \tilde{\varphi}(y_1, \ldots, x_N)\, dy_1 = \int_{-M}^{x_1} \partial_1 \tilde{\varphi}(y_1, \ldots, x_N)\, dy_1$$

for some $M > 0$. If M is large enough, we deduce that

$$|\varphi(\mathbf{x})|^2 \le \left(\int_{-M}^{M} |\partial_1 \tilde{\varphi}(y_1, \ldots, x_N)|\, dy_1 \right)^2 \le 2M \int_{-M}^{M} |\partial_1 \tilde{\varphi}(y_1, \ldots, x_N)|^2\, dy_1$$

and also

$$\int_\Omega |\varphi(\mathbf{x})|^2 \, d\mathbf{x} = \int_{(-M,M)^N} |\tilde{\varphi}(\mathbf{x})|^2 \, d\mathbf{x} \leq 2M \int_\Omega |\partial_1 \tilde{\varphi}(\mathbf{x})|^2 \, d\mathbf{x}.$$

This ends the proof. □

It is clear from the proof that this result still holds if Ω is bounded only in one direction. However, it does not hold i general for unbounded Ω.

It also fails outside $H_0^1(\Omega)$; see however Exercise 2.31, where inequalities of the kind (2.29) are established for functions in some specific subspaces of $H^1(\Omega)$ different from $H_0^1(\Omega)$.

From Theorem 2.5, we deduce that the bilinear form

$$((u, v)) := \sum_{|\alpha|=1} (\partial^\alpha u, \partial^\alpha v) \tag{2.30}$$

is in fact an inner product in $H_0^1(\Omega)$, with associated norm

$$\|\nabla v\| = \left[\sum_{|\alpha|=1} \|\partial^\alpha v\|^2 \right]^{1/2}.$$

Thanks to (2.29), this norm is equivalent in $H_0^1(\Omega)$ to the usual norm in $H^1(\Omega)$.

Let us now recall some embedding properties of Sobolev spaces:

Theorem 2.6 (Sobolev Embeddings) *Assume that* $1 \leq p \leq +\infty$ *and let* $p_\star$ *be defined as follows:*

$$\begin{cases} \dfrac{1}{p_\star} = \dfrac{1}{p} - \dfrac{1}{N} & \text{if } p < N, \\ p_\star \text{ is arbitrary in } [1, +\infty) & \text{if } p = N, \\ p_\star = +\infty & \text{if } p > N. \end{cases}$$

Then any $v \in W^{1,p}(\Omega)$ *belongs to* $L^{p_\star}(\Omega)$ *and the embedding*

$$W^{1,p}(\Omega) \hookrightarrow L^{p_\star}(\Omega) \tag{2.31}$$

is continuous and dense. Moreover, the embedding is compact in $L^s(\Omega)$, *for all* s *with* $1 \leq s < p_\star$.

More generally, if $m \geq 1$, $1 \leq p \leq +\infty$ and q is defined as

$$\begin{cases} \dfrac{1}{q} = \dfrac{1}{p} - \dfrac{m-n}{N}, & \text{if } (m-n)p < N, \\ q \in [1, +\infty), & \text{if } (m-n)p = N, \\ q = +\infty, & \text{if } (m-n)p > N, \end{cases} \tag{2.32}$$

then any $v \in W^{m,p}(\Omega)$ belongs to $W^{n,q}(\Omega)$ and the embedding

$$W^{m,p}(\Omega) \hookrightarrow W^{n,q}(\Omega) \tag{2.33}$$

is continuous and dense. Moreover, it is compact in $W^{n,s}(\Omega)$, for any n and s such that $n < m$ and $s = q$ or $n = m$ and $1 \leq s < q$.

Finally, if $\partial\Omega$ is regular enough, $mp > N$, k is the greatest integer such that $0 \leq k < m - N/p$ and

$$m - \frac{N}{p} = k + \alpha, \quad \text{with } \alpha \in (0, 1], \tag{2.34}$$

then any "class" in $W^{m,p}(\Omega)$ possesses exactly one representative that belongs to the space $C^{k,\alpha}(\overline{\Omega})$. Furthermore, the corresponding mapping is linear, continuous and one-to-one from $W^{m,p}(\Omega)$ into $C^{k,\alpha}(\overline{\Omega})$. Accordingly, it is usual to say in this case that

$$W^{m,p}(\Omega) \hookrightarrow C^{k,\alpha}(\overline{\Omega}), \tag{2.35}$$

with a continuous embedding.

The proof is given in Sect. 2.7.

It is usual to say that the embeddings (2.33) (resp. (2.35)) are of the Gagliardo-Nirenberg-Sobolev type (resp. of the Morrey type).

In the following chapters, in order to establish the existence or regularity of velocity fields, we will need several particular estimates. Some of them are contained in the following results:

Lemma 2.5 *Assume that $N = 2$. Then there exists a constant $C > 0$ such that*

$$\|v\|_{L^4} \leq C\|v\|_{L^2}^{1/2}\|\nabla v\|_{L^2}^{1/2} \quad \forall v \in H^1(\Omega). \tag{2.36}$$

Proof We will use the extension operator $E : H^1(\Omega) \mapsto H_0^1(G)$, where G is a neighborhood of $\overline{\Omega}$; see Exercise 2.22.

Accordingly, it will suffice to prove an inequality like (2.36) for the functions in $H_0^1(G)$:

$$\|v\|_{L^4(G)} \leq C\|v\|_{L^2(G)}^{1/2}\|\nabla v\|_{L^2(G)}^{1/2} \quad \forall v \in H_0^1(G). \tag{2.37}$$

By density, it can be assumed that $v \in \mathcal{D}(G)$. Then, if M is sufficiently large, one has:

$$|v(x_1, x_2)|^2 = 2 \int_{-M}^{x_1} \partial_1 v(y_1, x_2)\, v(y_1, x_2)\, dy_1$$
$$\leq 2 \left(\int_{-M}^{M} |\partial_1 v(y_1, x_2)|^2 \, dy_1 \right)^{1/2} \left(\int_{-M}^{M} |v(y_1, x_2)|^2 \, dy_1 \right)^{1/2}$$

for any $(x_1, x_2) \in G$.

Working similarly with $\partial_2 v$, we find that

$$|v(x_1, x_2)|^4 \leq 4 \left(\int_{-M}^{M} |\partial_1 v(y_1, x_2)|^2 \, dy_1 \right)^{1/2} \left(\int_{-M}^{M} |v(y_1, x_2)|^2 \, dy_1 \right)^{1/2}$$
$$\times \left(\int_{-M}^{M} |\partial_2 v(x_1, y_2)|^2 \, dy_2 \right)^{1/2} \left(\int_{-M}^{M} |v(x_1, y_2)|^2 \, dy_2 \right)^{1/2}$$

and, after integrating with respect to x_1 and x_2 and using the Cauchy-Schwarz inequality twice, the following appears:

$$\|v\|_{L^4(G)}^4 \leq \|\partial_1 v\|_{L^2(G)} \|\partial_2 v\|_{L^2(G)} \|v\|_{L^2(G)}^2.$$

This yields (2.37) and finishes the proof. □

Lemma 2.6 *Assume that $N = 3$. Then there exists constants $C > 0$ such that*

$$\|v\|_{L^4} \leq C \|v\|_{L^2}^{1/4} \|\nabla v\|_{L^2}^{3/4} \quad \forall v \in H^1(\Omega) \tag{2.38}$$

and

$$\|v\|_{L^3} \leq C \|v\|_{L^2}^{1/2} \|\nabla v\|_{L^2}^{1/2} \quad \forall v \in H^1(\Omega).$$

Proof Once more, we use the extension operator $E : H^1(\Omega) \mapsto H_0^1(G)$. Thus, we must prove that

$$\|v\|_{L^4(G)} \leq C \|v\|_{L^2(G)}^{1/4} \|\nabla v\|_{L^2(G)}^{3/4} \quad \forall v \in H_0^1(G) \tag{2.39}$$

and

$$\|v\|_{L^3(G)} \leq C \|v\|_{L^2(G)}^{1/2} \|\nabla v\|_{L^2(G)}^{1/2} \quad \forall v \in H_0^1(G). \tag{2.40}$$

Let v be given in $H_0^1(G)$. Then

$$\int_G |v|^4 = \int_G |v|^a |v|^b \le \left(\int_G |v|^{ar}\right)^{1/r} \left(\int_G |v|^{br'}\right)^{1/r'}$$

provided $a + b = 4$, $r \in (1, +\infty)$ and $r' = r/(r-1)$.

If we impose $ar = 2$ and $br' = 6$, we find $a = 1$, $b = 3$ and $r = r' = 2$. Therefore, we have

$$\|v\|^4_{L^4(G)} \le \|v\|_{L^2(G)} \|v\|^3_{L^6(G)} \le C \|v\|_{L^2(G)} \|\nabla v\|^3_{L^2(G)},$$

where we have applied (2.31) for $p = 2$ and $p^* = 6$ and the fact that, in the space $H_0^1(G)$, $v \mapsto \|\nabla v\|_{L^2(G)}$ is a norm equivalent to the usual norm $\|\cdot\|_{H^1(G)}$.

This proves (2.39). The proof of (2.40) is completely similar and is left to the reader. □

In fact, Lemma 2.5 and Lemma 2.6 are particular cases of the so called Galiardo-Nirenberg inequalities. For completeness, let us recall the following general result:

Lemma 2.7 *Let $\Omega \subset \mathbb{R}^N$ be a bounded connected open set satisfying the so called exterior cone condition.*[8]

Let the exponent $1 \le q \le +\infty$ and the nonnegative integers j and m be given, with $j < m$. Furthermore, let $1 \le r \le +\infty$, $1 \le p < +\infty$ and $\theta \in [0, 1]$ be given and assume that

$$\frac{1}{p} = \frac{j}{N} + \theta\left(\frac{1}{r} - \frac{m}{N}\right) + \frac{1-\theta}{q}, \quad \frac{j}{m} \le \theta \le 1.$$

If $r > 1$ and $m - j - N/r$ is a nonnegative integer, assume also that $\theta < 1$. Then, the following holds:

1. *For any $\sigma \ge 1$, there exists a constant $C_\sigma > 0$ such that, for all $u \in L^q(\Omega)$ with $D^m u \in L^r(\Omega)$, one has:*

$$\|D^j u\|_{L^p} \le C_\sigma \|D^m u\|^\theta_{L^r} \|u\|^{1-\theta}_{L^q} + C_\sigma \|u\|_{L^\sigma}.$$

2. *There exists $C > 0$ such that*

$$\|D^j u\|_{L^p} \le C \|u\|^\theta_{W^{m,r}} \|u\|^{1-\theta}_{L^q} \quad \forall u \in L^q(\Omega) \cap W^{m,r}(\Omega). \tag{2.41}$$

A complete proof of Lemma 2.7 is given in [20].

In addition to Lemmas 2.5 and 2.6, we also have another consequence of (2.41):

[8] This means that, for every $\mathbf{x}_0 \in \partial\Omega$, there exists a *cone* of vertex $\mathbf{x}_0$ contained in $\mathbb{R}^N \setminus \Omega$. If $\partial\Omega$ is for instance of class C^2, then this property is clearly satisfied at any boundary point.

Lemma 2.8 *Assume that $N \geq 1$. Then there exists a constant $C > 0$ such that*

$$\|u\| \leq C\|u\|_{L^1}^{2/(N+2)}\|\nabla u\|^{N/(N+2)} \quad \forall u \in H^1(\Omega).$$

This is called *Nash inequality*. It is deduced from (2.41) by taking $j = 0, m = 1$, $p = 2, q = 1, r = 2$ and $\theta = N/(N+2)$.

As an immediate consequence of the argument used in the proof of Theorem 2.6, we get:

Lemma 2.9 *Assume that $N = 3$, $1 \leq p < 3$ and $1/p_\star = 1/p - 1/3$. Then*

$$\|\phi\|_{L^{p_\star}} \leq C\|\nabla\phi\|_{L^p} \quad \forall \phi \in \mathcal{D}(\Omega),$$

where C depends on p but is independent of Ω.

The following technical results, taken from [26, 27], will be useful:

Lemma 2.10 *Assume that $N = 3$. The following assertions hold:*

(i) Let s and p satisfy $1 \leq s \leq p \leq +\infty$. If $p_\star$ is as in Theorem 2.6 and $1/r = 1/p_\star + 1/s$, then the bilinear mapping that to each couple (u, v) assigns the product uv is well defined and continuous from $W_0^{1,p}(\Omega) \times W_0^{1,s}(\Omega)$ into $W_0^{1,r}(\Omega)$.

(ii) If $6/5 \leq s \leq 3$ and $1/r = 1/6 + 1/s$, then $(\mathbf{u}, v) \mapsto v\mathbf{u}$ is a well defined continuous mapping from $H_0^1(\Omega)^3 \times W_0^{1,s}(\Omega)$ into $W_0^{1,r}(\Omega)^3$.

(iii) Let s and p satisfy $1 \leq s, p \leq +\infty$, $s > 1$ and assume that $1/r = 1/p_\star + 1/s$. For any $u \in W_0^{1,p}(\Omega)$ and any $g \in W^{-1,r'}(\Omega)$, consider the distribution μu, given by

$$\langle gu, \varphi\rangle_{W^{-1,s'}, W_0^{1,s}} = \langle g, u\varphi\rangle_{W^{-1,r'}, W_0^{1,r}}, \quad \forall \varphi \in W_0^{1,s}(\Omega). \tag{2.42}$$

Then, the bilinear mapping $(u, g) \mapsto gu$ is well defined and continuous from $W_0^{1,p}(\Omega) \times W^{-1,r}(\Omega)$ into $W^{-1,s}(\Omega)$.

(iv) If $s > 2$ and $1/r = 1/6 + 1/s$, then $(\mathbf{u}, g) \mapsto g\mathbf{u}$ is a well defined continuous mapping from $H_0^1(\Omega)^3 \times W^{-1,r}(\Omega)$ into $W^{-1,s}(\Omega)^3$.

Proof The previous properties are almost direct consequences of Hölder inequalities and Sobolev embeddings. For instance, let us see that the assertion in *(i)* holds for $p < 3$.

Let u and v be given, with $u \in W_0^{1,p}(\Omega)$ and $v \in W_0^{1,s}(\Omega)$. Then $u \in L^{p^*}(\Omega)$ and $v \in L^{s^*}(\Omega)$, where

$$\frac{1}{p^*} = \frac{1}{p} - \frac{1}{3} \text{ and } \frac{1}{s^*} = \frac{1}{s} - \frac{1}{3}.$$

Recall that

$$\frac{1}{p^*} + \frac{1}{s} = \frac{1}{r} \tag{2.43}$$

and, consequently,

$$\frac{1}{q} := \frac{1}{p^*} + \frac{1}{s^*} < \frac{1}{r}.$$

Therefore, $uv \in L^q(\Omega) \subset L^r(\Omega)$. On the other hand, for any i with$1 \leq i \leq 3$, the functions $\partial_i u\, v$ and $u\, \partial_i v$ belong to $L^r(\Omega)$, in view of (2.43). This shows that $uv \in W^{1,r}(\Omega)$.

We deduce that the bilinear mapping $(u, v) \mapsto uv$ is well defined from $W_0^{1,p}(\Omega) \times W_0^{1,s}(\Omega)$ into $W^{1,r}(\Omega)$.

Moreover,

$$\|uv\|_{W^{1,r}} \leq \|u\|_{W_0^{1,p}} \|uv\|_{W_0^{1,s}}$$

and this shows that the mapping is continuous.

Observe that a similar argument shows that the mapping $(u, v) \mapsto uv$ is well defined and continuous from $W^{1,p}(\Omega) \times W_0^{1,s}(\Omega)$ into $W_0^{1,r}(\Omega)$.

We also have that $uv \in W_0^{1,r}(\Omega)$ for all $u \in W_0^{1,p}(\Omega)$ and $v \in W_0^{1,s}(\Omega)$.

Indeed, note that u and v can be written as strong limits in their respective spaces of sequences in $\mathcal{D}(\Omega)$; therefore, the same can be said about uv.

This concludes the proof of *(i)* when $p < 3$.

Let us now prove the third assertion.

Note that, if $u \in W_0^{1,p}(\Omega)$ and $\mu \in W^{-1,r'}(\Omega)$, then (2.42) provides the definition of a distribution $u\mu \in W^{-1,s'}(\Omega)$. Furthermore, one has

$$\|u\mu\|_{W^{-1,s}} = \sup_{\varphi \in W_0^{1,s}(\Omega)} \frac{\langle \mu, u\varphi \rangle_{W^{-1,r'}, W_0^{1,r}}}{\|\varphi\|_{W_0^{1,s}}} \leq C \|u\|_{W_0^{1,p}} \|\mu\|_{W^{-1,r'}},$$

that is, the bilinear mapping $(u, \mu) \mapsto u\mu$ is continuous.

This proves part *(iii)*.

The remaining assertions are left to the reader; see Exercise 2.38. □

For any given $T > 0$, we can define as before the spaces $\mathcal{D}(0, T)$, $\mathcal{D}'(0, T)$, $L^p_{\text{loc}}(0, T)$, $W^{m,p}(0, T)$, $W_0^{m,p}(0, T)$ and $W^{-m,p'}(0, T)$.

For future use, it is convenient to reformulate Theorem 2.6 in this framework. Thus, we have the following results (more details can be found in [4]):

Lemma 2.11 *Let $T > 0$, $v \in L^1(0, T)$ and $M \in \mathbb{R}$ be given and let us set*

$$u(t) := M + \int_0^t v(s)\,ds \quad \forall t \in [0, T].$$

Then u is absolutely continuous in $[0, T]$. Moreover, (the class represented by) u belongs to $W^{1,1}(0, T)$ and satisfies $u' = v$ in the sense of distributions and also in $L^1(0, T)$, that is, a.e. in $(0, T)$.

Proof For any $t_0 \in [0, T)$ and any $h > 0$ small enough one has, in view of Lebesgue's Theorem, that

$$u(t_0 + h) - u(t_0) = \int_{t_0}^{t_0+h} v(s)\,ds = \int_0^T \mathbb{1}_{[t_0,t_0+h]}(s)v(s)\,ds$$

converges to zero as $h \to 0$. This, together with a similar argument for $h < 0$, shows that u is continuous in $[0, T]$.

Now, let $\phi \in \mathcal{D}(0, T)$ be given. One has

$$\begin{aligned} -\int_0^T u(t)\phi'(t)\,dt &= -M\int_0^T \phi'(t)\,dt - \int_0^T \left(\int_0^t v(s)\phi'(t)\,ds\right) dt \\ &= -\int_0^T \left(\int_0^t v(s)\phi'(t)\,ds\right) dt, \end{aligned}$$

since $\phi(0) = \phi(T) = 0$. From Fubini's Theorem, we get

$$\begin{aligned} -\int_0^T u(t)\phi'(t)\,dt &= -\int_0^T \left(\int_0^T \mathbb{1}_{[0\le s\le t]}v(s)\,ds\right)\phi'(t)\,dt \\ &= -\int_0^T \left(\int_0^T \mathbb{1}_{[0\le s\le t]}v(s)\,dt\right)\phi'(t)\,ds \\ &= -\int_0^T \left(v(s)\int_s^T \phi'(t)\,dt\right) ds \\ &= \int_0^T v(s)\phi(s)\,ds. \end{aligned}$$

Therefore, $v = u'$ in the sense of distributions and, since $v \in L^1(0, T)$, we finally have that u is absolutely continuous and belongs to $W^{1,1}(0, T)$. □

Lemma 2.12 *Let $T > 0$ and $u \in W^{1,1}(0, T)$ be given. Then u has a representative $\tilde{u}$ which is absolutely continuous in $[0, T]$. Moreover,*

$$\tilde{u}(t_2) = \tilde{u}(t_1) + \int_{t_1}^{t_2} u'(s)\,ds \quad \forall t_1, t_2 \in [0, T]. \tag{2.44}$$

Proof Let w be defined by

$$w(t) := \int_0^t u'(s)ds.$$

It follows from Lemma 2.11 that w is continuous, belongs to $W^{1,1}(0,T)$ and satisfies $w' = u'$ in the sense of distributions. Therefore, the distributional derivative of $u - w$ is zero, which implies $u - w = C$ a.e. for some C. Now, set $\tilde{u} := w + C$, so that $\tilde{u} = u$ a.e. Then, one has

$$\tilde{u}(t_2) - \tilde{u}(t_1) = \left(C + \int_0^{t_2} u'(s)\,ds\right) - \left(C + \int_0^{t_1} u'(s)\,ds\right) = \int_{t_1}^{t_2} u'(s)\,ds$$

for all $t_1, t_2 \in [0,T]$. This ends the proof. □

Combining this lemma and the Sobolev embedding in Theorem 2.6 with $N = 1$, $m = 1$ and $p \in [1, +\infty)$, we find that any $u \in W^{1,p}(0,T)$ has a unique representative $\tilde{u}$ that belongs to $C^{0,\alpha}([0,T])$ with $\alpha = 1 - 1/p$ and satisfies the identity (2.44). Furthermore, the linear embedding $u \mapsto \tilde{u}$ is continuous.

It is usual to identify u and $\tilde{u}$. Accordingly, as before, we can view $W^{1,p}(0,T)$ as a subspace of $C^{0,\alpha}([0,T])$.

In the sequel, for functions with values in $\mathbb{R}^N$, we will specify the components with appropriate indices. For example, by writing $\mathbf{v} \in W^{m,p}(\Omega)^N$ we mean that $\mathbf{v} = (v_1, \dots, v_N)$ and each component v_i belongs to $W^{m,p}(\Omega)$. The same notation will be kept for $\mathbb{R}^N$-valued distributions, i.e., N-tuples $\mathbf{S} = (S_1, \dots, S_N)$ with $S_i \in \mathcal{D}'(\Omega)$ for all $i = 1, \dots, N$.

Obviously, $W^{m,p}(\Omega)^N$ is a Banach space with respect to the standard product norm.

On the other hand, the scalar products in $H^m(\Omega)^N$ and $H_0^m(\Omega)^N$ are defined as in (2.18) or (2.30), with $\partial^\alpha u\, \partial^\alpha v$ replaced by the Euclidean inner product $\partial^\alpha \mathbf{u} \cdot \partial^\alpha \mathbf{v}$ for all α.

2.3 Functions and Distributions with Values in Banach Spaces

In order to solve evolution (i.e., time-dependent) PDE problems, it is very appropriate to view their candidates to solutions as functions defined in a time interval $[0,T]$, with values in a Banach space. In most cases, this will be a space of real-valued functions defined on an open set $\Omega \subset \mathbb{R}^N$.

This motivates the study and analysis of functions $f : G \mapsto B$, where $G \in \mathbb{R}^k$ is a non-empty open set and B is a Banach space and also of similar B-valued distributions.

Let the function $f : G \mapsto B$ be given. Recall that it is said that f is Böchner-measurable (or strongly measurable) if there exists a zero-measure set $Z \subset G$ such that $f|_{G\setminus Z}$ takes values in a separable space $B_0 \subset B$ and f is weakly measurable, that is, $\langle \ell, f \rangle_{B',B} : G \mapsto \mathbb{R}$ is measurable for every $\ell \in B'$.

This is equivalent to have $f(\mathbf{z}) = \lim_{n\to+\infty} f_n(\mathbf{z})$ for all $\mathbf{z}$, for some measurable functions $f_n : G \mapsto B$ that take values in a countable set. Furthermore, if the Banach space B is separable, then $f : G \mapsto B$ is Böchner-measurable if and only if it is weakly measurable; see [8] for more details.

If $f : G \mapsto B$ is Böchner-measurable, then $\mathbf{z} \mapsto \|f(\mathbf{z})\|_B$ is measurable and the Lebesgue integral

$$\int_0^T \|f(\mathbf{z})\|_B \, d\mathbf{z}$$

has always a sense (finite or not). If it is finite, it is usual to say that f is Bochner-integrable. This allows to define $\mathcal{L}^1(G; B)$ and then, after identification of functions that coincide a.e. in G, the set $L^1(G; B)$.

More generally, for any p with $1 \le p < +\infty$, we will denote by $L^p(G; B)$ the set of all (classes of) Bochner-measurable functions $f : G \mapsto B$ such that

$$\|f\|_{L^p(G;B)} := \left[\int_0^T \|f(\mathbf{z})\|_B^p \, dt\right]^{1/p} < +\infty.$$

On the other hand, we will denote by $L^\infty(0, T; B)$ the set of (classes of) Böchner-measurable functions $f : G \mapsto B$ that are essentially bounded, that is, satisfy

$$\|f\|_{L^\infty(0,T;B)} := \inf\{M : \|f(\mathbf{z})\|_B \le M \text{ a.e.}\} < +\infty.$$

The $L^p(G; B)$ are vector spaces for the usual linear operations. Furthermore, for any $p \in [1, +\infty]$, $\|\cdot\|_{L^p(G;B)}$ is a norm in $L^p(G; B)$ and, endowed with this norm, $L^p(G; B)$ is complete.

In particular, if $p = 2$ and $B = H$ is a Hilbert space with product $(\cdot\,,\cdot)_H$, $L^2(G; H)$ is also a Hilbert space for the scalar product

$$(f, g)_{L^2(G;H)} := \int_0^T (f(\mathbf{z}), g(\mathbf{z}))_H \, d\mathbf{z}.$$

Proposition 2.9 *Let $1 \le p < +\infty$. If the Banach space B is separable, then the same happens to $L^p(G; B)$.*

As before, we can also speak of locally p-integrable B-valued functions on G. This leads to the spaces $L^p_{\text{loc}}(G; B)$. For brevity, we omit the details.

The $L^p_{\text{loc}}(G; B)$ form an ordered family of Fréchet spaces for some appropriate semi-norms. Accordingly, it will be said that $f_n \to f$ strongly (resp. weakly) in $L^p_{\mathrm{loc}}(G; B)$ if, for any bounded open set $G_0 \subset\subset G$, the associated restrictions $f_n|_{G_0}$ converge to $f|_{G_0}$ strongly (resp. weakly) in $L^p(G_0; B)$.

Let us now introduce Banach-valued distributions:

Definition 2.5 A distribution on G with values in the Banach space B, also called a B-valued distribution on G, is a (sequentially) continuous linear mapping $S : \mathcal{D}(G) \mapsto B$, i.e., a linear mapping such that

$$\text{If } \phi_n \to 0 \text{ in } \mathcal{D}(G), \text{ then } S(\phi_n) \to 0 \text{ in } B.$$

The set of B-valued distributions on G is a vector space for the usual laws and will be denoted by $\mathcal{D}'(G; B)$.

As in the scalar case, it is usual to denote by $\langle S, \phi \rangle$ the value assigned by S to the test function ϕ.

In particular, it is also possible to consider B-valued distributions on $(0, T)$ for any $T \in (0, +\infty]$. The corresponding space is usually denoted $\mathcal{D}'(0, T; B)$.

Many of the properties mentioned above for real-valued distributions are also satisfied in this context. They will be mentioned in the following paragraphs.

Again, the space $\mathcal{D}'(G; B)$ can be endowed with a notion of convergence: we say that a sequence $\{S_n\}$ converges to S in $\mathcal{D}'(G; B)$ if $\langle S_n, \phi \rangle \to \langle S, \phi \rangle$ in B for any $\phi \in \mathcal{D}(G)$. In this case, we write

$$S_n \to S \text{ in } \mathcal{D}'(G; B).$$

There exists a vector topology on $\mathcal{D}'(G; B)$ compatible with this notion of convergence,[9] see [29] for details.

Once more, locally integrable functions may be viewed as distributions. More precisely, any function $f \in L^1_{\mathrm{loc}}(G; B)$ defines a distribution $S_f \in \mathcal{D}'(G; B)$ by

$$\langle S_f, \varphi \rangle := \int_0^T f(\mathbf{z})\varphi(\mathbf{z})\, d\mathbf{z} \quad \forall \varphi \in \mathcal{D}(G).$$

The distribution S_f is unique, as the following result shows:

Proposition 2.10 *Assume that $f, g \in L^1_{\mathrm{loc}}(G; B)$. Then, f and g define the same distributions if and only if f and g coincide a.e.*

Consequently, we can view $L^1_{\mathrm{loc}}(G; B)$ as a proper subspace of $\mathcal{D}'(G; B)$. It is furthermore clear that the mapping $f \mapsto S_f$ is sequentially continuous. Accordingly, we will write that $L^1_{\mathrm{loc}}(G; B) \hookrightarrow \mathcal{D}'(G; B)$ and will make no distinction in the notation, denoting in the sequel the distribution S_f by f.

---

[9] Indeed, it suffices to consider the topology induced by the family of seminorms $\{p^B_\phi\}_{\phi \in \mathcal{D}(\Omega)}$, where $p^B_\phi(S) := \|\langle S, \phi \rangle\|_B$ for any $S \in \mathcal{D}'(G; B)$ and any $\phi \in \mathcal{D}(\Omega)$.

Definition 2.6 Let $S \in \mathcal{D}'(G; B)$ be given and let α be a multi-index. The (distributional) derivative of S of order α is the B-valued distribution defined as follows:

$$\langle \partial^\alpha S, \phi \rangle = (-1)^{|\alpha|} \langle S, \partial^\alpha \phi \rangle \quad \forall \phi \in \mathcal{D}(G).$$

Let us give an example. Let us take $B = C^0([0, 1])$ and let $f \in L^1((-1, 1) \times (-1, 1); B)$ be given by

$$f(y_1, y_2) = \begin{cases} v & \text{if } y_1 < y_2, \\ w & \text{if } y_1 > y_2, \end{cases}$$

where the functions v and w are given as follows:

$$v(s) = e^s, \quad w(s) = \sin s \quad \forall s \in [0, 1].$$

Let us consider the B-valued distribution defined by f. Then, for instance, $\partial^{(1,0)} f$ is defined as follows:

$$\langle \partial^{(1,0)} f, \varphi \rangle = \left(\int_{-1}^{1} \varphi(y, y)\, dy \right) (v + w) \quad \forall \varphi \in \mathcal{D}((-1, 1) \times (-1, 1)).$$

The distributional derivative is a well-defined linear sequentially continuous mapping from $\mathcal{D}'(G; B)$ into itself.

Proposition 2.11 *Assume that $m \geq 1$ and $f : G \mapsto B$ is of class C^m. Then, for any multi-index $\alpha = (\alpha_1, \dots, \alpha_k)$ with $|\alpha| := \alpha_1 + \cdots \alpha_k \leq m$, the distribution defined by $\partial^\alpha f$ is the derivative of order α of S_f, that is:*

$$S_{\partial^\alpha f} = \partial^\alpha S_f.$$

Definition 2.6 and Proposition 2.11 allow us to speak of derivatives of all orders of functions in $L^p(G; B)$. In particular, note that the (first-order) derivative of $f \in L^p(0, T; B)$ in the sense of $\mathcal{D}'(0, T; B)$ is the distribution defined by

$$\langle \frac{df}{dt}, \phi \rangle := - \int_0^T f(t) \frac{d\phi}{dt}(t)\, dt, \quad \forall \phi \in \mathcal{D}(0, T).$$

As in the case of real-valued functions and distributions, we can define Sobolev spaces for Banach-valued functions:

Definition 2.7 Let $m \geq 1$ be an integer and assume that $1 \leq p \leq +\infty$. The Sobolev space $W^{m,p}(G; B)$ is the space of functions $f \in L^p(G; B)$ such that

$$\partial^\alpha f \in L^p(G; B) \text{ for } |\alpha| \leq m.$$

The $W^{m,p}(G; B)$ are Banach spaces for the norms

$$\|f\|_{W^{m,p}(G;B)} := \left(\sum_{|\alpha| \le m} \|\partial^\alpha f\|^p_{L^p(G;B)} \right)^{1/p} \quad \text{if } p < +\infty,$$
$$\|f\|_{W^{m,\infty}(G;B)} := \max_{|\alpha| \le m} \|\partial^\alpha f\|_{L^\infty(G;B)} \quad \text{if } p = +\infty.$$

If $p = 2$ and $B = H$ is a Hilbert space with scalar product $(\cdot, \cdot)_H$, then it is usual to denote $W^{m,2}(G; B)$ by $H^m(G; H)$. It is a Hilbert space as well, with scalar product

$$(f, g)_{H^m(G;H)} := \sum_{|\alpha| \le m} \int_G (\partial^\alpha f, \partial^\alpha g)_H \, d\mathbf{z}. \tag{2.45}$$

Proposition 2.12 *For any $p \in [1, +\infty)$, the Banach space $(L^p(G; B))'$ is isometrically isomorphic to $L^{p'}(G; B')$. More precisely, for any continuous linear form $\ell \in (L^p(G; B))'$, there exists exactly one $g \in L^{p'}(G; B')$ such that*

$$\langle \ell, f \rangle_{L^{p'}(G;B'), L^p(G;B)} = \int_G \langle g(\mathbf{z}), f(\mathbf{z}) \rangle_{B',B} \, d\mathbf{z} \quad \forall f \in L^p(G; B).$$

The mapping $\ell \mapsto g$ is an isometric isomorphism. Consequently, if $1 < p < +\infty$ and B is a reflexive Banach space, the same happens to $L^p(G; B)$.

As in the case of real-valued functions, it can be proved that any function $f \in W^{1,1}(0, T; B)$ has a unique representative that is absolutely continuous in $[0, T]$ and, consequently, it can be written that $W^{1,1}(0, T; B) \hookrightarrow C^0([0, T]; B)$, with a continuous embedding.

We will denote by $\mathcal{D}(G; B)$ the space of C^∞ functions $\varphi : G \mapsto B$ with compact support. Also, the closure of $\mathcal{D}(G; B)$ in the space $W^{m,p}(G; B)$ (resp. $H^m(G; H)$) will be denoted by $W_0^{m,p}(G; B)$) (resp. $H_0^m(G; H)$).

Many of the results in Sect. 2.2 can be extended to the spaces $W^{m,p}(G; B)$ and $W_0^{m,p}(G; B)$.

In particular, with arguments similar to those in Sect. 2.2, it can be proved that, for any $p \in [1, +\infty)$, the space $C^1([0, T]; B)$, formed by the restrictions to $[0, T]$ of the functions in $C^1(\mathbb{R}; B)$, is *dense* in $H^1(0, T; B)$.

Furthermore, $H^1(0, T; B) \hookrightarrow C^0([0, T]; B)$ with a continuous embedding, whence the *trace operator* γ is well defined on $H^1(0, T; B)$, with

$$\gamma v = \left(v\big|_{t=0}, v\big|_{t=T} \right) \quad \forall v \in H^1(0, T; B)$$

and $N(\gamma) = H_0^1(0, T; B)$.

For any $p \in [1, +\infty)$, the dual of $W_0^{m,p}(G; B)$ is denoted by $W^{-m,p'}(G; B')$. It can be identified to a space of B'-valued distributions, as the following proposition shows:

Proposition 2.13 *For any integer $m \geq 0$ and any $p \in [1, +\infty)$, one has*

$$W^{-m,p'}(G; B') \cong \{S \in \mathcal{D}'(G; B') : S = \sum_{|\alpha| \leq m} \partial^\alpha v_\alpha \text{with the} v_\alpha \in L^{p'}(0, T; B')\},$$

where the identification is as follows: $S \in W^{-m,p'}(G; B')$ if and only if there exist functions $v_\alpha \in L^{p'}(G; B')$ such that

$$\langle S, u \rangle_{W^{-m,p'}(G;B'), W_0^{m,p}(G;B)} = \sum_{|\alpha| \leq m} (-1)^{|\alpha|} \int_G \langle v_\alpha, \partial^\alpha u \rangle_{B',B} \, d\mathbf{z}$$

for all $u \in W_0^{m,p}(G; B)$.

Recall that $\Omega \subset \mathbb{R}^N$ is a bounded connected open set with Lipschitz-continuous boundary.

Note that, whenever $S \in \mathcal{D}'(\Omega; B)$ and all the first-order derivatives $\partial_i S$ vanish, S must be a constant; see Exercise 2.47.

In the sequel, we will denote by $\mathcal{V}$ the space

$$\mathcal{V} := \{\mathbf{v} \in \mathcal{D}(\Omega)^N : \nabla \cdot \mathbf{v} = 0\}.$$

We have a very important result that characterizes the gradient of a distribution:

Theorem 2.7 (De Rham's Lemma) *Let B be a Banach space. Suppose that*

$$\mathbf{f} = (f_1, \ldots, f_N) \text{ with } f_i \in \mathcal{D}'(\Omega; B), \quad 1 \leq i \leq N.$$

Then $\mathbf{f} = \nabla p$ for some $p \in \mathcal{D}'(\Omega; B)$ if and only if

$$\langle \mathbf{f}, \mathbf{v} \rangle := \sum_{i=1}^{N} \langle f_i, v_i \rangle = 0 \quad \forall \mathbf{v} \in \mathcal{V}. \tag{2.46}$$

Furthermore, for any $m \geq 0$ and any $r \in [1, +\infty)$, we can assign to each $\mathbf{f} \in W^{m-1,r}(\Omega; B)^N$ satisfying (2.46) a distribution $p \in W^{m,r}(\Omega; B)$ such that $\mathbf{f} = \nabla p$. This can be done in such a way such that the mapping $\mathbf{f} \mapsto p$ is well defined and continuous from $W^{m-1,r}(\Omega; B)^N \cap \{\mathbf{f} : \langle \mathbf{f}, \mathbf{v} \rangle = 0 \ \forall \mathbf{v} \in \mathcal{V}\}$ into $W^{m,r}(\Omega; B)$.

Remark 2.1 A constructive proof of this result due to J. Simon is reproduced in Sect. 2.8. Other proofs can be found in the literature; see for example [11, 13, 17, 31, 32] and the more recent contribution [2]. □

Remark 2.2 Later, we will apply this Theorem with $\mathbf{f}$ in $H^{m-1}(\Omega; B)$ satisfying (2.46) for various m and various B; see for instance Theorem 2.11 and Proposition 5.1. □

In the following chapters, we will almost permanently need the spaces

$$H := \text{ the closure of } \mathcal{V} \text{ in } L^2(\Omega)^N, \tag{2.47}$$

$$V := \text{ the closure of } \mathcal{V} \text{ in } H_0^1(\Omega)^N. \tag{2.48}$$

Obviously, H and V are separable Hilbert spaces for the norms of $L^2(\Omega)^N$ and $H_0^1(\Omega)^N$, respectively.

Theorem 2.8 *The spaces H and V satisfy the following properties:*

$$H = \{\mathbf{v} \in L^2(\Omega)^N : \nabla \cdot \mathbf{v} = 0, \quad \mathbf{v} \cdot \mathbf{n} = 0 \text{ on } \partial\Omega\} \tag{2.49}$$

and

$$V = \{\mathbf{v} \in H_0^1(\Omega)^N : \nabla \cdot \mathbf{v} = 0\}. \tag{2.50}$$

Moreover, the orthogonal complement $H^\perp$ *of* H *in* $L^2(\Omega)^N$ *is*

$$H^\perp = \{\mathbf{v} \in L^2(\Omega)^N : \exists p \in H^1(\Omega) \text{ such that } \mathbf{v} = \nabla p\}. \tag{2.51}$$

Proof In view of Theorem 2.4, any $\mathbf{v} \in L^2(\Omega)^N$ such that $\nabla \cdot \mathbf{v} \in L^2(\Omega)$ possesses a normal trace in the Hilbert space $H^{-1/2}(\partial\Omega)$.

Thus, to any $\mathbf{v}$ with $\nabla \cdot \mathbf{v} = 0$, it is correct to impose that $\mathbf{v} \cdot \mathbf{n}|_{\partial\Omega} = 0$. This shows that the right-hand side of (2.49) makes sense.

Note that (2.51) is an immediate consequence of Theorem 2.7.

Indeed, let us denote by $\tilde{H}$ and $\tilde{V}$ the spaces on the right-hand sides of (2.49) and (2.50), respectively.

It is clear that $H \subset \tilde{H}$. To prove that $\tilde{H} \subset H$, let us suppose that there exists $\mathbf{v}_0 \in \tilde{H} \setminus H$. Then, as an immediate consequence of the *Hahn-Banach Theorem,* there must exist $\mathbf{z} \in L^2(\Omega)^N$ such that

$$(\mathbf{z}, \mathbf{v}_0) = 1 \text{ and } (\mathbf{z}, \mathbf{v}) = 0 \ \forall \mathbf{v} \in H.$$

From Theorem 2.7, we find that $\mathbf{z} = \nabla p$ for some $p \in H^1(\Omega)$. But then we are led to an absurdity: $(\mathbf{z}, \mathbf{v}_0) = 0$.

Consequently, $H = \tilde{H}$.

The proof that $V = \tilde{V}$ is similar. □

2.4 Some Additional Technical Results

This section deals with a collection of results that will play important roles in the theoretical analysis in Chaps. 3 and 5.

First, let us recall some estimates of the Gronwall kind. They have been taken from [12, 26] and will help to get estimates of the (exact or approximate) solutions in the following chapters.

Lemma 2.13 *Let g and k be nonnegative functions satisfying $g \in W^{1,1}(0,T)$, $k \in L^1(0,T)$ and*

$$\begin{cases} \dfrac{d}{dt} g^2 \le kg \ \text{ a.e. in } (0,T), \\ g(0) \le g_0, \end{cases}$$

where $g_0 \in \mathbb{R}_+$. Then

$$g(t) \le g_0 + \frac{1}{2}\int_0^t k(s)\,ds \quad \forall t \in [0,T]. \tag{2.52}$$

Proof Since $g \in W^{1,1}(0,T)$, it is absolutely continuous and its derivative makes sense a.e. in $[0,T]$. The same happens to g^2 and, moreover,

$$\frac{d}{dt}g^2 = 2g\frac{dg}{dt} \le kg,$$

whence

$$\frac{dg}{dt} \le \frac{k}{2} \quad \text{a.e. in } \{t \in [0,T] : g(t) > 0\}.$$

Here, we have used that $k \ge 0$.

Let $t \in [0,T]$ be given. Then

$$\begin{aligned} g(t) &= g(0) + \int_0^t \frac{dg}{dt}(s)\,ds \\ &\le g_0 + \int_A \frac{dg}{dt}(s)\,ds + \int_B \frac{dg}{dt}(s)\,ds, \end{aligned}$$

where A (resp. B) denotes the set of times $t \in [0,T]$ such that $g(t) > 0$ (resp. $g(t) = 0$). Consequently,

$$g(t) \le g_0 + \frac{1}{2}\int_0^t k(s)\,ds + \int_B \max\left(\frac{dg}{dt}(s), 0\right)\,ds.$$

But it is clear that $\frac{dg}{dt} = 0$ a.e. in B.

Hence, we get the inequality (2.52) and the lemma is proved. □

Lemma 2.14 *Let g be a nonnegative function in $W^{1,1}(0,T)$ and let k be given in $L^1(0,T)$. Assume that*

$$\begin{cases} \dfrac{dg}{dt} \leq F(g) + k(t) \ \text{ a.e. in } (0,T), \\ g(0) \leq g_0, \end{cases}$$

where $F : \mathbb{R} \mapsto \mathbb{R}$ is measurable and sublinear, that is, such that

$$|F(z)| \leq C_1 + C_2|z| \quad \forall z \in \mathbb{R}.$$

Then, for every $\varepsilon > 0$ there exists t_ε with $0 < t_\varepsilon \leq T$ (depending only on ε, C_1, C_2, T, g_0 and k), such that

$$g(t) \leq g_0 + \varepsilon \quad \forall t \in [0, t_\varepsilon].$$

Proof For all $t \in [0,T]$, we have

$$\begin{aligned} g(t) &= g(0) + \int_0^t F(g(s))\,ds + \int_0^t k(s)\,ds \\ &\leq g_0 + \int_0^t |F(g(s))|\,ds + \int_0^t |k(s)|\,ds \\ &\leq g_0 + C_1T + \|k\|_{L^1(0,T)} + C_2 \int_0^t |g(s)|\,ds, \end{aligned}$$

whence

$$g(t) \leq (g_0 + C_1T + \|k\|_{L^1(0,T)})e^{C_2t} \leq C_3.$$

Therefore,

$$g(t) \leq g_0 + \int_0^t (C_1 + C_2C_3 + |k(s)|)\,ds \quad \forall t \in [0,T],$$

and the assertion holds. □

Lemma 2.15 *Let the functions ϕ, ψ and f be Lipschitz-continuous and defined for all $t \geq 0$. Assume that $\phi, \psi \geq 0$ and*

$$\begin{cases} \dfrac{d\phi}{dt} + \psi \leq G(\phi) + f(t) \ \text{ a.e. in } [0,+\infty), \\ \phi(0) = \phi_0, \end{cases} \tag{2.53}$$

where $G : \mathbb{R}_+ \mapsto \mathbb{R}_+$ is a Lipschitz-continuous nondecreasing function. Then

$$\phi(t) \leq F(t;\phi_0) \quad \forall t \in [0, T(\phi_0)), \tag{2.54}$$

where $F(\cdot\,;\phi_0)$ is the maximal to the right solution to the initial-value problem

$$\begin{cases} \dfrac{dF}{dt} = G(F) + f(t) \ \text{ a.e. in } (0, T(\phi_0)), \\ F(0) = \phi_0. \end{cases}$$

Moreover, $\int_0^t \psi(\tau)\,d\tau \le \tilde{F}(t;\phi_0)$ for all $t \in [0, T(\phi_0))$, where

$$\tilde{F}(t;\phi_0) := \phi_0 + \int_0^t (G(F(\tau;\phi_0)) + f(\tau))\,d\tau.$$

Proof Set $v := F(\cdot\,;\phi_0) - \phi$. We can write that $v(0) = 0$ and

$$\frac{dv}{dt} \ge G(F(t;\phi_0)) - G(\phi(t)) = H(t)v \ \text{ in } [0, T(\phi_0)),$$

where H is given by

$$H(t) = \begin{cases} \dfrac{G(F(t;\phi_0)) - G(\phi(t))}{F(t;\phi_0) - \phi(t)} & \text{if } F(t;\phi_0) - \phi(t) \ne 0, \\ 0 & \text{otherwise.} \end{cases}$$

We deduce that $v \ge 0$ and the inequality (2.54) holds.

On the other hand, we also have from (2.53) that

$$\phi(t) + \int_0^t \psi(s)\,ds \le \int_0^t (G(\phi(s)) + f(s))\,ds + \phi_0 \le \tilde{F}(t;\phi_0) \ \ \forall t \in [0, T(\phi_0))$$

and this ends the proof. □

The following technical result will also be used:

Lemma 2.16 *Assume that $h \in L^1(0, +\infty)$, $h \ge 0$ and*

$$\int_{t_0}^t h(\tau)\,d\tau \le a_1(t - t_0) + a_2 \quad \forall t, t_0 \text{ with } 0 \le t_0 \le t, \tag{2.55}$$

where $a_1, a_2 \ge 0$. Then

$$\sup_{t \ge 0} \left(e^{-t} \int_0^t e^{\tau} h(\tau)\,d\tau \right) < +\infty. \tag{2.56}$$

Proof First, suppose that $0 \le t \le 1$. Then,

$$e^{-t} \int_{t_0}^t e^{\tau} h(\tau)\,d\tau \le \int_0^1 h(\tau)\,d\tau \le a_1 + a_2.$$

Now, if $t > 1$, let us introduce $n \geq 1$ and $r \in [0, 1)$ be such that $t = n + r$. We have:

$$\int_0^t e^\tau h(\tau)\,d\tau = \int_0^{n+r} e^\tau h(\tau)\,d\tau = \sum_{j=1}^{n} \int_{j-1}^{j} e^\tau h(\tau)\,d\tau + \int_n^{n+r} e^\tau h(\tau)\,d\tau$$
$$\leq \sum_{j=1}^{n} e^j \int_{j-1}^{j} h(\tau)\,d\tau + e^{n+1} \int_n^{n+1} h(\tau)\,d\tau.$$

From (2.55), we see that

$$\int_{j-1}^{j} h(\tau)\,d\tau \leq a_1 + a_2 \text{ for } j = 1, \ldots, n+1.$$

Therefore,

$$e^{-t} \int_0^t e^\tau h(\tau)\,d\tau = e^{-n-r} \int_0^{n+r} e^\tau h(\tau)\,d\tau \leq e^{-n-r} \sum_{j=1}^{n+1} e^j \int_{j-1}^{j} h(\tau)\,d\tau$$
$$\leq (a_1 + a_2) e^{-n-r} \sum_{j=1}^{n+1} e^j = (a_1 + a_2) e^{-n-r} \frac{e^{n+2} - e}{e - 1}$$
$$= (a_1 + a_2) \frac{e^{2-r} - e^{-n-r+1}}{e - 1} \leq (a_1 + a_2) \frac{e^2}{e - 1}$$

and, finally, (2.56) holds. □

Let X be a Banach space and let the mapping $\Phi : K \subset X \mapsto X$ be given. Recall that it is said that Φ is *compact* if the image $\Phi(B)$ of any bounded set $B \subset K$ is precompact in X.

Let us recall *Schauder's Fixed-Point Theorem.* This is one of the most important tools used in the theoretical analysis of nonlinear PDEs.

Theorem 2.9 *Let X be a Banach space, and let $K \subset X$ be a non-empty bounded closed convex set. Then any continuous compact mapping $\Phi : K \mapsto K$ has at least one fixed-point.*

For a proof, see for instance [36].

There are in fact several relevant variants of this result. For example, the compactness of Φ can be exchanged with the compactness of K; for details, see [18, 23]. However, without compactness assumptions on Φ and K, the result does not hold; it is not difficult to find counter-examples of mappings defined for instance in the closed unit ball of the Hilbert space ℓ^2 of square-summable sequences.

The existence of weak solutions to the variable density Navier-Stokes equations will be proved in Chap. 3. There, we will need the so called *Nikolskii spaces,* that are defined as follows.

Let B be a Banach space. Given a function $f : (0, T) \mapsto B$ and a (small) number $h > 0$, we denote by $\tau_h f$ the function

$$(\tau_h f)(t) := f(t + h) \quad \forall t \in (-h, T - h).$$

Definition 2.8 Let B be a Banach space. For any q and s with $1 \leq q \leq +\infty$ and $0 < s < 1$, the associated Nikolskii space $N^{s,q}(0, T; B)$ is

$$N^{s,q}(0, T; B) := \{f \in L^q(0, T; B) : \sup_{h>0} h^{-s} \|\tau_h f - f\|_{L^q(0,T-h;B)} < +\infty\}.$$

The $N^{s,q}(0, T; B)$ are Banach spaces with respect to the norms

$$\|f\|_{N^{s,q}(0,T;B)} := \|f\|_{L^q(0,T;B)} + \sup_{0<h<T} h^{-s} \|\tau_h f - f\|_{L^q(0,T-h;B)}.$$

In view of this definition, it can be said that the functions in $N^{s,q}(0, T; B)$ possess a kind of fractional derivative (of order s) that belongs to $L^q(0, T; B)$. In fact, for $q = +\infty$, we are speaking of Hölder-continuous B-valued functions with exponent s. Also, Nikolskii spaces are similar to the so called Besov spaces; see [33].

We will use in the sequel various consequences of the following general *compactness criterion* in the spaces $L^p(0, T; B)$, due to J. Simon [25]:

Theorem 2.10 *Assume that $1 \leq p < +\infty$ and B is a Banach space. Let $\mathcal{F} \subset L^p(0, T; B)$ be given. Then $\mathcal{F}$ is relatively compact in $L^p(0, T; B)$ if and only if the following two conditions are satisfied:*

$$\{\int_a^b f(t)\, dt : f \in \mathcal{F}\} \text{ is relatively compact in } B \text{ for all } a, b \in [0, T], \tag{2.57}$$

$$\sup_{f \in \mathcal{F}} \|\tau_h f - f\|_{L^1(0,T-h;B)} \to 0 \text{ as } h \to 0. \tag{2.58}$$

If $p = +\infty$, the same result holds if we replace $L^\infty(0, T; B)$ by $C^0([0, T]; B)$.

Proof First, let us assume that $\mathcal{F}$ is relatively compact in $L^p(0, T; B)$.

Let $a, b \in [0, T]$ be given. Since the mapping $f \mapsto \int_a^b f(t)\, dt$ is continuous, it is clear that $\{\int_a^b f(t)\, dt : f \in \mathcal{F}\}$ is relatively compact in B.

Now, let $\varepsilon > 0$ be given. There exist $f_1, \dots, f_n \in L^p(0, T; B)$ such that $\mathcal{F} \subset \bigcup_{i=1}^n B(f_i; \varepsilon/3)$. For every $i = 1, \dots, n$, there exists $h_i > 0$ such that

$$0 < h \leq h_i \Rightarrow \|\tau_h f_i - f_i\|_{L^p(0,T-h;B)} \leq \varepsilon/3.$$

This is a straightforward consequence of Lebesgue's Theorem.

Then, it is clear that, if $0 < h \le h_0 := \min_i h_i$, one has

$$\|\tau_h f - f\|_{L^p(0,T-h;\,B)} \le \varepsilon \quad \forall f \in \mathcal{F}.$$

Consequently, one has (2.57)–(2.58).

Conversely, let us assume that (2.57) and (2.58) are satisfied. Let $a > 0$ be given and let us introduce $M_a(f)$, with

$$M_a(f)(t) := \frac{1}{a}\int_t^{t+a} f(s)\,ds \ \ \forall t \in [0, T-a], \ \ \forall f \in \mathcal{F}.$$

Let us see that, for every $a > 0$, $\mathcal{F}_a := \{M_a(f) : f \in \mathcal{F}\}$ is relatively compact in $C^0([0, T-a]; B)$.

It is evident that $\mathcal{F}_a$ is pointwise bounded in B, that is, $\|M_a(f)(t)\|_B \le C_a(t)$ for all $f \in \mathcal{F}$ and $t \in [0, T-a]$. On the other hand, $\mathcal{F}_a$ is equi-continuous since, for any $f \in \mathcal{F}$ and any $t_1, t_2 \in [0, T-a]$ with $t_1 \le t_2$, one has

$$\begin{aligned}\|M_a(f)(t_1) - M_a(f)(t_2)\|_B &= \frac{1}{a}\Big\|\int_{t_1}^{t_1+a} \left(\tau_{t_2-t_1} f - f\right)(s)\,ds\Big\|_B \\ &\le \frac{1}{a}\|\tau_{t_2-t_1} f - f\|_{L^1(0,T-(t_2-t_1);\,B)}.\end{aligned}$$

Therefore, in view of Ascoli's Theorem, the relative compactness of $\mathcal{F}_a$ is ensured.

Note that

$$(M_a(f) - f)(t) = \frac{1}{a}\int_0^a (\tau_h f - f)(t)\,dh \quad \text{(an identity in } L^p(0, T-a;\, B)\text{)}.$$

Consequently,

$$\begin{aligned}\int_0^{T-a} \|(M_a(f) - f)(t)\|_B\,dt &\le \frac{1}{a}\int_0^{T-a}\int_0^a \|(\tau_h(f) - f)(t)\|_B\,dh\,dt \\ &= \frac{1}{a}\int_0^a\int_0^{T-a} \|(\tau_h(f) - f)(t)\|_B\,dt\,dh,\end{aligned}$$

whence

$$\|M_a(f) - f\|_{L^1(0,T-a;\,B)} \le \sup_{h\in[0,a]} \|\tau_h f - f\|_{L^1(0,T-a;\,B)}.$$

This proves that, for any $T' < T$, the set $\{f|_{[0,T']} : f \in \mathcal{F}\}$ is relatively compact in $L^1(0, T'; B)$.

Obviously, the change of variable $t \to T - t$ allows to obtain the same property for the family of restrictions $f|_{[T',T]}$ for all $T' > 0$.

Therefore, (2.58) holds. □

It is usual to find the following situation: X, B and Y are three Banach spaces with $X \hookrightarrow B \hookrightarrow Y$, the embedding $X \hookrightarrow B$ is compact and the embedding $B \hookrightarrow Y$ is continuous.

In this particular situation, we have:

Lemma 2.17 *If $1 \le p < +\infty$ and $\mathcal{F}$ is bounded in $L^p(0, T; X)$ and relatively compact in $L^p(0, T; Y)$, then it is also relatively compact in $L^p(0, T; B)$.*

Proof We will use the following:

$$\forall \delta > 0 \ \exists C_\delta > 0 \ \text{ such that } \ \|v\|_B \le C_\delta \|v\|_Y + \delta \|v\|_X, \quad \forall v \in X. \tag{2.59}$$

Let us assume that $\|v\|_{L^p(0,T;X)} \le M$ for all $v \in \mathcal{F}$ and let $\{v_n\}$ be a sequence in $\mathcal{F}$. Then $\{v_n\}$ possesses a Cauchy subsequence in $L^p(0, T; Y)$ that will be indexed again by n.

In view of (2.59), for each $\delta > 0$ there exists $C_\delta > 0$ such that

$$\begin{aligned} \|v_n - v_m\|_{L^p(0,T;B)} &\le C_\delta \|v_n - v_m\|_{L^p(0,T;Y)} + \delta \|v_n - v_m\|_{L^p(0,T;X)} \\ &\le C_\delta \|v_n - v_m\|_{L^p(0,T;Y)} + 2\delta M. \end{aligned}$$

Using these inequalities, it is not difficult to show that $\{v_n\}$ is also a Cauchy sequence in $L^p(0, T; B)$.

This ends the proof. □

The $p = +\infty$ version of this lemma is as follows:

Lemma 2.18 *If $\mathcal{F}$ is bounded in $L^\infty(0, T; X)$ and relatively compact in the space $C^0([0, T]; Y)$, then $\mathcal{F}$ is also relatively compact in $C^0([0, T]; B)$.*

The proof is similar and is left to the reader (see Exercise 2.51).

Note that Lemmas 2.17 and 2.18 are particularly interesting for the following reason: the space Y can be very large; consequently, compactness in $L^p(0, T; Y)$ or $C^0([0, T]; Y)$ may be relatively easy to establish in practice.

In other words, with the help of any of these results, we can deduce the compactness of a sequence in a "small" space of B-valued functions from boundedness in a smaller space of X-valued functions and compactness in a "large" space.

This will be seen in the following chapters in several specific situations.

The following two corollaries are consequences of Theorem 2.10:

Corollary 2.2 *Assume that the Banach spaces X, B and Y are as before. Let $\delta : \mathbb{R}_+ \mapsto \mathbb{R}_+$ be a function such that*

$$\delta(h) \to 0 \quad \text{as} \quad h \to 0$$

and let us assume that $1 < p < +\infty$. *Let* W *be the space*

$$W := \{v \in L^p(0, T; X) : \sup_{0<h<T} \delta(h)^{-1} \|\tau_h(v) - v\|_{L^p(0,T-h;Y)} < +\infty\},$$

endowed with its natural norm. Then, W *is compactly embedded in* $L^p(0, T; B)$.

Corollary 2.3 *With* X, B *and* Y *as before, the following embeddings are compact:*

1. $L^p(0, T; X) \cap \{v : v_t \in L^r(0, T; Y)\} \hookrightarrow L^p(0, T; B)$ *for all* $p \in [1, +\infty)$ *and* $r \in (1, +\infty]$.
2. $L^\infty(0, T; X) \cap \{v : v_t \in L^r(0, T; Y)\} \hookrightarrow C^0([0, T]; B)$ *for all* $r \in (1, +\infty]$.
3. $L^q(0, T; X) \cap N^{s,q}(0, T; Y) \hookrightarrow L^q(0, T; B)$ *for all* $s \in (0, 1]$ *and* $q \in [1, +\infty]$.
4. *Furthermore, if* $\mathcal{B}$ *is a bounded set in* $L^\infty(0, T; X)$, $k \in L^1(0, T)$ *is a fixed function,* $r > 1$ *and* K *is a constant, any set of the form*

$$\mathcal{F} = \mathcal{B} \cap \{v : \|v_t\|_Y \le k + \psi \text{ a.e.}, \|\psi\|_{L^r(0,T)} \le K\}$$

is relatively compact in $C^0([0, T]; B)$.

For any Banach space X, we will denote by $C^0_w([0, T]; X)$ the space of X-valued weakly continuous functions on $[0, T]$, that is, the space of functions $v : [0, T] \mapsto X$ such that

$$t \mapsto \langle \ell, v(t)\rangle_{X',X} \text{ is continuous}$$

for every $\ell \in X'$.

In this space, we can introduce a notion of convergence: it will be said that a sequence $\{v_n\}$ converges to v in $C^0_w([0, T]; X)$ if, for every $\ell \in X'$, one has:

$$\langle \ell, v_n\rangle_{X',X} \to \langle \ell, v\rangle_{X',X} \text{ strongly in } C^0([0, T]) \text{ as } n \to +\infty.$$

A relevant result is the following:

Lemma 2.19 *Let* X *and* Y *be Banach spaces such that* X *is reflexive and* $X \hookrightarrow Y$ *with a continuous embedding. Assume that* $f \in L^\infty(0, T; X)$ *and, regarded as a* Y*-valued function, it is continuous, i.e.* f *possesses a version in* $C^0([0, T]; Y)$. *Then, regarded as a* X*-valued function,* f *is weakly continuous, that is,* f *possesses a version in* $C^0_w([0, T]; X)$.

See Exercise 2.55 for the proof.

Assume that $N = 3$. In order to work with unbounded domains in the next chapters, it is convenient to introduce the space

$$K^1(\Omega) := \{v \in L^6(\Omega) : \nabla v \in L^2(\Omega)^3\}.$$

This is a Banach space for the norm

$$v \mapsto \|v\|_{K^1} := \|v\|_{L^6} + \|\nabla v\|_{L^2}.$$

In general, $H^1(\Omega)$ and $K^1(\Omega)$ do not coincide. Indeed, if Ω is unbounded, there is no reason for a function in $K^1(\Omega)$ to belong to $L^2(\Omega)$. However, these spaces are the same if Ω is bounded, due to Theorem 2.6.

We will denote by $K_0^1(\Omega)$ the closure of $\mathcal{D}(\Omega)$ in $K^1(\Omega)$. Obviously, $K_0^1(\Omega)$ is also a Banach space for the norm $\|\cdot\|_{K^1}$ and

$$H_0^1(\Omega) \hookrightarrow K_0^1(\Omega),$$

with a continuous embedding. Moreover, in $K_0^1(\Omega)$ the seminorm

$$u \mapsto \|\nabla u\|_{L^2}$$

is in fact a norm, which is equivalent to $\|\cdot\|_{K^1}$ (see Lemma 2.9 below).

At the end of Chap. 3, we will need the analog of the working space V in an unbounded domain. It will be denoted by $W(\Omega)$ and is defined as follows:

$$W(\Omega) := \{\mathbf{v} \in K_0^1(\Omega)^3 : \nabla \cdot \mathbf{v} = 0 \ \text{in} \ \Omega\}.$$

It is not difficult to check that, for unbounded Ω, Lemma 2.10 is still true if we replace in items *(ii)* and *(iv)* the space $H_0^1(\Omega)^3$ by $K_0^1(\Omega)^3$.

We will end this section with an interpolation result in L^p spaces:

Proposition 2.14 *Let us assume that $\Omega \subset \mathbb{R}^N$ is a non-empty open set and $v \in L^{p_0}(0,T;L^{q_0}(\Omega)) \cap L^{p_1}(0,T;L^{q_1}(\Omega))$, where $1 \le p_i, q_i \le +\infty$. Then, for any $\theta \in [0,1]$, one has $v \in L^{p_\theta}(0,T;L^{q_\theta}(\Omega))$, where p_θ and q_θ are defined as follows:*

$$\frac{1}{p_\theta} = \frac{1-\theta}{p_0} + \frac{\theta}{p_1} \quad \textit{and} \quad \frac{1}{q_\theta} = \frac{1-\theta}{q_0} + \frac{\theta}{q_1}.$$

Furthermore, the embedding

$$L^{p_0}(0,T;L^{q_0}(\Omega)) \cap L^{p_1}(0,T;L^{q_1}(\Omega)) \hookrightarrow L^{p_\theta}(0,T;L^{q_\theta}(\Omega))$$

is continuous.

Proof Obviously, we can assume that $0 < \theta < 1$. Let us also suppose that the p_i and q_i are finite.

For any measurable function $v = v(\mathbf{x}, t)$, one has

$$
\begin{aligned}
\int_0^T \left(\int_\Omega |v|^{q_\theta}\right)^{p_\theta/q_\theta} &= \int_0^T \left(\int_\Omega |v|^{a+b}\right)^{p_\theta/q_\theta} \\
&\le \int_0^T \left(\left(\int_\Omega |v|^{ar}\right)^{1/r} \left(\int_\Omega |v|^{br'}\right)^{1/r'}\right)^{p_\theta/q_\theta} \\
&= \int_0^T \|v\|_{L^{ar}}^{ap_\theta/q_\theta} \|v\|_{L^{br'}}^{bp_\theta/q_\theta} \\
&\le \left(\int_0^T \|v\|_{L^{ar}}^{ap_\theta s/q_\theta}\right)^{1/s} \left(\int_0^T \|v\|_{L^{br'}}^{bp_\theta s'/q_\theta}\right)^{1/s'}
\end{aligned}
$$

whenever $q_\theta = a + b$ $(a, b \ge 0)$, $1 < r, s < +\infty$ and r' and s' are the conjugate exponents of r and s.

With $a = (1-\theta)q_\theta$, $b = \theta q_\theta$, $r = ((1-\theta)q_1 + \theta q_0)/(1-\theta)q_1$ and $s = p_0/(1-\theta)p_\theta$, one has:

$$
ar = q_0, \quad ap_\theta s/q_\theta = p_0, \quad br' = p_1 \text{ and } bp_\theta s'/q_\theta = p_1
$$

and the inequality becomes

$$
\int_0^T \|v\|_{L^{q_\theta}}^{p_\theta} \le \left(\int_0^T \|v\|_{L^{q_0}}^{p_0}\right)^{p_\theta/p_0} \left(\int_0^T \|v\|_{L^{q_1}}^{p_1}\right)^{p_\theta/p_1}.
$$

This proves the result when $1 \le p_i, q_i < +\infty$. A similar argument leads to the proof of the result when some of the $p_i = +\infty$ and/or some of the $q_i = +\infty$. □

2.5 The Stokes Operator

The stationary Stokes problem in Ω with homogeneous Dirichlet conditions on the boundary $\partial\Omega$ is the following:

$$
\begin{cases}
-\Delta \mathbf{u} + \nabla p = \mathbf{f} & \text{in } \Omega, \\
\nabla \cdot \mathbf{u} = 0 & \text{in } \Omega, \\
\mathbf{u} = 0 & \text{on } \partial\Omega,
\end{cases}
\tag{2.60}
$$

where $\mathbf{f}$ is given, for instance, in $L^2(\Omega)^N$.

The unknowns are $\mathbf{u}$ and p. If they are smooth, for any $\mathbf{v} \in \mathcal{V}$ one has

$$
(\mathbf{f}, \mathbf{v}) = (-\Delta \mathbf{u}, \mathbf{v}) + (\nabla p, \mathbf{v}) = (\nabla \mathbf{u}, \nabla \mathbf{v}),
$$

since $(\nabla p, \mathbf{v}) = -\int_\Omega p\,(\nabla \cdot \mathbf{v}) = 0$. Consequently,

$$
(\nabla \mathbf{u}, \nabla \mathbf{v}) = (\mathbf{f}, \mathbf{v}).
$$

By continuity, this identity is still valid for any $\mathbf{v} \in V$. Moreover, if $\mathbf{u}$ is smooth and vanishes at the boundary, we have that $\mathbf{u} \in H_0^1(\Omega)^N$ and therefore $\mathbf{u} \in V$.

These considerations lead us to define the weak solution to problem (2.60):

Definition 2.9 We say that $\mathbf{u}$ is a weak solution for the Stokes problem (2.60) if

$$\begin{cases} \mathbf{u} \in V, \\ (\nabla \mathbf{u}, \nabla \mathbf{v}) = (\mathbf{f}, \mathbf{v}) \quad \forall \mathbf{v} \in V. \end{cases} \tag{2.61}$$

From the *Lax-Milgram Theorem*, noting that

$$(\nabla \mathbf{u}, \nabla \mathbf{v}) = (\mathbf{u}, \mathbf{v})_{H_0^1} \quad \forall \mathbf{u}, \mathbf{v} \in V,$$

it becomes clear that, for each $\mathbf{f} \in L^2(\Omega)^N$, the Stokes problem (2.61) admits exactly one weak solution.

Furthermore, using De Rham's Lemma 2.7, one can recover the pressure p from a weak solution $\mathbf{u}$ of the Stokes problem. More precisely, $\mathbf{u}$ is a weak solution of the Stokes problem if and only if $\mathbf{u} \in H_0^1(\Omega)^N$ and there exists $p \in \mathcal{D}'(\Omega)$ such that

$$\begin{cases} -\Delta \mathbf{u} + \nabla p = \mathbf{f} \quad \text{in } \mathcal{D}'(\Omega), \\ \nabla \cdot \mathbf{u} = 0 \quad \text{in } \mathcal{D}'(\Omega), \\ \gamma(u_i) = 0, \quad i = 1, \dots, N. \end{cases} \tag{2.62}$$

The following result shows that, if Ω is regular enough, the Stokes system also exhibits elliptic regularity:

Theorem 2.11 *Assume that Ω is bounded and the boundary $\partial\Omega$ is regular enough (for instance, of class C^2). Let $\mathbf{f} \in L^r(\Omega)^N$ be given with $2 \le r < +\infty$ and suppose that $\mathbf{u} \in H_0^1(\Omega)^N$ and $p \in L^2(\Omega)$ solve* (2.62). *Then $\mathbf{u} \in W^{2,r}(\Omega)^N$ and $p \in W^{1,r}(\Omega)$. Moreover, there exists a positive constant C such that*

$$\|\mathbf{u}\|_{W^{2,r}} + \|\nabla p\|_{L^r} \le C \|\mathbf{f}\|_{L^r}.$$

Sketch of the Proof This result was first established by Cattabbriga [6]. The proof is lengthy and needs several previous technical results. For brevity, we will only present the main ideas when $r = 2$; for more details, see for instance [7, Theorem 3.7] and [4, Theorem IV.6.3].

In a first step, we can establish *interior* regularity, that is, the H^2 regularity of $\mathbf{u}$ and the H^1 regularity of p in any open set Ω' with $\overline{\Omega'} \subset \Omega$.

This can be achieved by introducing a cut-off function $\psi \in \mathcal{D}(\Omega)$ with $\psi = 1$ in Ω' and considering the system satisfied by the couple $(\mathbf{w}, q) := (\psi \mathbf{u}, \psi p)$, that is

$$\begin{cases} -\Delta \mathbf{w} + \nabla q = \tilde{\mathbf{f}}, \quad \mathbf{x} \in \Omega, \\ \nabla \cdot \mathbf{w} = \tilde{g}, \quad \mathbf{x} \in \Omega, \\ \mathbf{w} = 0, \quad \mathbf{x} \in \partial\Omega, \end{cases} \tag{2.63}$$

where $\tilde{\mathbf{f}}$ and $\tilde{g}$ are respectively given by $\tilde{\mathbf{f}} = \varphi\mathbf{f} - (\Delta\psi)\mathbf{u} - 2(\nabla\psi \cdot \nabla)\mathbf{u} + p\nabla\psi$ and $\tilde{g} = \psi g + \nabla\psi \cdot \mathbf{v}$.

For example, it can be shown that to prove that $(\partial_1 \mathbf{w}, \partial_1 q) \in H^1(\Omega)^N \times L^2(\Omega)$ is equivalent to establish uniform estimates in this product space of the divided differences

$$\delta_h \mathbf{w} := \frac{1}{h}\left(\mathbf{w}(x_1 + h, \ldots, x_N) - \mathbf{w}(x_1, \ldots, x_N)\right)$$

and

$$\delta_h q := \frac{1}{h}\left(q(x_1 + h, \ldots, x_N) - q(x_1, \ldots, x_N)\right),$$

but this can be done in a relatively easy way by manipulating (2.63), looking at the problem satisfied by $(\delta_h \mathbf{w}, \delta_h q)$ and observing that $\delta_h \tilde{\mathbf{f}}$ and $\delta_h \tilde{g}$ are uniformly bounded in $L^2(\Omega)^N$ and $L^2(\Omega)$ respectively.

Therefore, the so called interior $H^2 \times H^1$ regularity holds.

In other words, one has $(\mathbf{w}, q) \in H^2(\Omega')^N \times H^1(\Omega')$ for any open set $\Omega' \subset\subset \Omega$.

Then, we can prove regularity "up to the boundary".

Thus, let us assume that $\mathbf{x}_0 \in \partial\Omega$, $O', O \subset \mathbb{R}^N$ are two small open neighborhoods of $\mathbf{x}_0$ such that $\overline{O'} \subset O$, let us introduce a cut-off function $\psi \in \mathcal{D}(O)$ with $\psi = 1$ in O' and let us consider the system satisfied by the couple $(\mathbf{z}, h) := (\psi\mathbf{u}|_{\Omega\cap O}, \psi p|_{\Omega\cap O})$, which is similar to (2.63).

After an appropriate change of variables $\mathbf{x} \to \boldsymbol{\xi}$, it is possible to reformulate this problem in a set of the form

$$G = D \cap \{\boldsymbol{\xi} \in \mathbb{R}^N : \xi_N > 0\},$$

where D is a neighborhood of $\mathbf{0}$.

Now, again with an argument relying on uniform estimates of difference quotients (although involving more technical calculations), it can be proved that $(\mathbf{z}, h)$ belongs to $H^2(\Omega \cap O)^N \times H^1(\Omega \cap O)$.

In order to end the proof, we use a partition of unit in Ω, that is, a collection of non-empty open sets $\{\Omega_0, \Omega_1, \ldots, \Omega_I\}$ and functions $\{\varphi_0, \varphi_1, \ldots, \varphi_I\}$ such that

$$\begin{cases} \Omega \subset \bigcup_{i=0}^{I} \Omega_i, \quad \overline{\Omega_0} \subset \Omega, \quad \partial\Omega \subset \bigcup_{i=1}^{I} \Omega_i, \\ \mathbf{x}^i \in \Omega_i, \quad \varphi_i \in \mathcal{D}(\mathbb{R}^N) \text{ and } \operatorname{Supp}\varphi_i \subset \Omega_i \quad \forall i = 0, 1, \ldots, I, \end{cases}$$

the $\mathbf{x} \in \Omega_i$ can be parametrized by local coordinates $y_j = Y_j(x_1, \ldots, x_N)$, where the Y_j are of class C^2,

$$\mathbf{x} \in \Omega_i \cap \Omega \ (\text{resp}\,.\ \mathbf{x} \in \Omega_i \setminus \overline{\Omega}) \ \Leftrightarrow \ y_N > 0 \ (\text{resp}\,.\ y_N < 0)$$

and $\sum_{i=0}^{I} \varphi_i = 1$ in Ω.

Decomposing $\mathbf{u}$ and q as sums of functions with supports in the Ω_i, working in each Ω_i with $i \geq 1$ in local coordinates and taking into account the previous estimates, the proof can be achieved. □

Remark 2.3 It is interesting to try to deduce minimal or sharp regularity assumptions on $\partial\Omega$ for which Theorem 2.11 holds. This question was analyzed in [3]; there, it was proved that $W^{2,\infty}$ regularity suffices. On the other hand, a similar result can be proved when $\mathbf{f} \in L^r(\Omega)^N$ with $1 < r < 2$. □

Let us introduce the orthogonal projector $P_H : L^2(\Omega)^N \mapsto H$. By definition, P_H satisfies the following:

$$P_H\mathbf{v} \in H \text{ and } \mathbf{v} - P_H\mathbf{v} \in H^{\perp} \quad \forall \mathbf{v} \in L^2(\Omega)^N.$$

Of course, $P_H\mathbf{v} = \mathbf{v}$ if and only if $\mathbf{v} \in H$.

Using the operator P_H, we see in view of Theorem 2.11 that, when $\mathbf{f} \in L^2(\Omega)^N$ and $\partial\Omega$ is of class C^2, the Stokes problem (2.60) can be also written in the form

$$\mathbf{u} \in V \cap H^2(\Omega)^N, \quad P_H(-\Delta\mathbf{u}) = P_H\mathbf{f}.$$

We will end this chapter by introducing the so called *Stokes operator* and recalling briefly some of its properties. They are direct consequences of the general spectral theory for compact operators in Hilbert spaces.

The details can be found, together with many other sharper results, for instance, in [2, 7, 14, 15].

Definition 2.10 Let $D(A) := V \cap H^2(\Omega)$. The Stokes operator $A : D(A) \subset H \mapsto H$ is defined by

$$A\mathbf{v} = P(-\Delta\mathbf{v}) \quad \forall \mathbf{v} \in D(A).$$

It is not difficult to check that, for any $\mathbf{w} \in D(A)$ and any $\mathbf{v} \in V$, one has

$$(A\mathbf{w}, \mathbf{v}) = \int_\Omega (-\Delta\mathbf{w}) \cdot \mathbf{v}\, d\mathbf{x} = (\mathbf{w}, \mathbf{v})_{H_0^1}. \tag{2.64}$$

On the other hand, the following important result holds:

Lemma 2.20 *The Stokes operator $A : D(A) \subset H \mapsto H$ is definite positive and self-adjoint. Furthermore, it has a compact inverse $A^{-1} : H \mapsto H$. Consequently, A possesses a sequence $\{\lambda_n\}$ of eigenvalues satisfying*

$$0 < \lambda_1 \leq \lambda_2 \leq \ldots, \quad \lambda_n \to +\infty \text{ as } n \to +\infty.$$

The associated eigenfunctions $\mathbf{w}^n$ with $\|\mathbf{w}^n\| = 1$ form a complete orthonormal system for H.

Let us denote by V_m the m-dimensional space spanned by the first m eigenfunctions $\mathbf{w}^1, \dots, \mathbf{w}^m$, corresponding to the eigenvalues $\lambda_1, \dots, \lambda_m$. Let us denote by P_m be the associated orthogonal projector, that is,

$$P_m : L^2(\Omega)^N \mapsto V_m,$$

with

$$P_m \mathbf{f} = \sum_{n=1}^{m} (\mathbf{f}, \mathbf{w}^n)\mathbf{w}^n, \quad \forall \mathbf{f} \in L^2(\Omega)^N.$$

The following results hold:

Lemma 2.21 *The functions $\lambda_m^{-1/2}\mathbf{w}^m$ form a complete orthonormal system in V. On the other hand, for any $\mathbf{f} \in V$, $P_m\mathbf{f} \to \mathbf{f}$ in V as $m \to +\infty$.*

Lemma 2.22 *Let us assume that $\partial\Omega$ is of class C^2. Then the eigenfunctions $\mathbf{w}^m$ belong to $W^{2,r}(\Omega)^N$ for all $r \in [1, +\infty)$.*

2.6 Appendix A: Spaces of p-Integrable Functions on the Boundary

We will only present the details when $N = 2$ or $N = 3$, although the definition can also be achieved in higher dimensions. As usual, we will denote the points $\mathbf{x} \in \mathbb{R}^N$ in the form

$$\mathbf{x} = (x', x_N), \quad \text{with } x' = (x_1, \dots, x_{N-1}).$$

Let $G \subset \mathbb{R}^m$ be a non-empty open set and let the function $\psi : G \mapsto \mathbb{R}^\ell$ be given. It will be said that ψ is *bi-Lipschitzian* (or bi-Lipschitz-continuous) if there exist constants $L_1, L_2 > 0$ such that

$$L_1|\mathbf{y}^1 - \mathbf{y}^2| \le |\psi(\mathbf{y}^1) - \psi(\mathbf{y}^2)| \le L_2|\mathbf{y}^1 - \mathbf{y}^2| \quad \forall \mathbf{y}^1, \mathbf{y}^2 \in G.$$

Thus, let us assume that Ω is bounded and its boundary $\partial\Omega$ is *Lipschitz-continuous* and leaves Ω at one side.

By this we mean that there exist an open set Ω_0, points $\mathbf{x}^1, \dots, \mathbf{x}^I \in \partial\Omega$ and associated bounded open sets Ω_i and functions $\varphi_0, \varphi_1, \dots, \varphi_I$ such that

$$\begin{cases} \Omega \subset \bigcup_{i=0}^{I} \Omega_i, \quad \overline{\Omega_0} \subset \Omega, \quad \partial\Omega \subset \bigcup_{i=1}^{I} \Omega_i, \\ \mathbf{x}^i \in \Omega_i, \quad \varphi_i \in \mathcal{D}(\mathbb{R}^N) \text{ and } \operatorname{Supp}\varphi_i \subset \Omega_i \quad \forall i = 0, 1, \dots, I, \end{cases}$$

the $\mathbf{x} \in \Omega_i$ can be parametrized by Lipschitz-continuous local coordinates $y_j = Z^i_j(x_1, \ldots, x_N)$ $(j = 1, \ldots, N)$, with

$$\begin{cases} |Z^i_N(x', x_N) - Z^i_N(x', y_N)| \geq k|x_N - y_N| \quad \forall (x', x_N), (x', y_N) \in \Omega_i, \quad k > 0, \\ \mathbf{x} \in \Omega_i \cap \Omega \ (\text{resp} . \ \mathbf{x} \in \Omega_i \setminus \overline{\Omega}) \ \Leftrightarrow \ y_N > 0 \ (\text{resp} . \ y_N < 0) \end{cases}$$

and $\sum_{i=0}^{I} \varphi_i = 1$ in Ω; see for instance [1, 24, 34] for more details.

Under these assumptions, thanks to the version of the *Implicit Function Theorem* for Lipschitz-continuous functions (see for instance [22]), it is clear that, for every $i = 1, \ldots I$, the points on $\partial\Omega \cap \operatorname{Supp} \varphi_i$ can be written in the form

$$x_j = Y_{j,i}(s), \quad s \in [0, 1]^{N-1}, \quad j = 1, \ldots, N$$

or, in short, $\mathbf{x} = \mathbf{Y}^i(s)$, for some bi-Lipschitzian functions $\mathbf{Y}^i : [0, 1]^{N-1} \mapsto \mathbb{R}^N$.

Note that other similar definitions dealing with the regularity of $\partial\Omega$ ca be given. For example, it will be said that $\partial\Omega$ is of class C^m if the same property is satisfied with functions Z^i_j that are m times continuously differentiable in the Ω_i.

With the previous open sets and local coordinates, we can introduce a measure on $\partial\Omega$. It suffices to do as follows:

1. For any i with $1 \leq i \leq I$ and any set of the form $R := \{\mathbf{Y}^i(s) : s \in \Pi_k[a_k, b_k]\}$ where $0 \leq a_k < b_k \leq 1$, we set

$$\begin{aligned} \lambda^*_{\partial\Omega}(R) &= \int_{\Pi_k[a_k,b_k]} \varphi_i(\mathbf{Y}^i(s))\, A(\mathbf{Y}^i)(s)\, ds \\ &= \int_{[0,1]^{N-1}} (\varphi_i \mathbb{1}_R)(\mathbf{Y}^i(s))\, A(\mathbf{Y}^i)(s)\, ds, \end{aligned}$$

where we have introduced the notation

$$A(\mathbf{Y}^i)(s) := \left| \frac{d\mathbf{Y}^i}{ds}(s) \right| \quad \text{if } N = 2$$

and

$$A(\mathbf{Y}^i)(s) := \left| \left(\frac{\partial \mathbf{Y}^i}{\partial s_1} \times \frac{\partial \mathbf{Y}^i}{\partial s_2} \right)(s) \right| \quad \text{if } N = 3.$$

Observe that $A(\mathbf{Y}^i)(s)$ is defined and bounded a.e. in $[0, 1]^{N-1}$ (in the Lebesgue sense).

2. Then, for any non-empty $B \subset \partial\Omega$ we denote by $\lambda^*_{\partial\Omega}(B)$ the infimum of all sums $\sum_k \lambda^*_{\partial\Omega}(R_k)$ associated with countable coverings $\{R_k\}$ of B, where the R_k are as above. We complete this definition by setting $\lambda^*_{\partial\Omega}(\emptyset) = 0$.

3. It is not difficult to check that $\lambda^*_{\partial\Omega}$ is an exterior measure. Accordingly, we say that a set $B \subset \partial\Omega$ is $\lambda_{\partial\Omega}$-measurable if and only if

$$\lambda^*_{\partial\Omega}(Z) = \lambda^*_{\partial\Omega}(Z \cap B) + \lambda^*_{\partial\Omega}(Z \setminus B)$$

for any $Z \subset \partial\Omega$. In view of Caratheodory's Theorem, the family of $\lambda_{\partial\Omega}$-measurable sets is a σ-algebra $\mathcal{A}_{\partial\Omega}$ and the restriction to $\mathcal{A}_{\partial\Omega}$ of the exterior measure $\lambda^*_{\partial\Omega}$ is a complete measure on $\partial\Omega$ that will be denoted by $\lambda_{\partial\Omega}$.

Note that any $\psi \in C^0(\partial\Omega)$ can be integrated on $\partial\Omega$ with respect to this measure. The corresponding integral is given as follows:

$$\langle \lambda_{\partial\Omega}, \psi \rangle = \sum_{i=1}^{I} \int_{[0,1]^{N-1}} (\varphi_i \psi)(\mathbf{Y}^i(s))\, A(\mathbf{Y}^i)(s)\, ds.$$

It can be seen that this quantity is independent of the family of open sets Ω_i, the partition of unit $\{\varphi_i\}$ and the local coordinates $y_j = Z_{j,i}(x_1, x_2)$ we have chosen.

Now, just as we did in the case of the Lebesgue measure, we can introduce simple and integrable functions on $\partial\Omega$ with respect to the measure $\lambda_{\partial\Omega}$. We can also introduce classes of integrable functions, we can define the Banach spaces $L^p(\partial\Omega)$, etc.

For brevity, if $f \in L^1(\partial\Omega)$, the integral of f on $\partial\Omega$ with respect to the measure-$\lambda_{\partial\Omega}$ will be denoted

$$\int_{\partial\Omega} f(\mathbf{x})\, d\Gamma(\mathbf{x}) \quad \text{or simply} \quad \int_{\partial\Omega} f\, d\Gamma.$$

It is easy to check that, for smooth boundaries $\partial\Omega$ and regular functions $f : \partial\Omega \mapsto \mathbb{R}$, we are redefining the usual line and surface integrals. For instance, if $\Omega = (a_1, b_1) \times (a_2, b_2)$, one has

$$\begin{aligned}\int_{\partial\Omega} f\, d\Gamma = &\int_{a_1}^{b_1} f(x_1, a_2)\, dx_1 + \int_{a_2}^{b_2} f(b_1, x_2)\, dx_2 \\ &- \int_{a_1}^{b_1} f(x_1, b_2)\, dx_1 - \int_{a_2}^{b_2} f(a_1, x_2)\, dx_1.\end{aligned}$$

Also, if $\Omega = \{\mathbf{x} \in \mathbb{R}^3 : |\mathbf{x}| < R,\ x_3 > 0\}$ for some $R > 0$,

$$\begin{aligned}\int_{\partial\Omega} f\, d\Gamma = R^2 &\int_0^{2\pi}\int_0^{\pi/2} f(R\cos\theta\ \cos\phi,\, R\cos\theta\ \sin\phi,\, R\sin\theta)\cos\theta\, d\theta\, d\phi \\ &+ \int_0^{2\pi}\int_0^{R} f(\rho\ \cos\phi,\, \rho\ \sin\phi, 0)\rho\, d\rho\, d\phi.\end{aligned}$$

2.7 Appendix B: Proof of Sobolev Embeddings

The proof is composed of several steps. In order to avoid technicalities, we will only indicate the main ideas. More details and comments can be found, for instance, in [9].

Step 1: Suppose first that $1 \leq p < N$.
We first note that, if $W^{1,p}(\mathbb{R}^N) \subset L^q(\mathbb{R}^N)$, we necessarily have $q = p^*$. This is a clear consequence of scaling; see Exercise 2.32. Let us see that the converse is also true.

We will use that $\mathcal{D}(\mathbb{R}^N)$ is dense in $W^{1,p}(\mathbb{R}^N)$, a property that can be established arguing as in the proof of Lemma 2.3.

Suppose, for the moment, that $p = 1$. Then $p^* = N/(N-1)$ and we want to prove that

$$\int_{\mathbb{R}^N} |v|^{N/(N-1)}\, d\mathbf{x} \leq C \left(\int_{\mathbb{R}^N} |\nabla v|\, d\mathbf{x} \right)^{N/(N-1)} \quad \forall v \in W^{1,1}(\mathbb{R}^N). \tag{2.65}$$

For any $v \in \mathcal{D}(\mathbb{R}^N)$, one has

$$v(\mathbf{x}) = \int_{-\infty}^{x_1} \partial_1 v(y_1, x_2, \dots, x_N)\, dy_1 \quad \forall \mathbf{x} \in \mathbb{R}^N$$

and similar identities holds for $i = 2, \dots, N$.

Let us introduce the notation

$$S_j(x_{j+1}, \dots, x_N) := \int \dots \int_{\mathbb{R}^j} |v(\mathbf{x})|^{N/(N-1)}\, dx_1 \dots dx_j$$

and

$$G_j(x_{j+1}, \dots, x_N) := \int \dots \int_{\mathbb{R}^j} |\nabla v(\mathbf{x})|\, dx_1 \dots dx_j$$

for $j = 1, \dots, N$. Then

$$|v(\mathbf{x})|^{N/(N-1)} \leq \prod_{i=1}^{N} \left(\int_{\mathbb{R}} |\nabla v|_{x_i = y_i}|\, dy_i \right)^{1/(N-1)}$$

and, consequently,

$$\begin{aligned}
S_1(x_2,\ldots,x_N) &\leq \left(\int_{\mathbb{R}} |\nabla v|_{x_1=y_1}|\,dy_1\right)^{1/(N-1)} \int_{\mathbb{R}} \prod_{i=2}^{N} \left(\int_{\mathbb{R}} |\nabla v|_{x_i=y_i}|\,dy_i\right)^{1/(N-1)} dx_1 \\
&\leq \left[\int_{\mathbb{R}} |\nabla v|_{x_1=y_1}|\,dy_1 \prod_{i=2}^{N} \int_{\mathbb{R}}\left(\int_{\mathbb{R}} |\nabla v|_{x_i=y_i}|\,dy_i\right)dx_1\right]^{1/(N-1)} . \\
&= \left[G_1(x_2,\ldots,x_N) \prod_{i=2}^{N} \int_{\mathbb{R}}\left(\int_{\mathbb{R}} |\nabla v|_{x_i=y_i}|\,dy_i\right)dx_1\right]^{1/(N-1)} .
\end{aligned}$$

Now, we can estimate the integral of $|v|^{N/(N-1)}$ with respect to x_1 and x_2 in a very similar way:

$$\begin{aligned}
S_2(x_3,\ldots,x_N) &= \int_{\mathbb{R}} S_1(x_2,\ldots,x_N)\,dx_2 \\
&\leq \int_{\mathbb{R}} \left[G_1(x_2,\ldots,x_N) \prod_{i=2}^{N} \int_{\mathbb{R}}\left(\int_{\mathbb{R}} |\nabla v|_{x_i=y_i}|\,dy_i\right)dx_1\right]^{1/(N-1)} dx_2 \\
&\leq \left[G_2(x_3,\ldots,x_N)^2 \prod_{i=3}^{N} \iint_{\mathbb{R}^2}\left(\int_{\mathbb{R}} |\nabla v|_{x_i=y_i}|\,dy_i\right)dx_1\,dx_2\right]^{1/(N-1)} .
\end{aligned}$$

After some computations, we see that

$$S_{N-1}(x_N) \leq \left[G_{N-1}(x_N)^{N-1} \int_{\mathbb{R}^N} |\nabla v(\mathbf{y})|\,d\mathbf{y}\right]^{1/(N-1)},$$

whence we obtain the following desired inequality:

$$\begin{aligned}
\int_{\mathbb{R}^N} |v|^{N/(N-1)}\,d\mathbf{x} &= \int_{\mathbb{R}} S_{N-1}(x_N)\,dx_N \\
&\leq \left(\int_{\mathbb{R}} G_{N-1}(x_N)\,dx_N\right)\left(\int_{\mathbb{R}^N} |\nabla v|\,d\mathbf{y}\right)^{1/(N-1)} \\
&= \left(\int_{\mathbb{R}^N} |\nabla v|\,d\mathbf{x}\right)^{N/(N-1)} .
\end{aligned}$$

Now, let us assume that $1 < p < N$.

Again, let $v \in \mathcal{D}(\mathbb{R}^N)$ be given. For any $\gamma > 1$, one has $|v|^\gamma \in W^{1,1}(\mathbb{R}^N)$ and $\nabla |v|^\gamma = \gamma |v|^{\gamma-1}(\operatorname{sign} v)\,\nabla v$ a.e. (see Exercise 2.33). Thus, as a consequence of (2.65), one has:

$$\int_{\mathbb{R}^N} |v|^{\gamma N/(N-1)}\, d\mathbf{x} \le \int_{\mathbb{R}^N} |\nabla |v|^\gamma|\, d\mathbf{x} \le \gamma \int_{\mathbb{R}^N} |v|^{\gamma-1}|\nabla v|\, d\mathbf{x}$$
$$\le \gamma \left(\int_{\mathbb{R}^N} |v|^{(\gamma-1)p'}\, d\mathbf{x} \right)^{1/p'} \left(\int_{\mathbb{R}^N} |\nabla v|^p\, d\mathbf{x} \right)^{1/p}.$$

With $\gamma = p(N-1)/(N-p)$, we get $\gamma N/(N-1) = (\gamma-1)p' = p^*$ and therefore

$$\left(\int_{\mathbb{R}^N} |v|^{p^*}\, d\mathbf{x} \right)^{1/p^*} \le \frac{(N-1)p}{N-p} \left(\int_{\mathbb{R}^N} |\nabla v|^p\, d\mathbf{x} \right)^{1/p}. \tag{2.66}$$

Hence, we have (2.31) when $1 \le p < N$, with Ω replaced by $\mathbb{R}^N$.

Note that a slight modification of this argument also gives (2.31) when $p = N$ with p^* arbitrary in $[1, +\infty)$ and Ω replaced by $\mathbb{R}^N$; see Exercise 2.34.

Step 2: Now, let $\Omega \subset \mathbb{R}^N$ be a non-empty bounded connected open set with Lipschitz-continuous boundary.

Note that there exist an extension operator $E : W^{1,p}(\Omega) \mapsto W^{1,p}(\mathbb{R}^N)$, an open set $G \subset \mathbb{R}^N$ and a constant $C > 0$ such that, for every $v \in W^{1,p}(\Omega)$, Ev has compact support in G,

$$\|Ev\|_{L^p(\mathbb{R}^N)} \le C\|v\|_{L^p(\Omega)} \quad \text{and} \quad \|\nabla Ev\|_{L^p(\mathbb{R}^N)} \le C\|\nabla v\|_{L^p(\Omega)} \tag{2.67}$$

(see Exercise 2.22).

Let $v \in W^{1,p}(\Omega)$ be given. There exists a sequence $\{v_n\}$ in $\mathcal{D}(\mathbb{R}^N)$ with $v_n \to Ev$ strongly in $W^{1,p}(\mathbb{R}^N)$. Obviously, $\{v_n\}$ is a Cauchy sequence in $W^{1,p}(\mathbb{R}^N)$ and also in $L^{p^*}(\mathbb{R}^N)$. Moreover, $v_n \to Ev$ strongly in $L^{p^*}(\mathbb{R}^N)$. Consequently, (2.66) is satisfied by each v_n and

$$\|v\|_{L^{p^*}(\Omega)} \le \|Ev\|_{L^{p^*}(\mathbb{R}^N)} \le \frac{(N-1)p}{N-p}\|\nabla Ev\|_{L^p(\mathbb{R}^N)} \le C\|v\|_{W^{1,p}(\Omega)}.$$

This yields the continuous embedding (2.31) when $1 \le p \le N$.

Step 3: Let us first assume that $N < p < +\infty$ and let the function v belong to $C^1(\mathbb{R}^N)$ and have compact support.

For any bounded open set $G \subset \mathbb{R}^N$ and any $f \in L^1(G)$, we will use the notation

$$⨍_G f\, d\mathbf{x} := \frac{1}{|G|} \int_G f\, d\mathbf{x},$$

where $|G|$ is the measure of G.

For any $\mathbf{x} \in \mathbb{R}^N$ and $r > 0$, one has

$$\begin{aligned}
\int_{B(\mathbf{x};r)} |v(\mathbf{y}) - v(\mathbf{x})|\, d\mathbf{y} &= \int_0^r \left(\int_{\partial B(0;1)} |v(\mathbf{x} + \tau\mathbf{w}) - v(\mathbf{x})| \tau^{N-1}\, d\Gamma(\mathbf{w}) \right) d\tau \\
&= \int_0^r \left(\int_{\partial B(0;1)} \left| \int_0^\tau \frac{d}{ds} v(\mathbf{x} + s\mathbf{w})\, ds \right| \tau^{N-1}\, \Gamma(\mathbf{w}) \right) d\tau \\
&\le \int_0^r \tau^{N-1} \left(\int_{\partial B(0;1)} \left(\int_0^r |\nabla v(\mathbf{x} + s\mathbf{w})|\, ds \right) d\Gamma(\mathbf{w}) \right) d\tau \\
&= \frac{r^N}{N} \int_{\partial B(0;1)} \left(\int_0^r |\nabla v(\mathbf{x} + s\mathbf{w})|\, ds \right) d\Gamma(\mathbf{w}) \\
&= \frac{r^N}{N} \int_{B(\mathbf{x};r)} \frac{|\nabla v(\mathbf{y})|}{|\mathbf{y} - \mathbf{x}|^{N-1}}\, d\mathbf{y}.
\end{aligned}$$

Consequently,

$$⨍_{B(\mathbf{x};r)} |v(\mathbf{y}) - v(\mathbf{x})|\, d\mathbf{y} \le C \int_{B(\mathbf{x};r)} \frac{|\nabla v(\mathbf{y})|}{|\mathbf{y} - \mathbf{x}|^{N-1}}\, d\mathbf{y}. \tag{2.68}$$

Note that the right-hand side of (2.68) is bounded as follows:

$$\int_{B(\mathbf{x};r)} \frac{|\nabla v(\mathbf{y})|}{|\mathbf{y} - \mathbf{x}|^{N-1}}\, d\mathbf{y} \le C(N, p) r^{1-N/p} \|\nabla v\|_{L^p(\mathbb{R}^N)}, \tag{2.69}$$

since $1 < p' < N/(N-1)$. Taking into account that

$$|v(\mathbf{x})| \le ⨍_{B(\mathbf{x};r)} |v(\mathbf{y}) - v(\mathbf{x})|\, d\mathbf{y} + ⨍_{B(\mathbf{x};r)} |v(\mathbf{y})|\, d\mathbf{y}$$

for any $\mathbf{x} \in \mathbb{R}^N$, we easily deduce from (2.68) and (2.69) that

$$\sup_{\mathbb{R}^N} |v| \le C \|v\|_{W^{1,p}(\mathbb{R}^N)}. \tag{2.70}$$

Let us now assume that $\mathbf{x}, \mathbf{y} \in \mathbb{R}^N$ satisfies $|\mathbf{x} - \mathbf{y}| = r > 0$ and v is as before and let us take $D = B(\mathbf{x}; r/2) \cup B(\mathbf{y}; r/2)$. Using the inequality

$$|v(\mathbf{x}) - v(\mathbf{y})| \le ⨍_D |v(\mathbf{x}) - v(\mathbf{z})|\, d\mathbf{z} + ⨍_D |v(\mathbf{z}) - v(\mathbf{y})|\, d\mathbf{z},$$

and (2.68) twice, we see that

$$|v(\mathbf{x}) - v(\mathbf{y})| \le \frac{C(N, p)}{2^{1-N/p}} |\mathbf{x} - \mathbf{y}|^{1-N/p} \|\nabla v\|_{L^p(\mathbb{R}^N)} \quad \forall \mathbf{x}, \mathbf{y} \in \mathbb{R}^N. \tag{2.71}$$

Since the functions v considered in this step are dense in $W^{1,p}(\mathbb{R}^N)$, in view of (2.70) and (2.71), we have (2.35) for $m = 1$, $k = 0$ and α as in (2.34), with Ω replaced by $\mathbb{R}^N$.

Now, let us consider the particular case where $p = +\infty$.

Let v be given in $W^{1,\infty}(\mathbb{R}^N)$. Then, for every bounded open set $G \subset \mathbb{R}^N$, the restriction to G belongs to $W^{1,q}(G)$ for all finite q.

Therefore, it also belongs to $C^{0,\alpha}(\overline{G})$ for all $\alpha \in [0, 1)$ and, in particular, $v \in C^0(\mathbb{R}^N)$.

Let $\{\zeta_\varepsilon\}$ be a regularizing sequence in $\mathbb{R}^N$ and let us set

$$v_\varepsilon(\mathbf{x}) := \int_{\mathbb{R}^N} v(\mathbf{z})\zeta_\varepsilon(\mathbf{x} - \mathbf{z})\, d\mathbf{z} \quad \forall \mathbf{x} \in \mathbb{R}^N.$$

The functions v_ε are well defined, belong to $C^\infty(\mathbb{R}^N)$ and satisfy $v_\varepsilon(\mathbf{x}) \to v(\mathbf{x})$ as $\varepsilon \to 0$ for every $\mathbf{x} \in \mathbb{R}^N$.

On the other hand,

$$\partial_i v_\varepsilon(\mathbf{x}) = \int_{B(\mathbf{x};\varepsilon)} v(\mathbf{z})\, \partial_i \zeta_\varepsilon(\mathbf{x} - \mathbf{z})\, d\mathbf{z} = \int_{B(\mathbf{x};\varepsilon)} \partial_i v(\mathbf{z})\, \zeta_\varepsilon(\mathbf{x} - \mathbf{z})\, d\mathbf{z}$$

for all $\mathbf{x} \in \mathbb{R}^N$ and every $i = 1, \dots, N$ and

$$|v_\varepsilon(\mathbf{x}) - v_\varepsilon(\mathbf{y})| = \left| \int_0^1 \frac{d}{ds} v_\varepsilon(\mathbf{y} + s(\mathbf{x} - \mathbf{y}))\, ds \right| \le \|\nabla v_\varepsilon\|_{L^\infty(\mathbb{R}^N)} |\mathbf{x} - \mathbf{y}| \tag{2.72}$$

for all $\mathbf{x}, \mathbf{y} \in \mathbb{R}^N$. Consequently, $\|\nabla v_\varepsilon\|_{L^\infty(\mathbb{R}^N)} \le \|\nabla v\|_{L^\infty(\mathbb{R}^N)}$ and, taking limits in (2.72), we deduce that

$$|v(\mathbf{x}) - v(\mathbf{y})| \le \|\nabla v\|_{L^\infty(\mathbb{R}^N)} |\mathbf{x} - \mathbf{y}| \quad \forall \mathbf{x}, \mathbf{y} \in \mathbb{R}^N.$$

This proves that (2.35) also holds with Ω replaced by $\mathbb{R}^N$ when $p = +\infty$ for $m = 1$, $k = 0$ and $\alpha = 1$.

Again, for an open set Ω as in Step 2, the extension operator E allows to deduce the embedding (2.35) with $m = 1$, that is,

$$W^{1,p}(\Omega) \hookrightarrow C^{0,1-N/p}(\overline{\Omega}).$$

It is not difficult to extend the results in Steps 1 to 3 to the other cases considered in this theorem with $m \ge 1$. The details are left to the reader (see Exercise 2.36).

Step 4: Let us prove the compactness of some embeddings.

For instance, let us consider the case where $1 \le p \le N$ and $1 \le q < p^*$.

Again, let the extension mapping $E : W^{1,p}(\Omega) \mapsto W^{1,p}(\mathbb{R}^N)$ and the open set $G \subset \mathbb{R}^N$ be such that $\operatorname{Supp} Ev \subset G$ and (2.67) holds for every $v \in W^{1,p}(\Omega)$.

Let $\{v_n\}$ be a bounded sequence in $W^{1,p}(\Omega)$. Let us see that $\{v_n\}$ possesses a subsequence that converges strongly in $L^q(\Omega)$.

Let us introduce the extensions Ev_n and the (regularized) functions v_n^ε, with

$$v_n^\varepsilon(\mathbf{x}) := \int_{\mathbb{R}^N} \zeta_\varepsilon(\mathbf{x}-\mathbf{y})\, Ev_n(\mathbf{y})\, d\mathbf{y} \quad \forall \mathbf{x} \in \mathbb{R}^N.$$

Then the following holds:

- The functions $v_n^\varepsilon - Ev_n$ converge in $L^q(G)$ uniformly in n to 0 as $\varepsilon \to 0$, that is,

$$\sup_{n\geq 1} \|v_n^\varepsilon - Ev_n\|_{L^q(G)} \to 0 \text{ as } \varepsilon \to 0. \tag{2.73}$$

  (Using that $q < p^*$ and proving first that $\sup_{n\geq 1} \|v_n^\varepsilon - Ev_n\|_{L^1(G)} \to 0$).
- For every $\varepsilon > 0$, the v_n^ε belong to a compact set of $C^0(\overline{G})$, i.e.,

$$\{v_n^\varepsilon : n \geq 1\} \text{is uniformly bounded and equi-continuous} \tag{2.74}$$

  (Using that $|v_n^\varepsilon(\mathbf{x})| \leq C/\varepsilon^N$ and $|\nabla v_n^\varepsilon(\mathbf{x})| \leq C/\varepsilon^{N+1}$).
- Finally, from (2.73) and (2.74), we see that

$$\{v_n\} \text{possesses a Cauchy subsequence in } L^q(\Omega). \tag{2.75}$$

Hence, we can extract from any bounded sequence in $W^{1,p}(\Omega)$ a strongly convergent subsequence in $L^q(\Omega)$ and this shows that the embedding $W^{1,p}(\Omega) \hookrightarrow L^q(\Omega)$ is compact.

The proofs of the remaining compact embeddings asserted in Theorem 2.6 are left as an exercise.

2.8 Appendix C: Proof of de Rham's Lemma

As already mentioned, we will follow a constructive approach due to J. Simon. For brevity, we will only indicate the main ideas; more details are given in [26, 28].

If $\mathbf{f}$ is a gradient, we obviously have (2.46). Indeed, assume that $\mathbf{f} = \nabla p$ for some $p \in \mathcal{D}'(\Omega; B)$. Then, if $\mathbf{v} \in \mathcal{V}$, one has

$$\langle \mathbf{f}, \mathbf{v} \rangle = \sum_{i=1}^{N} \langle \partial_i p, v_i \rangle = -\langle p, \nabla \cdot \mathbf{v} \rangle = 0.$$

Let us prove the reciprocal.

Step 1: First, note that (2.46) implies

$$\partial_i f_j = \partial_j f_i \quad \forall i, j = 1, \ldots, N, \tag{2.76}$$

that is, $\nabla \times \mathbf{f} = 0$.

Indeed, for any $\varphi \in \mathcal{D}(\Omega)$, the function $\mathbf{v}$, with

$$v_i = -\partial_j \varphi, \ v_j = \partial_i \varphi \ \text{ and } \ v_k = 0 \ \text{ for } \ k \neq i, j,$$

belongs to $\mathcal{V}$. Consequently,

$$0 = \langle \mathbf{f}, \mathbf{v} \rangle = \langle \partial_j f_i - \partial_i f_j, \varphi \rangle$$

and this gives (2.76).

Step 2: In the sequel, for any set $D \subset \mathbb{R}^N$, we will use the notation

$$\Omega_D := \{\mathbf{x} \in \mathbb{R}^N : \mathbf{x} + D \in \Omega\}.$$

Let the compact set $K \subset \mathbb{R}^N$ be given. Then, for any $S \in \mathcal{D}'(\Omega; B)$ and any $\mu \in \mathcal{D}'(\mathbb{R}^N)$ with $\operatorname{Supp} \mu \subset K$, we introduce the *$\mu$-ponderation* or *$\mu$-weighted average* of S as follows:

$$\begin{cases} S \odot \mu \in \mathcal{D}'(\Omega_K; B), \\ \langle S \odot \mu, \varphi \rangle = \langle S, \langle \mu, \tilde{\tau}_{\bullet} \varphi \rangle \rangle \ \ \forall \varphi \in \mathcal{D}(\Omega_K). \end{cases}$$

Here, $\tilde{\tau}_{\bullet}\varphi$ stands for the *lower translation* of φ, i.e.

$$\tilde{\tau}_{\mathbf{y}} \varphi(\mathbf{x}) := \varphi(\mathbf{x} - \mathbf{y}) \quad \forall \mathbf{x}, \mathbf{y} \in \mathbb{R}^N.$$

This definition is correct. Indeed, under the previous assumptions on μ, the mapping $\varphi \mapsto \langle \mu, \tilde{\tau}_{\bullet} \varphi \rangle$ is well defined and sequentially continuous from $\mathcal{D}(\Omega_K)$ into $\mathcal{D}(\Omega)$.

Note that, if $S \in L^1_{\text{loc}}(\Omega; B)$ and $\mu \in L^1(K)$, one has

$$\langle S \odot \mu, \varphi \rangle = \int_{\Omega_K} \left(\int_K S(\mathbf{x} + \mathbf{y}) \mu(\mathbf{y}) \, d\mathbf{y} \right) \varphi(\mathbf{x}) \, d\mathbf{x} \quad \forall \varphi \in \mathcal{D}(\Omega_K),$$

that is, $S \odot \mu \in L^1_{\text{loc}}(\Omega; B)$ and

$$(S \odot \mu)(\mathbf{x}) = \int_K S(\mathbf{x} + \mathbf{y}) \mu(\mathbf{y}) \, d\mathbf{y} \ \text{ a.e. in } \Omega_K.$$

Let $\mathcal{D}'_K(\mathbb{R}^N)$ stand for the space of distributions $\mu \in \mathcal{D}'(\mathbb{R}^N)$ with $\operatorname{Supp}\mu \subset K$. Then the following holds:

$$\begin{cases} (S,\mu) \mapsto S \odot \mu \text{is sequentially continuous in } \mathcal{D}'(\Omega; B) \times \mathcal{D}'_K(\mathbb{R}^N) \\ and\ \partial_j(S \odot \mu) = \partial_j S \odot \mu = -S \odot \partial_j \mu \text{for all } j = 1, \ldots, N. \end{cases} \tag{2.77}$$

Furthermore, if $\mu \in \mathcal{D}(\mathbb{R}^N)$ and $\operatorname{Supp}\mu \subset K$, one has $S \odot \mu \in C^\infty(\Omega_K; B)$.

For any $r > 0$, any vector $\mathbf{b} \in \mathbb{R}^N$ with $|\mathbf{b}| = 1$ and any β with $0 < \beta \leq \pi$, let us introduce the portion of cone

$$\Lambda(r, \mathbf{b}, \beta) := \{\mathbf{x} \in \mathbb{R}^N : \mathbf{x} \cdot \mathbf{b} \geq |\mathbf{x}| \cos\beta,\ |\mathbf{x}| \leq r\}.$$

Due to the Lipschitz-continuity assumption on $\partial\Omega$, it is not difficult to prove that there exist $\beta \in (0, \pi]$, $r > 0$, an integer $I \geq 1$, vectors $\mathbf{b}^i$ and associated $\Lambda_i := \Lambda(r, \mathbf{b}^i, \alpha)$, functions $\alpha_i \in C^\infty(\Omega)$ and connected open sets ω_i $(1 \leq i \leq I)$ satisfying

$$\begin{cases} \bigcup_{i=1}^{I} \Omega_{\Lambda_i} = \Omega, \quad \operatorname{Supp}\alpha_i \cap \Omega \subset \omega_i \subset \Omega_{\Lambda_i}, \\ \forall \mathbf{x}, \mathbf{y} \in \omega_i \ \ \exists P(\mathbf{x}, \mathbf{y}) \subset \omega_i \text{ (a path from } \mathbf{x} \text{ to } \mathbf{y}\text{) of bounded length,} \\ 0 \leq \alpha_i \leq 1, \ \sum_{i=1}^{I} \alpha_i = 1 \text{ in } \Omega, \\ K(\gamma) := \sup_{\mathbf{x}\in\Omega} \sup_i |\partial^\gamma \alpha_i(\mathbf{x})| < +\infty. \end{cases} \tag{2.78}$$

(see the proof of Lemma 12 in [28], p. 212).

Step 3: Let $\Lambda = \Lambda(r, \mathbf{b}, \beta)$ be a cone like in the previous step and let ω be a non-empty open connected set satisfying $\omega \subset \Omega_\Lambda$. We are going to construct a distribution $p \in \mathcal{D}'(\omega; B)$ such that

$$\mathbf{f} = \nabla p \ \text{ in } \omega. \tag{2.79}$$

Thus, let us introduce the functions ζ_j $(1 \leq j \leq N)$ and ζ_0, with

$$\zeta_j(\mathbf{x}) := -\frac{1}{s(N)} \sigma(|\mathbf{x}|^{-1}\mathbf{x}) \frac{\theta(|\mathbf{x}|)x_j}{|\mathbf{x}|^N}$$

and

$$\zeta_0(\mathbf{x}) := -\frac{1}{s(N)} \sigma(|\mathbf{x}|^{-1}\mathbf{x}) \frac{\theta'(|\mathbf{x}|)}{|\mathbf{x}|^{N-1}}$$

for $\mathbf{x} \neq 0$. Here, we assume the following:

- σ is a C^∞ nonnegative nonzero function defined on the unit sphere S^{N-1} satisfying $\sigma(\mathbf{s}) = 0$ for $\mathbf{s} \cdot \mathbf{b} \le \cos\beta$,
- $s(N) = \int_{S^{N-1}} \sigma \, d\Gamma$,
- $\theta \in C^\infty([0, +\infty))$, $\theta(x) = 1$ for $x \le r/2$ and $\theta(x) = 0$ for $x \ge r$.

Let $\mathbf{a} \in \omega$ be fixed and let us set

$$h(\mathbf{x}) := \int_{P(\mathbf{a},\mathbf{x})} \mathbf{f} \odot \zeta_0 \cdot d\boldsymbol{\ell} \quad \text{and} \quad p := h + \sum_{k=1}^{N} f_k \odot \zeta_k \quad \text{in } \omega, \tag{2.80}$$

where $P(\mathbf{a}, \mathbf{x})$ is any C^1 path from $\mathbf{a}$ to $\mathbf{x}$ inside ω and $d\boldsymbol{\ell}$ is the usual integration element along $P(\mathbf{a}, \mathbf{x})$. Then h and p are well defined B-valued distributions in ω (in fact, $h \in C^1(\omega; B)$) and one has (2.79). Indeed,

$$\partial_j h(\mathbf{x}) = \lim_{s \to 0} \frac{1}{s} \int_{[\mathbf{x}, \mathbf{x}+s\mathbf{e}^j]} \mathbf{f} \odot \zeta_0 \cdot d\boldsymbol{\ell} = (\mathbf{f} \odot \zeta_0)_j(\mathbf{x}) \quad \forall \mathbf{x} \in \omega,$$

while

$$\partial_j (f_k \odot \zeta_k) = \partial_j f_k \odot \zeta_k = -f_j \odot \partial_k \zeta_k.$$

Therefore, $\partial_j p = f_j \odot (\zeta_0 - \sum_{k=1}^N \partial_k \zeta_k)$ for every j. But it is not difficult to see that

$$\zeta_0 - \sum_{k=1}^{N} \partial_k \zeta_k = \delta \quad \text{in } \mathcal{D}'(\mathbb{R}^N).$$

Consequently, $\partial_j p = f_j$ for all j and the previous assertion follows.

Step 4: Let $\beta \in (0, \pi]$, $r > 0$, I, the $\mathbf{b}^i$, the α_i and the ω_i satisfy (2.78). For every $i = 1, \ldots, I$, there exists $p_i \in \mathcal{D}'(\omega_i; B)$ satisfying

$$\mathbf{f} = \nabla p_i \quad \text{in } \omega_i.$$

It can be assumed that the p_i belong to $\mathcal{D}'(\mathbb{R}^N; B)$ and satisfy $\operatorname{Supp} p_i \cap \Omega \subset \Omega_{\Lambda_i}$. Indeed, it will suffice to multiply p_i by an appropriate cut-off function identical to 1 in ω_i.

Let us see that there exist $c_1, \ldots, c_I$ in B and a new distribution $p \in \mathcal{D}'(\mathbb{R}^N; B)$ such that, in each ω_i, one has $p = p_i + c_i$. Obviously, this will imply that

$$\nabla p = \mathbf{f} \quad \text{in } \Omega. \tag{2.81}$$

This can be proved as follows:

1. First, we take $c_1 = 0$ and $p_1^* = p_1$ in ω_1.

2. Let i be fixed with $1 \leq i \leq I-1$ and let us set $\Omega_i := \omega_1 \cup \cdots \cup \omega_i$. Assume that there exist $c_1, \ldots, c_i \in B$ and $p_i^* \in \mathcal{D}'(\mathbb{R}^N; B)$ such that

$$p_i^* = p_j + c_j \text{ in } \omega_j \text{ for } j = 1, \ldots, i.$$

Let us also assume (for the moment) that

$$c_{i+1} := p_i^* - p_{i+1} \text{ is constant in } \Omega_i \cap \omega_{i+1}. \tag{2.82}$$

Then, if we put

$$p_{i+1}^* = \begin{cases} p_i^* & \text{in } \Omega_i, \\ p_{i+1} + c_{i+1} & \text{in } \mathbb{R}^N \setminus \Omega_i, \end{cases}$$

we find a distribution $p_{i+1}^* \in \mathcal{D}'(\mathbb{R}^N; B)$ such that $p_{i+1}^* = p_j + c_j$ in ω_j for $1 \leq j \leq i+1$.

Thus, we see that, after a finite amount of cases, the result holds with $p := p_I^*$. It remains to prove that, under the induction assumption, one has (2.82).

First, note that, in a neighborhood of every $\mathbf{x} \in \Omega_i \cap \omega_{i+1}$, $p_i^* - p_{i+1}$ is a constant, since $\nabla p_i^* = \nabla p_{i+1} = \mathbf{f}$.

Let $\mathbf{x}^1$, $\mathbf{x}^2$ be two different points in $\Omega_i \cap \omega_{i+1}$ and set

$$c := \left(p_i^*(\mathbf{x}^2) - p_{i+1}(\mathbf{x}^2)\right) - \left(p_i^*(\mathbf{x}^1) - p_{i+1}(\mathbf{x}^1)\right).$$

It can be assumed that they belong to different connected components (otherwise, $c = 0$).

Let ζ_ε be a regularizing function (recall (2.5)–(2.6)). If ε is sufficiently small and L_{12} and L_{21} are paths in $\Omega_i \cap \omega_{i+1}$ respectively from $\mathbf{x}^1$ to $\mathbf{x}^2$ and from $\mathbf{x}^2$ to $\mathbf{x}^1$, the following holds:

$$\begin{aligned} c &= ((p_i^* - p_{i+1}) \odot \zeta_\varepsilon)(\mathbf{x}^2) - ((p_i^* - p_{i+1}) \odot \zeta_\varepsilon)(\mathbf{x}^1) \\ &= \left((p_i^* \odot \zeta_\varepsilon)(\mathbf{x}^2) - (p_i^* \odot \zeta_\varepsilon)(\mathbf{x}^1)\right) + \left((p_{i+1} \odot \zeta_\varepsilon)(\mathbf{x}^1) - (p_{i+1} \odot \zeta_\varepsilon)(\mathbf{x}^2)\right) \\ &= \int_{L_{12}} \nabla(p_i^* \odot \zeta_\varepsilon) \cdot d\ell + \int_{L_{21}} \nabla(p_{i+1} \odot \zeta_\varepsilon) \cdot d\ell \\ &= \int_{L_{12}} (\mathbf{f} \odot \zeta_\varepsilon) \cdot d\ell + \int_{L_{21}} (\mathbf{f} \odot \zeta_\varepsilon) \cdot d\ell \\ &= \int_{L} (\mathbf{f} \odot \zeta_\varepsilon) \cdot d\ell, \end{aligned}$$

where $L = L_{12} \cup L_{21}$ is a closed path in Ω_i.

At this point, observe that this last integral can be written in the form

$$\langle S, \widetilde{\alpha \mathbf{f}} \odot \zeta_\varepsilon \rangle = \langle \widetilde{\alpha \mathbf{f}}, S \odot \check{\zeta}_\varepsilon \rangle,$$

where α is a cut-off function, $\widetilde{\alpha\mathbf{f}}$ denotes the extension of $\alpha\mathbf{f}$ to $\mathcal{D}'(\mathbb{R}^N; B)^N$,

$$\langle S, \boldsymbol{\varphi} \rangle := \int_L \boldsymbol{\varphi} \cdot d\boldsymbol{\ell} \quad \forall \boldsymbol{\varphi} \in \mathcal{D}(\mathbb{R}^N)^N$$

and $\check{\zeta}_\varepsilon(\mathbf{x}) := \zeta_\varepsilon(-\mathbf{x})$.

If we choose α such that $\alpha = 1$ in the support of $S \odot \check{\zeta}_\varepsilon$ (which is always possible), we have

$$\langle \widetilde{\alpha\mathbf{f}}, S \odot \check{\zeta}_\varepsilon \rangle = \langle \mathbf{f}, \alpha S \odot \check{\zeta}_\varepsilon \rangle = \langle \mathbf{f}, S \odot \check{\zeta}_\varepsilon \rangle = 0,$$

since $S \odot \check{\zeta}_\varepsilon \in \mathcal{V}$.

This proves that $c = 0$.

Step 5: Up to now, we have proved that if $\mathbf{f}$ satisfies (2.46), there exist distributions p such that $\mathbf{f} = \nabla p$. Since Ω is connected, p is unique up to an additive constant.

With the notation in Step 4, we have

$$p = \sum_{i=1}^{I} \alpha_i p = \sum_{i=1}^{I} \alpha_i (p_i + c_i) \text{ in } \Omega.$$

We can redefine p, for instance, by choosing a ball $B_0 \subset \Omega$ and a function $\xi \in \mathcal{D}(B_0)$ satisfying $\int_{B_0} \xi = 1$ and changing p by $p - \langle p, \xi \rangle$. This allows to speak of a distribution in $\mathcal{D}'(\Omega; B)$ assigned to $\mathbf{f}$ through (2.81) without ambiguity.

Let us assume that $\mathbf{f} \in W^{m-1,r}(\Omega; B)^N \cap \{\mathbf{f} : \langle \mathbf{f}, \mathbf{v} \rangle = 0 \ \forall \mathbf{v} \in \mathcal{V}\}$, where $m \geq 0$ is an integer and $p \in [1, +\infty)$.

Using (2.80), it can be proved that $p_i \in W^{m,r}(\omega_i; B)$ for all i. Indeed, if $m = 0$, the f_k can be written as sums of first-order derivatives of some functions $g_{kj} \in L^r(\Omega; B)$. Thus, the $f_k \odot \zeta_0$ can be written as a sum of ponderations of the g_{kj} and derivatives of ζ_0 and, consequently, also belong to $L^r(\omega_i; B)$.

On the other hand, the $f_k \odot \zeta_k$ with $1 \leq k \leq N$ can be written as sums of terms the form $\partial_j g_{kj} \odot \zeta_k = g_{kj} \odot \partial_j \zeta_k$. Note that the function $\Phi := \partial_j \zeta_k$ satisfies the assumptions of Theorem 2 in [30] (p. 35), since

$$|\Phi(\mathbf{x})| \leq C|\mathbf{x}|^{-N} \text{ and } \int_{\{|\mathbf{x}|>2|\mathbf{y}|\}} |\Phi(\mathbf{x}-\mathbf{y}) - \phi(\mathbf{x})| \, d\mathbf{y} \leq C$$

for all $\mathbf{x} \in B(\mathbf{0}; r)$ and

$$\int_{\{|\mathbf{x}|=s\}} \Phi(\mathbf{x}) \, d\Gamma = 0$$

for all $s \in (0, r)$. Therefore, all the $g_{kj} \odot \partial_j \zeta_k$ belong to $L^r(\omega; B)$ and this shows that $p_i \in L^r(\omega_i; B)$.

If $m \geq 1$, a very similar argument shows that $p_i \in W^{m,r}(\omega_i; B)$.

This argument proves that the function p constructed in the previous step belongs to $W^{m,r}(\Omega; B)$ provided $\mathbf{f} \in W^{m-1,r}(\Omega; B)^N$. It is also clear from the discussion above that the mapping $\mathbf{f} \mapsto p$ is continuous.

This ends the proof.

2.9 Exercises of Chapter 2

Exercise 2.1* Let $\mathcal{L}^1(\Omega)$ be the set of measurable and integrable functions on Ω, endowed with the usual algebraic operations. Prove that $\mathcal{L}^1(\Omega)$ is a linear subspace of $\mathcal{L}(\Omega)$ and $v \mapsto \int_\Omega v$ is a linear form on $\mathcal{L}^1(\Omega)$.

Exercise 2.2* Prove that, if $u, v \in \mathcal{L}(\Omega)$, the integral of one of these functions makes sense and $u = v$ *almost everywhere* (a.e.) in Ω, then the integral of the other one is also meaningful and both integrals coincide.

HINT: First, prove the result for positive functions. Then, extend the assertion to any couple of measurable functions.

Exercise 2.3* Prove that, if $u \in \mathcal{L}(\Omega)$, $u \geq 0$ a.e. in Ω and $\int_\Omega u = 0$, then $u = 0$ a.e.

HINT: Use the definition of $\int_\Omega u$.

Exercise 2.4* Prove that, if $u \in \mathcal{L}^1(\Omega)$, one has (2.1).

HINT: First, use that, if $u, v \in \mathcal{L}^1(\Omega)$ and $u \leq v$, then $\int_\Omega u \leq \int_\Omega v$. Then, prove (2.1).

Exercise 2.5* Using the Hölder inequality (2.3), prove the Minkowski inequality (2.2).

HINT: Start form the inequality $\int_\Omega |u + v|^p \leq \int_\Omega |u + v|^{p-1}(|u| + |v|)$ and apply (2.3).

Exercise 2.6* Prove the Hölder inequality (2.3).

HINT: First, assume that $\|u\|_{L^p} = \|v\|_{L^p} = 1$ and use *Young's inequality: $ab \leq \frac{1}{p}a^p + \frac{1}{p'}b^{p'}$* for all $a, b \geq 0$. Young's inequality is a consequence of the concavity of the logarithm: for $a, b > 0$, $\log(ab) = \frac{1}{p}\log a^p + \frac{1}{p'}\log b^{p'} \leq \log(\frac{1}{p}a^p + \frac{1}{p'}b^{p'})$.

Exercise 2.7* Prove that for any $p \in (1, +\infty)$, as a topological space, $L^p(\Omega)$ is *separable,* i.e., there exists a set $E \subset L^p(\Omega)$ that is *dense* and *countable.*

HINT: If Ω is bounded, the polynomial functions with rational coefficients are dense in $C^0(\overline{\Omega})$ and this space is dense in $L^p(\Omega)$. If Ω is unbounded and we set $\Omega(n) := \Omega \cap B(0; n)$ for all large n, the family of functions of the form $q 1_{\Omega(n)}$, where q is a polynomial function on $\mathbb{R}^N$ with rational coefficients are dense in $L^p(\Omega)$.

Exercise 2.8** Prove that $L^\infty(\Omega)$ is a non-separable Banach space.

HINT: Let $\{\Omega_n\}$ be a sequence of non-empty open subsets of Ω with $\Omega_n \cap \Omega_m = \emptyset$ for $n \neq m$. Let E be the set of (classes of) functions $e \in L^\infty(\Omega)$ that satisfy $e|_{\Omega_n} = 1$ or $e|_{\Omega_n} = 0$ a.e. for all n. Then E is not countable and any two different functions $e, e' \in E$ satisfy $\|e - e'\|_{L^\infty} = 1$.

Exercise 2.9* Prove that, for every $p \in [1, +\infty]$, the set $L^p_{\text{loc}}(\Omega)$ is a linear space for the usual algebraic operations. Prove also that it is a *Fréchet space* (i.e., a complete metrizable locally convex space) for the family of seminorms

$$\pi_{p,\Omega_0}(v) = \|v\|_{L^p(\Omega_0)} \quad \forall v \in L^p_{\text{loc}}(\Omega) \ \forall \Omega_0 \in \mathcal{K}(\Omega),$$

where $\mathcal{K}(\Omega)$ denotes the family of the compact subsets of Ω. Finally, prove that, if $p < +\infty$, $L^p_{\text{loc}}(\Omega)$ is separable.

HINT: Choose an increasing sequence of bounded open sets ω_j with $\cup_{j \geq 1} \omega_j = \Omega$ and consider the countable family $\{\pi_{p,\omega_j} : j \geq 1\}$. Prove that, if the $f_n \in L^p_{\text{loc}}(\Omega)$ and $\pi_{p,\omega_j}(f_n - f_m) \to 0$ as $n, m \to +\infty$ for all $j \geq 1$, there exist $f \in L^p_{\text{loc}}(\Omega)$ such that $\pi_{p,\omega_j}(f_n - f) \to 0$ as $n \to +\infty$ for all $j \geq 1$. Also, use Exercise 2.7 to find a dense countable set in $L^p_{\text{loc}}(\Omega)$.

Exercise 2.10* Prove that the $L^p_{\text{loc}}(\Omega)$ are algebraically and topologically ordered, that is,

$$L^p_{\text{loc}}(\Omega) \hookrightarrow L^q_{\text{loc}}(\Omega) \quad \forall p, q \in [1, +\infty] \text{ with } p \geq q,$$

where the embeddings are (sequentially) continuous.

HINT: Use Hölder's inequality to deduce that $\pi_{q,K}(f) \leq C_K \pi_{p,K}(f)$ for any $f \in L^p_{\text{loc}}(\Omega)$ and any $K \in \mathcal{K}(\Omega)$.

Exercise 2.11** Prove the Clarkson inequalities in Lemma 2.2.

HINT: For $p \geq 2$, it is enough to prove that $a^p + b^p \leq (a^2 + b^2)^{p/2}$ for any $a, b \geq 0$ (to get this, assume that for instance $b > 0$, divide by b^p and analyze the behavior of the resulting function of a/b); then, use this inequality with $a = \frac{1}{2}|f(\mathbf{x}) + g(\mathbf{x})|$ and $b = \frac{1}{2}|f(\mathbf{x}) - g(\mathbf{x})|$. For $1 < p < 2$, it suffices to prove the second inequality (if it is written for $f + g$ and $f - g$, we get the first one); to get it, assume that Ω has unit Lebesgue measure, introduce the functions

$$\alpha(s) := \mathbb{1}_{[0,1/2]} - \mathbb{1}_{[1/2,1]} \text{ and } \beta(s) := \mathbb{1}_{[0,1/4] \cup [1/2,3/4]} - \mathbb{1}_{[1/4,1/2] \cup [3/4,1]}$$

and note that

$$\begin{aligned} \|f + g\|^p_{L^p} + \|f - g\|^p_{L^p} &= \int_0^1 \left(\int_\Omega |\alpha(s) f(\mathbf{x}) + \beta(s) g(\mathbf{x})|^p \right) ds \\ &\leq \int_\Omega (|f(\mathbf{x})|^2 + |g(\mathbf{x})|^2)^{p/2}. \end{aligned}$$

Exercise 2.12* Prove Theorem 2.2 for $p = 1$ in the case of an unbounded open set Ω.

HINT: Use an increasing family of bounded open sets $\Omega_n \subset \Omega$ such that $\Omega = \cup_{n\geq 1}\Omega_n$ and an auxiliary "multiplier" $h \in L^2(\Omega)$ to pass from $L^2(\Omega)$ to $L^1(\Omega)$.

Exercise 2.13* Prove that $(L^\infty(\Omega))'$ is not $L^1(\Omega)$.

HINT: Find a linear form on $L^\infty(\Omega)$ that is not defined by any function in $L^1(\Omega)$. To this purpose, consider an appropriate linear form on $C_c^0(\Omega)$ whose values cannot be given by the integral of the product by a function in $L^1(\Omega)$ and apply Hahn-Banach Theorem. Note that the same result can be proved in an indirect way: if we had $(L^\infty(\Omega))'$ isomorphic to $L^1(\Omega)$, then $(L^\infty(\Omega))'$ would necessarily be separable; but then $L^\infty(\Omega)$ would also be separable, which is not the case.

Exercise 2.14* Prove that the distribution S defined in (2.15) "does not belong" to $L^1_{\rm loc}(\Omega)$.

HINT: Prove that, if $S = S_f$ ifor some $f \in L^1_{\rm loc}(\Omega)$, then $f = 0$ a.e., whence we get $S = 0$, which is absurd.

Exercise 2.15* Prove Proposition 2.3.

Exercise 2.16* Prove (2.16).

Exercise 2.17* Prove that, for any distribution S and any multi-index α, $\partial^\alpha S$ is a new distribution.

Exercise 2.18* Prove Proposition 2.4.

HINT: Use integration by parts to show that $\int_\Omega \partial^\alpha f\, \varphi = (-1)^{|\alpha|}\int_\Omega f\, \partial^\alpha \varphi$ for all $\varphi \in \mathcal{D}(\Omega)$.

Exercise 2.19* Let H be the distribution defined by (2.17). Prove that the derivative $\partial^{(1,0)}H$ is given by (2.15).

HINT: Use integration by parts in the sets $\{\mathbf{x} \in \mathbb{R}^2 : x_1 > 0\}$ and $\{\mathbf{x} \in \mathbb{R}^2 : x_1 < 0\}$.

Exercise 2.20* Prove that, for any multi-index α, the differential operator $\partial^\alpha : \mathcal{D}'(\Omega) \mapsto \mathcal{D}'(\Omega)$ is well defined, linear and sequentially continuous.

Exercise 2.21* Prove that the spaces $W^{m,p}(\Omega)$ are complete.

HINT: Use that the $L^p(\Omega)$ are complete.

Exercise 2.22** Prove that, if $\Omega \subset \mathbb{R}^N$ is a bounded connected open set with Lipschitz-continuous boundary and $1 \leq p < +\infty$, there exists a bounded linear mapping $E : W^{1,p}(\Omega) \mapsto W^{1,p}(\mathbb{R}^N)$, an open set $G \subset \mathbb{R}^N$ and a constant $C > 0$ such that $Ev = v$ in Ω, $\text{Supp}\,(Ev) \subset G$,

$$\|Ev\|_{L^p(\mathbb{R}^N)} \leq C\|v\|_{L^p(\Omega)} \quad \text{and} \quad \|\nabla Ev\|_{L^p(\mathbb{R}^N)} \leq C\|\nabla v\|_{L^p(\Omega)}$$

for all $v \in W^{1,p}(\Omega)$.

HINT: Introduce an appropriate partition of unit $\{\varphi_0, \varphi_1, \dots, \varphi_I\}$ to work separatedly in an open set $\Omega_0 \subset\subset \Omega$ and in a finite amount of neighborhoods of pieces of the boundary. Then, for any $v \in W^{1,p}(\mathbb{R}^N)$, write that $v = \varphi_0 v + \sum_{i=1}^{I} \varphi_i v$ and, in each Ω_i with $i \geq 1$, work in local coordinates and extend by reflection.

Exercise 2.23* Prove that the functions (2.23) belong to $C^\infty(\overline{\Omega})$ and satisfy $v_\varepsilon \to v$ strongly in $H^1(\Omega)$ as $\varepsilon \to 0$.

HINT: First, note that the v_ε are restrictions to Ω of functions in $C^\infty(\mathbb{R}^N)$. Then, write $\|v_\varepsilon - v\|_{L^2}^2$ as a double integral and apply Lebesgue's Theorem. Also, proceed similarly with the $\|\partial_i v_\varepsilon - \partial_i v\|_{L^2}^2$.

Exercise 2.24** Assume that Ω is bounded and $\partial\Omega$ is Lipschitz-continuous. Prove that the classical restriction mapping $v \mapsto v|_{\partial\Omega}$ is linear and continuous from $C^1(\overline{\Omega})$, endowed with the norm of $H^1(\Omega)$, into $L^2(\partial\Omega)$.

HINT: First, prove the result when Ω is a N-rectangle; for instance, the following is obtained: $\int_{a_1}^{b_1} |v(0,x')|^2 \, dx' \leq C(\|v\|^2 + \|\partial_1 v\|^2)$, where the notation is self-explanatory. Then, for a general Ω, use local coordinates to get a similar estimate of the L^2 norm of $v|_{\partial\Omega}$ in a neighborhood of an arbitrary point $\mathbf{x}_0 \in \partial\Omega$. Finally, use a partition of unit to deduce a global estimate.

Exercise 2.25** Complete the details of the proof of the identity $N(\gamma) = H_0^1(\Omega)$ in Theorem 2.3. Also, prove that $R(\gamma)$ is a proper Hilbert subspace of $L^2(\partial\Omega)$ and γ possesses a continuous inverse on $R(\gamma)$.

Exercise 2.26* Prove Proposition 2.8. Prove also that the indicated embeddings are compact.

Exercise 2.27* Prove that, endowed with the bilinear form in (2.26), $E(\Omega)$ is a Hilbert space.

Exercise 2.28* Prove Lemma 2.4.

HINT: Use regularization by convolution.

Exercise 2.29* Prove that (2.28) implies that $\mathbf{v} \mapsto \mathbf{v} \cdot \mathbf{n}|_{\partial\Omega}$ is a linear continuous mapping from $C^1(\overline{\Omega})^N$, endowed with the norm of $E(\Omega)$, into $H^{-1/2}(\partial\Omega)$.

Exercise 2.30** Let $S \in \mathcal{D}'(\Omega)$ be given. Prove that, if the distributional derivatives of S vanish, then $S = C$ a.e. for some C.

HINT: First, prove that a function $\varphi \in \mathcal{D}(\Omega)$ is, for every $i = 1, \dots, N$, the primitive of a function $\psi_i \in \mathcal{D}(\Omega)$ if and only if $\int_\Omega \varphi = 0$. Then, choose $\alpha \in \mathcal{D}(\Omega)$ such that $\int_\Omega \alpha = 1$ and prove that any $\varphi \in \mathcal{D}(\Omega)$ can be written in the form $\varphi = (\int_\Omega \varphi)\alpha + \varphi^0$, with $\int_\Omega \varphi^0 = 0$. Finally, deduce that $S = C$, with $C = \langle S, \alpha \rangle$.

Exercise 2.31** Let $\Omega \subset \mathbb{R}^N$ be bounded. Prove that there exist constants $C(\Omega)$ such that one has

$$\|v\| \leq C(\Omega)\|\nabla v\| \quad \forall v \in X(\Omega), \tag{2.83}$$

when $X(\Omega)$ is one of the following spaces:

- $X(\Omega) = \{v \in H^1(\Omega) : \int_\Omega v\,d\mathbf{x} = 0\}$ (in this case, (2.83) is called the *Poincaré-Wirtinger* inequality).
- $X(\Omega) = \{v \in H^1(\Omega) : \gamma v = 0 \text{ on } \Gamma_0\}$ (Γ_0 is an open subset of $\partial\Omega$).
- $X(\Omega) = \{v \in H^1(\Omega) : \int_{\Gamma_0} \gamma v\,d\Gamma = 0\}$.

HINT: In all cases, argue by contradiction: if (2.83) is not satisfied, there must exist a sequence $\{v_n\}$ in $X(\Omega)$ such that $\|v_n\| = 1$ for all n and $\|\nabla v_n\| \to 0$ as $n \to +\infty$. But this is impossible.

Exercise 2.32* Prove that, if $1 \le p < N$ and $W^{1,p}(\mathbb{R}^N) \hookrightarrow L^q(\mathbb{R}^N)$, then one has necessarily

$$\frac{1}{q} = \frac{1}{p} - \frac{1}{N}.$$

HINT: Use the embedding estimates for a nonzero function $v \in W^{1,p}(\mathbb{R}^N)$ and the rescaled family of functions $\{v_\varepsilon\}$, with $v_\varepsilon(\mathbf{x}) := v(\varepsilon\mathbf{x})$.

Exercise 2.33* Prove that, if $G \subset \mathbb{R}^k$ is an open set, $v \in \mathcal{D}(G)$ and $\gamma \ge 1$, then $|v|^\gamma \in W^{1,p}(G)$ for all $p \in [1, +\infty]$ and $\partial_j |v|^\gamma = \gamma |v|^{\gamma-1}$ Sign (v) a.e. in G.

HINT: Use a family of regular approximations $G_\varepsilon : \mathbb{R} \mapsto \mathbb{R}$ to the function $s \mapsto |s|^\gamma$ and deduce the result taking limits as $\varepsilon \to 0$.

Exercise 2.34* Prove that, $W^{1,N}(\mathbb{R}^N) \subset L^{p_*}(\mathbb{R}^N)$ for all $p_* \in [1, +\infty)$.

Exercise 2.35* Prove that, if $\Omega \subset \mathbb{R}^N$ is a bounded connected open set, $\partial\Omega$ is regular enough, $mp > N$, k is the largest integer satisfying $0 \le k < m - N/p$ and $m - N/p = k + \alpha$ with $\alpha \in (0, 1]$, then $W^{m,p}(\Omega) \hookrightarrow C^{k,\alpha}(\overline{\Omega})$.

HINT: Consider the $(m-1)$-th order derivatives of a function $v \in W^{m,p}(\Omega)$ and argue as in Step 3 of the proof of Theorem 2.6.

Exercise 2.36* Prove (2.32)–(2.35) for $m \ge 2$ using these results for $m = 1$.

Exercise 2.37* Prove (2.73), (2.74) and (2.75).

Exercise 2.38* Prove Lemma 2.10 (i) when $p \ge N$. Prove also parts (iii) and (iv).

Exercise 2.39* Let B be a Banach space. Prove that the $L^p_{\text{loc}}(0, T; B)$ form an ordered family of Fréchet spaces for some appropriate semi-norms.

HINT: Argue as in Exercises 2.9 and 2.10.

Exercise 2.40** Prove Proposition 2.10. Deduce that the mapping $f \mapsto S_f$ from $L^1_{\text{loc}}(0, T; B)$ into the space of B-valued distributions $\mathcal{D}'((0, T); B)$ is well defined, one-to-one and sequentially continuous.

HINT: Use that any $f \in L^1_{\text{loc}}(0, T; B)$ takes values a.e. in a separable subspace of B and, also, that any separable Banach can be embedded in a Banach space X such that X' is separable.

Exercise 2.41* Prove that the distributional derivative is a well defined linear sequentially continuous mapping from $\mathcal{D}'(0, T; B)$ into itself.

Exercise 2.42* Prove Proposition 2.11.
HINT: Argue as in Exercise 2.18.

Exercise 2.43* Prove Proposition 2.12.
HINT: Argue as in the proof of Theorem 2.2 and Exercise 2.12.

Exercise 2.44** Prove that, as in the case of real-valued functions, one has

$$W^{1,1}(0, T; B) \hookrightarrow C^0([0, T]; B)$$

and any $f \in W^{1,1}(0, T; B)$ has a unique representative that is absolutely continuous in $[0, T]$.
HINT: Argue as in the proof of Lemma 2.12.

Exercise 2.45* Prove Proposition 2.13.
HINT: Argue as in the proof of Proposition 2.6.

Exercise 2.46* Prove that (2.46) implies (2.76).

Exercise 2.47* Prove that, if the open set $G \subset \mathbb{R}^k$ is connected, any distribution $S \in \mathcal{D}'(G; B)$ whose derivatives $\partial_i S$ vanish is necessary a constant.
HINT: Use regularization by convolution. More precisely, for any $S \in \mathcal{D}'(G; B)$, introduce the distributions $S_\varepsilon := S * \zeta_\varepsilon$, where $\{\zeta_\varepsilon\}$ is a regularizing sequence and

$$\langle S * \zeta_\varepsilon, \varphi \rangle = \langle S, \langle \zeta_\varepsilon, \tilde{\tau}_{\bullet}\varphi \rangle \rangle \quad \forall \varphi \in \mathcal{D}(G).$$

Here, for any $\mathbf{y} \in G$, $\tilde{\tau}_{\mathbf{y}}\varphi$ stands for the function $\mathbf{x} \mapsto \varphi(\mathbf{x} + \mathbf{y})$ (we implicitly assume that φ is extended by zero to the whole $\mathbb{R}^k$). Prove that $S_\varepsilon \in C^\infty(G; B)$ and $\partial_j S_\varepsilon = \partial_j S * \zeta_\varepsilon$ for all $\varepsilon > 0$. Also, prove that $S_\varepsilon \to S$ in $\mathcal{D}'(G; B)$ as $\varepsilon \to 0$. Deduce the result for S_ε and then for S.

Exercise 2.48* Prove (2.77).

Exercise 2.49** Using the constructive argument in the proof of Theorem 2.7, prove that the mapping $\mathbf{f} \mapsto p$ satisfies the continuity properties asserted in the statement.

Exercise 2.50* Prove (2.59).
HINT: Argue by contradiction and use the compactness of the embedding $X \hookrightarrow B$.

Exercise 2.51* Prove Lemma 2.18.
HINT: Use (2.57) and adapt the argument of the proof of Lemma 2.17.

Exercise 2.52* Give a proof of Lemma 2.19. Prove also that the embedding $L^\infty(0, T; X) \cap C^0([0, T]; Y) \hookrightarrow C^0_w([0, T]; X)$ is sequentially continuous, i.e. that the following assertion holds: if $f_n \to f$ strongly in $C^0([0, T]; Y)$ and strongly in $L^\infty(0, T; X)$, then $\langle h, f_n\rangle_{X',X} \to \langle h, f\rangle_{X',X}$ strongly in $C^0([0, T])$ for every $h \in X'$.

HINT: Let $f \in L^\infty(0, T; X)$ be given and let us also denote by f its Y-valued continuous version. Let $\ell \in X'$ and $t_0 \in [0, T]$ be given and assume that $t_n \to t_0$. Using that X is reflexive and the uniqueness of the limit, prove that $\langle \ell, f(t_n)\rangle_{X',X} \to \langle \ell, f(t_0)\rangle_{X',X}$ as $n \to +\infty$.

Exercise 2.53* Prove Proposition 2.14 when some p_i or q_i is infinity..

Exercise 2.54* Prove Corollary 2.3.

Exercise 2.55* Prove Lemma 2.19.

HINT: First, check that, if $t_n \to t_0$ in $[0, T]$, the $f(t_n)$ are uniformly bounded in X and one has $f(t_n) \to f(t_0)$ strongly in Y. Deduce that $f(t_n) \to f(t_0)$ weakly in Y.

Exercise 2.56* Prove (2.64).

HINT: Use the properties of the orthogonal projector P_H.

Exercise 2.57* Prove Lemma 2.20.

HINT: Use the density of $D(A)$ in H and the compactness of the embedding $V \hookrightarrow H$.

Exercise 2.58* Prove Lemma 2.21.

Exercise 2.59* Prove Lemma 2.22.

HINT: Use Theorem 2.11.

References

1. Adams, R.A.: Sobolev Spaces. Pure and Applied Mathematics, vol. 65. Academic Press [A subsidiary of Harcourt Brace Jovanovich, Publishers], New York-London (1975)
2. Amrouche, C., Girault, V.: On the existence and regularity of the solutions of Stokes problem in arbitrary dimension. Proc. Jpn. Acad. Ser. A Math. Sci. **67**(5), 171–175 (1991)
3. Bello, J.A.: L^r regularity for the Stokes and Navier-Stokes problems. Ann. Mat. Pura Appl. (4) **170**, 187–206 (1996)
4. Boyer, F., Fabrie, P.: Mathematical Tools for the Study of the Incompressible Navier-Stokes Equations and Related Models. Applied Mathematical Sciences, vol. 183. Springer, Berlin (2013)
5. Brézis, H.: Functional Analysis, Sobolev Spaces and Partial Differential Equations. Universitext. Springer, New York (2011)
6. Cattabriga, L.: Su un problema al contorno relativo al sistema di equazioni di Stokes. Rend. Sem. Mat. Univ. Padova **31**, 308–340 (1961)
7. Constantin, P., Foias, C.: Navier-Stokes Equations. Chicago Lectures in Mathematics. University of Chicago Press, Chicago (1988)
8. Diestel, J., Uhl, Jr. J.J.: Vector Measures. Mathematical Surveys, No. 15. American Mathematical Society, Providence (1977). With a foreword by B. J. Pettis
9. Evans, L.C.: Partial Differential Equations. Graduate Studies in Mathematics, vol. 19. American Mathematical Society, Providence (1998)

10. Felix, L., Denjoy, A., Montel, P.: Henri Lebesgue, le savant, le professeur, l'homme. Enseign. Math. (2) **3**, 1–18 (1957)
11. Girault, V., Raviart, P.-A.: Finite Element Methods for Navier-Stokes Equations. Springer Series in Computational Mathematics, vol. 5. Springer, Berlin (1986). Theory and algorithms
12. Heywood, J.G.: The Navier-Stokes equations: on the existence, regularity and decay of solutions. Indiana Univ. Math. J. **29**(5), 639–681 (1980)
13. Ladyzhenskaya, O.A.: The Mathematical Theory of Viscous Incompressible Flow. Mathematics and Its Applications, vol. 2. Gordon and Breach Science Publishers, New York-London-Paris (1969)
14. Ladyzhenskaya, O.A.: The sixth millennium problem: Navier-Stokes equations, existence and smoothness. Uspekhi Mat. Nauk **58**(2(350)), 45–78 (2003)
15. Ladyzhenskaya, O.A., Seregin, G.A.: On partial regularity of suitable weak solutions to the three-dimensional Navier-Stokes equations. J. Math. Fluid Mech. **1**(4), 356–387 (1999)
16. Lang, S.: Real and Functional Analysis. Graduate Texts in Mathematics, vol. 142, 3rd edn. Springer, New York (1993)
17. Lions, J.-L.: Quelques méthodes de résolution des problèmes aux limites non linéaires. Dunod, Paris; Gauthier-Villars, Paris (1969)
18. Pata, V.: Fixed Point Theorems and Applications. Unitext, vol. 116. Springer, Cham (2019). La Matematica per il 3+2
19. Paumier, A.-S., Barany, M.J., Lützen, J.: From Nancy to Copenhagen to the world: the internationalization of Laurent Schwartz and his theory of distributions. Historia Math. **44**(4), 367–394 (2017)
20. Roskovec, T.G., Fiorenza, A., Formica, M.R., Soudský, F.: Detailed proof of classical Gagliardo-Nirenberg interpolation inequality with historical remarks. Z. Anal. Anwend. **40**(2), 217–236 (2021)
21. Schwartz, L.: Théorie des distributions. Tomes I et II. Publ. Inst. Math. Univ. Strasbourg, vols. 9–10. Hermann & Cie., Paris (1950–1951)
22. Shannon, C.: On Lipschitz implicit function theorems in Banach spaces and applications. J. Math. Anal. Appl. **494**(2):Paper No. 124589, 16 (2021)
23. Shapiro, J.H.: A Fixed-Point Farrago. Universitext. Springer, Cham (2016)
24. Simader, C., Naumann, J.: Measure and Integration on Lipschitz Manifolds. Humboldt-Univ., Universität zu Berlin, Institut für Mathematik (2007)
25. Simon, J.: Compact sets in the space $L^p(0, T; B)$. Ann. Mat. Pura Appl. (4) **146**, 65–96 (1987)
26. Simon, J.: Existencia de solución del problema de Navier-Stokes con densidad variable. (Spanish) [Existence of solution for the variable density Navier-Stokes problem]. Lectures at the University of Sevilla, Spain (1989)
27. Simon, J.: Nonhomogeneous viscous incompressible fluids: existence of velocity, density and pressure. SIAM J. Math. Anal. **21**(5), 1093–1117 (1990)
28. Simon, J.: Démonstration constructive d'un théorème de G. de Rham. C. R. Acad. Sci. Paris Sér. I Math. **316**(11), 1167–1172 (1993)
29. Simon, J.: Banach, Fréchet, Hilbert and Neumann spaces. Mathematics and Statistics Series. ISTE, London; John Wiley & Sons, Inc., Hoboken (2017). Analysis for PDEs set. Vol. 1
30. Stein, E.M.: Singular Integrals and Differentiability Properties of Functions. Princeton Mathematical Series, No. 30. Princeton University Press, Princeton (1970)
31. Tartar, L.: Topics in Nonlinear Analysis. Publications Mathématiques d'Orsay 78, vol. 13. Université de Paris-Sud, Département de Mathématiques, Orsay (1978)
32. Témam, R.: Navier-Stokes Equations. AMS Chelsea Publishing, Providence (2001). Theory and numerical analysis, Reprint of the 1984 edition
33. Triebel, H.: Theory of Function Spaces. Monographs in Mathematics, vol. 78. Birkhäuser Verlag, Basel (1983)
34. Vo-Khac, K.: Introduction aux méthodes mathématiques modernes de la physique. Espaces préhilbertiens; série de Fourier; spectre des opérateurs. Distributions, convolutions et transformation de Fourier; espaces de Sobolev. Equations différentielles; transformation de Laplace; fonctions orthogonales. Equations aux dérivées partielles; méthode de Galerkine-Fourier.

Centre de Documentation Universitaire, Paris (1968). Faculté des Sciences d'Orléans, Cours de Mathématiques C2 de la Maîtrise de Physique

35. Yushkevich, A.P.: Some remarks on the history of the theory of generalized solutions of partial differential equations and generalized functions. Istor.-Mat. Issled. (34), 256–267 (1993)
36. Zeidler, E.: Nonlinear Functional Analysis and Its Applications. IV. Springer, New York (1988). Applications to mathematical physics, Translated from the German and with a preface by Juergen Quandt

Chapter 3
The Existence of a Global Weak Solution

In this chapter, we focus in proving the existence of a global weak solution to the three-dimensional variable density incompressible Navier-Stokes equations, both for bounded and unbounded domains.

In general terms, a weak solution is a triplet $\{\mathbf{u}, \rho, p\}$ such that the PDEs are satisfied in a slightly stronger sense than the sense of distributions (to be specified below).

The existence result for bounded domains is due to Simon [18, 19]. A previous pioneering paper concerning only the case of strictly positive density is due to Kazhikhov [13]; see also [14]. The generalization to unbounded domains is due to Fernández-Cara and Guillén-González [10]. We will follow these references closely; see also [1, 17] for other improvements and presentations.

In the following sections, the arguments that lead to the existence of a weak solution are presented with detail. We believe that this presentation may be useful to understand what one has to do not only in the present setting but also when considering other nonlinear PDEs.

3.1 The Fundamental Result in a Bounded Domain

In order to fix ideas, unless otherwise indicated, it will be assumed that $\Omega \subset \mathbb{R}^3$ is a non-empty bounded connected open set, with C^2 boundary $\partial\Omega$ that leaves Ω at one side (the definition is given in Sect. 2.6).

We set $Q := \Omega \times (0, T)$ and $\Sigma := \partial\Omega \times (0, T)$. Recall that the spaces H and V satisfy

$$H = \{\mathbf{u} \in L^2(\Omega)^N : \nabla \cdot \mathbf{u} = 0, \quad \mathbf{u} \cdot \mathbf{n} = 0 \text{ on } \partial\Omega\}$$

and

$$V = \{\mathbf{u} \in H_0^1(\Omega)^N : \nabla \cdot \mathbf{u} = 0\}.$$

P. Braz e Silva et al., *Analysis and Control of the Variable Density Incompressible Navier-Stokes Equations*, MS&A 22, https://doi.org/10.1007/978-3-032-14510-9_3

Our aim in this section is to prove the following result (for the definition of the Nikolskii spaces $N^{s,q}(0, T; B)$, see Chap. 2, Definition 2.8):

Theorem 3.1 *Let $T > 0$ be given. Assume that $\mathbf{u}_0 \in H$, $\rho_0 \in L^\infty(\Omega)$ with $\rho_0 \geq 0$ and $\mathbf{f} \in L^1(0, T; L^2(\Omega)^3)$. Then, there exist $\mathbf{u} \in L^2(0, T; V)$, $\rho \in L^\infty(Q)$ and $p \in W^{-1,\infty}(0, T; L^2(\Omega))$ such that*

$$\rho\mathbf{u} \in L^\infty(0, T; L^2(\Omega)^3) \cap N^{1/4,2}(0, T; W^{-1,3}(\Omega)^3), \tag{3.1}$$

$$\inf_\Omega \rho_0 \leq \rho(x, t) \leq \sup_\Omega \rho_0 \quad \text{a.e. in } Q, \tag{3.2}$$

the equations

$$\frac{\partial \rho\mathbf{u}}{\partial t} + \nabla \cdot (\rho\mathbf{u} \otimes \mathbf{u}) + \nabla p = \mu\Delta\mathbf{u} + \rho\mathbf{f}, \tag{3.3}$$

$$\nabla \cdot \mathbf{u} = 0, \tag{3.4}$$

$$\frac{\partial \rho}{\partial t} + \nabla \cdot (\rho\mathbf{u}) = 0 \tag{3.5}$$

are satisfied in Q (in the distributional sense), the boundary condition

$$\mathbf{u} = 0 \tag{3.6}$$

is satisfied as an equality in $L^2(0, T; H^{1/2}(\partial\Omega)^3)$ and the following initial conditions hold:

$$\left(\int_\Omega \rho\mathbf{u} \cdot \mathbf{v}\right)\Big|_{t=0} = \int_\Omega \rho_0\mathbf{u}_0 \cdot \mathbf{v} \quad \forall \mathbf{v} \in V, \tag{3.7}$$

$$\rho|_{t=0} = \rho_0 \ \text{ in } \Omega. \tag{3.8}$$

Let $\{\mathbf{u}, \rho, p\}$ be a solution to (3.3)–(3.8) furnished by Theorem 3.1. Note that, in view of (3.1) and (3.3), for any $\mathbf{v} \in V$ the real-valued function

$$t \mapsto \left(\int_\Omega \rho\mathbf{u} \cdot \mathbf{v}\right)(t)$$

is well defined and absolutely continuous in $[0, T]$, since it belongs to $L^\infty(0, T)$ and possesses a distributional time derivative in $L^1(0, T)$. This gives a sense to (3.7).

On the other hand, $\rho \in L^\infty(Q)$ and, thanks to (3.4), possesses a time derivative in $L^\infty(0, T; H^{-1}(\Omega))$. Consequently, ρ can be viewed as a continuous $H^{-1}(\Omega)$-valued function on $[0, T]$ and a weakly continuous $L^r(\Omega)$-valued function for any $r \in (1, +\infty)$ and (3.8) is also meaningful, at least as an equality in $L^r(\Omega)$.

The PDEs in (3.3) and (3.5) are nonlinear. This means that, even if with $\mathbf{f} = 0$, a linear combination of solutions is not a solution.

Any triplet $\{\mathbf{u}, \rho, p\}$ given by Theorem 3.1 is called a *weak solution* to (3.3)–(3.8). Note that it satisfies the incompressibility condition $\nabla \cdot \mathbf{u} = 0$ and the homogeneous boundary condition (3.6) in the following sense:[1]

$$\mathbf{u} \in L^2(0, T; V) \text{ and, consequently, } \mathbf{u}(\cdot, t) \in V \text{ a.e. in } (0, T).$$

For more regular initial data with $\rho_0 \geq \alpha > 0$ a.e. in Ω, it can be shown that $\mathbf{u} \in L^2(0, T; V) \cap L^\infty(0, T; H)$ and Eq. (3.3) is satisfied in a stronger sense; see the considerations on regularity below, in Sect. 3.2 (see also the arguments in [17]).

Now, we prove Theorem 3.1. To this end, we will apply classical techniques. More precisely, we will introduce a family of *semi-Galerkin* approximations, we will obtain appropriate estimates, we will deduce the existence of convergent subsequences and, finally, we will take limits and conclude that a weak solution exists.

The following arguments (suitably modified) can serve to prove the existence of weak solutions to many other linear and nonlinear PDE problems. To this respect, several related comments will be given at the end of the chapter.

For clarity, the proof will be divided in several steps.

Step 1: The Approximated Solutions

In view of the results in Chap. 2, we can work with a "base" $\{\mathbf{w}^1, \ldots, \mathbf{w}^m, \ldots\}$ of V, that is, a set of linearly independent functions whose linear combinations are dense in V, with $\mathbf{w}^m \in C^1(\overline{\Omega})^3$ for all $m \geq 1$,

$$(\mathbf{w}^i, \mathbf{w}^j) = \delta_{ij} \text{ and } (\mathbf{w}^i, \mathbf{w}^j) = \lambda_i \delta_{ij} \quad \forall i, j \geq 1.$$

Let $V^m := [\mathbf{w}^1, \ldots, \mathbf{w}^m]$ be the space spanned by $\mathbf{w}^1, \ldots, \mathbf{w}^m$, let $\{\mathbf{f}^m\}$ be a sequence in $C^0([0, T]; L^2(\Omega)^3)$ such that $\mathbf{f}^m \to \mathbf{f}$ in $L^1(0, T; L^2(\Omega)^3)$ and let the $\mathbf{u}_0^m \in V^m$ and $\rho_0^m \in C^1(\overline{\Omega})$ be such that

$$\frac{1}{m} + \inf_\Omega \rho_0 \leq \rho_0^m \leq \frac{1}{m} + \sup_\Omega \rho_0 \text{ in } \Omega$$

[1] Recall that, in PDE theory, it is a common practice to introduce weak solutions even for simple linear problems. The reason is that, except in a very small and not significant amount of cases, this is the best solution concept for which existence can be ensured.

Indeed, if $f = f(\mathbf{x})$ is continuous and bounded in the open set Ω, in general, there is no C^2 solution to the problem

$$\begin{cases} -\Delta u = f(\mathbf{x}), & \mathbf{x} \in \Omega, \\ u = 0, & \mathbf{x} \in \partial\Omega \end{cases}$$

(see [11] for a counter-example).

Of course, it is also clear that in a differential problem where the coefficients are discontinuous it is not appropriate to look for pointwise (classical) solutions and we have to weaken or relax the definition.

for all $m \geq 1$,

$$\mathbf{u}_0^m \to u_0 \text{ in } H \text{ and } \rho_0^m \to \rho_0 \text{ weakly-}* \text{ in } L^\infty(\Omega). \tag{3.9}$$

It will be said that $\{\mathbf{u}^m, \rho^m\}$ is an approximate solution to the variable density Navier-Stokes problem (3.3)–(3.8) in Q if $\rho^m \in L^\infty(Q)$, $\mathbf{u}^m \in C^1([0, T]; V^m)$,

$$\frac{\partial \rho^m}{\partial t} + \mathbf{u}^m \cdot \nabla \rho^m = 0 \text{ in } Q, \tag{3.10}$$

$$\begin{cases} \int_\Omega \left[\rho^m (\frac{\partial \mathbf{u}^m}{\partial t} + (\mathbf{u}^m \cdot \nabla)\mathbf{u}^m) \cdot \mathbf{v} + \mu \nabla \mathbf{u}^m : \nabla \mathbf{v} \right] = \int_\Omega \rho^m \mathbf{f}^m \cdot \mathbf{v} \\ \text{in } (0, T) \ \forall \mathbf{v} \in V^m \end{cases} \tag{3.11}$$

and the following initial conditions are satisfied:

$$\rho^m|_{t=0} = \rho_0^m, \tag{3.12}$$

$$\mathbf{u}^m|_{t=0} = \mathbf{u}_0^m. \tag{3.13}$$

Our first goal is to prove the existence of the approximate solutions $\{\mathbf{u}^m, \rho^m\}$. After this, we will derive some uniform estimates which appropriate to our needs.

Remark 3.1 Obviously, if ρ^m is given, (3.11) can be viewed as a nonlinear ODE system of dimension m. On the other hand, if $\mathbf{u}^m$ is given, (3.10) is a first order linear PDE. This is why (3.10)–(3.13) is called a *semi-Galerkin approximation.*[2] This is a very appropriate way to approximate the original problem. Indeed, assuming that $\mathbf{u}^m$ is known, the Cauchy problem (3.10), (3.12) can be easily solved by the *method of characteristics;* then, the associated ρ^m can be put in (3.11) and it is expected that the initial-value problem (3.11), (3.13) possesses a unique solution, at least local in time. To this respect, see Lemma 3.1 below. □

Remark 3.2 Equations (3.10) and (3.11) can also be equivalently written in *conservative* form:

$$\frac{\partial \rho^m}{\partial t} + \nabla \cdot (\rho^m \mathbf{u}^m) = 0 \tag{3.14}$$

and

$$\int_\Omega \left[(\frac{\partial \rho^m \mathbf{u}^m}{\partial t} + \nabla \cdot (\rho^m \mathbf{u}^m \otimes \mathbf{u}^m) \cdot \mathbf{v} + \mu \nabla \mathbf{u}^m : \nabla \mathbf{v} \right] = \int_\Omega \rho^m \mathbf{f}^m \cdot \mathbf{v} \tag{3.15}$$

in $(0, T)$ for all $\mathbf{v} \in V^m$. □

[2] In a related *Galerkin* process, we would try to find finite-dimensional approximations of ρ as well.

Substep 1.1: The Linearized Approximated Problem

Let us fix a positive integer $m \geq 1$ and consider the following linear problems:

Given $\mathbf{w} \in C^0([0, T]; V^m)$, *find* $\rho \in L^\infty(Q) \cap C^0([0, T]; W^{-1,\infty}(\Omega))$ *such that*

$$\begin{cases} \dfrac{\partial \rho}{\partial t} + \mathbf{w} \cdot \nabla \rho = 0 \ \text{ in } \ Q, \\ \rho(\mathbf{x}, 0) = \rho_0^m(\mathbf{x}) \ \text{ in } \ \Omega \end{cases} \tag{3.16}$$

and then find $\mathbf{u} \in C^1([0, T]; V^m)$ *such that*

$$\begin{cases} \displaystyle\int_\Omega \left[\rho(\frac{\partial \mathbf{u}}{\partial t} + (\mathbf{w} \cdot \nabla)\mathbf{u}) \cdot \mathbf{v} + \mu \nabla \mathbf{u} : \nabla \mathbf{v} \right] = \int_\Omega \rho \mathbf{f}^m \cdot \mathbf{v} \ \text{ in } \ (0, T) \ \forall \mathbf{v} \in V^m, \\ \mathbf{u}|_{t=0} = \mathbf{u}_0^m. \end{cases} \tag{3.17}$$

We will prove that, for every $\mathbf{w} \in C^0([0, T]; V^m)$, there exists exactly one solution $\{\mathbf{u}, \rho\}$ to (3.16) and (3.17).

This will allow to consider the mapping $\mathbf{w} \mapsto \mathbf{u}$. Of course, a fixed-point for this mapping will be a solution to the approximated problem (3.10)–(3.13).

First of all, it is clear that, for each $\mathbf{w} \in C^0([0, T]; V^m)$, the corresponding transport problem (3.16) is uniquely solvable in $C^1(\overline{Q})$. This is a consequence of the following result:

Lemma 3.1 *Let* $\mathbf{z} \in C^0([0, T]; C^1(\overline{\Omega})^3)$ *and* $\eta_0 \in C^1(\overline{\Omega})^3$ *be given, with*

$$\nabla \cdot \mathbf{z} = 0 \quad \text{in } Q, \qquad \mathbf{z} = 0 \quad \text{on } \Sigma,$$

$$0 < \alpha \leq \eta_0 \leq \beta \quad \text{in } \Omega.$$

Then, there exists a unique solution $\eta \in C^1(\overline{Q})$ *to the problem*

$$\begin{cases} \dfrac{\partial \eta}{\partial t} + \mathbf{z} \cdot \nabla \eta = 0 \ \text{ in } \ Q, \\ \eta(\mathbf{x}, 0) = \eta_0(\mathbf{x}) \ \text{ in } \ \Omega. \end{cases} \tag{3.18}$$

Moreover, the function η *satisfies*

$$\alpha \leq \eta \leq \beta \quad \text{in } Q. \tag{3.19}$$

Proof As in Sect. 1.2, let us consider for each $\mathbf{x} \in \overline{\Omega}$ the characteristic system

$$\begin{cases} \dfrac{\partial \mathbf{Y}}{\partial t} = \mathbf{z}(\mathbf{Y}(\mathbf{x}, t), t), \quad t \in (0, T), \\ \mathbf{Y}(\mathbf{x}, 0) = \mathbf{x}. \end{cases} \tag{3.20}$$

Note that the maximal to the right solution to (3.20) is defined in the whole $[0, T]$ and in particular $\mathbf{Y}(\mathbf{x}, t) \equiv \mathbf{x}$ for all $\mathbf{x} \in \partial\Omega$. Moreover, for every $t \in [0, T]$, the mapping $S_t : \overline{\Omega} \mapsto \overline{\Omega}$ given by

$$S_t(\mathbf{x}) = \mathbf{Y}(\mathbf{x}, t) \quad \forall \mathbf{x} \in \overline{\Omega}$$

is a C^1 diffeomorphism on $\overline{\Omega}$.

Then, from the classical theory of first-order PDEs, it is found that the unique solution η to problem (3.16) is given by

$$\eta(\mathbf{y}, t) = \eta_0(S_t^{-1}(\mathbf{y})) \quad \forall \mathbf{y} \in \overline{\Omega}, \;\; \forall t \in [0, T];$$

for details, see [15].

As a consequence, we immediately get the announced existence, uniqueness and regularity of η and also that (3.19) holds. □

In fact, we have more information than (3.19) on η: the mass distribution of η is independent of t; in other words, for every $k \in \mathbb{R}_+$ the function

$$t \mapsto |\{\mathbf{x} \in \Omega : \eta(\mathbf{x}, t) \leq k\}|$$

is constant.

The following well-posedness result follows from the continuous dependence with respect to $\mathbf{w}$ of the solution to (3.20) (see [15] for a detailed proof):

Lemma 3.2 *Let $\mathbf{z}$ and the $\mathbf{z}_n$ $(n \geq 1)$ be functions satisfying the hypotheses in Lemma 3.1 and such that $\mathbf{z}_n \to \mathbf{z}$ in $C^0([0, T]; C^1(\overline{\Omega}))$. Let us denote by η_n the solution to problem* (3.18) *with $\mathbf{z}$ replaced by $\mathbf{z}_n$. Then, $\eta_n \to \eta$ in $C^0(\overline{Q})$ as $n \to +\infty$.*

In view of Lemmas 3.1 and 3.2, problem (3.17) has a unique solution $\rho \in C^1(\overline{Q})$ satisfying

$$\frac{1}{m} + \inf_{\Omega} \rho_0 \leq \rho(\mathbf{x}, t) \leq \frac{1}{m} + \sup_{\Omega} \rho_0 \;\; \text{in} \;\; Q \tag{3.21}$$

ad the mapping $\mathbf{w} \mapsto \rho$ is continuous from $C^0([0, T]; C^1(\overline{\Omega}))$ into $C^1(\overline{\Omega})$.

Now, given $\mathbf{w}$ and the associated ρ, let us look for a solution $\mathbf{u}$ to (3.17) of the form

$$\mathbf{u}(\mathbf{x}, t) = \sum_{j=1}^{m} \phi_j(t)\mathbf{w}^j(\mathbf{x}), \tag{3.22}$$

where the $\phi_j \in C^1([0, T])$ for $j = 1, \ldots, m$.

For each i, let us take $\mathbf{v} = \mathbf{w}^i$ in (3.17). In view of (3.22), we see that $\mathbf{u}$ solves (3.17) if and only if the ϕ_j satisfy

$$\begin{cases} \sum_{j=1}^{m} \rho_{ij}(t)\dfrac{d\phi_j}{dt} + \sum_{j=1}^{m} b_{ij}(t)\phi_j = d_i(t), \quad t \in (0,T), \ 1 \le i \le m, \\ \phi_j(0) = \text{the j-th component of } \mathbf{u}_0^m \text{ for } 1 \le j \le m, \end{cases} \tag{3.23}$$

where the coefficients ρ_{ij}, b_{ij} and d_i are given by

$$\rho_{ij} := \int_\Omega \rho\, \mathbf{w}^j \cdot \mathbf{w}^i, \quad b_{ij} := \int_\Omega \rho(\mathbf{w}\cdot\nabla)\mathbf{w}^j \cdot \mathbf{w}^i + \mu\lambda_i\delta_{ij}$$

and

$$d_i := \int_\Omega \rho\, \mathbf{f}^m \cdot \mathbf{w}^i \in C^0([0,T])$$

and consequently belong to $C^0([0,T])$.

The matrix $R_m := \{\rho_{ij}\}_{i,j=1}^m$ is symmetric and positive definite in $[0,T]$, since $\{\mathbf{w}^i\}$ is an orthonormal system in H. More precisely, one has:

$$\sum_{ij} \rho_{ij}(t)\xi_i\xi_j = \int_\Omega \rho(x,t)|\sum_{i=1}^m \xi_i \mathbf{w}^i(x)|^2 \ge \left(\frac{1}{m} + \inf_\Omega \rho_0\right)\sum_{i=1}^m |\xi_i|^2 \quad \forall \xi \in \mathbb{R}^m.$$

In particular, R_m is invertible and (3.23) can be rewritten in the form

$$\begin{cases} \dfrac{d\phi}{dt} = -R_m^{-1}B_m\phi - R_m^{-1}D_m \quad \text{in } (0,T), \\ \phi_j(0) = \text{the } j\text{-th component of } \mathbf{u}_0^m \text{ for } 1 \le j \le m, \end{cases} \tag{3.24}$$

where $B_m = \{b_{ij}\}_{i,j=1}^m$, D_m is the column vector with entries d_i and ϕ is the column vector with entries ϕ_j.

This Cauchy problem is uniquely solvable. Therefore, we conclude that problem (3.17) is uniquely solvable as well.

Moreover, its solution depends continuously of $\mathbf{w}$ and ρ. Indeed, if $\mathbf{w}$ and the $\mathbf{w}_n$ satisfy the conditions respectively satisfied by $\mathbf{z}$ and the $\mathbf{z}_n$ in Lemma 3.2 and we denote by $\{\mathbf{u}, \rho\}$ and $\{\mathbf{u}_n, \rho_n\}$ the solutions to the associated linearized problems, it is an easy consequence of (3.24) that $\mathbf{u}_n \to \mathbf{u}$ in $C^1([0,T]; V^m)$.

Substep 1.2: Some Estimates

Let $\{\mathbf{u}, \rho\}$ be the solution to (3.16) and (3.17) corresponding to $\mathbf{w} \in C^0([0,T]; V^m)$. We will derive in this step some estimates for $\mathbf{u}$ that are independent of $\mathbf{w}$. Actually, these estimates are fundamental for many purposes in the existence theory for (3.10)–(3.13); it will be seen that they also play a key role for (3.3)–(3.8).

Taking $\mathbf{v} \in V^m$, multiplying (3.10) by $\frac{1}{2}\mathbf{u} \cdot \mathbf{v}$, integrating over Ω and adding the resulting equation to (3.11), we get:

$$\int_\Omega \left[\left(\rho \frac{\partial \mathbf{u}}{\partial t} + \frac{1}{2} \frac{\partial \rho}{\partial t} \mathbf{u} + (\rho \mathbf{w} \cdot \nabla)\mathbf{u} + \frac{1}{2}(\nabla \cdot (\rho \mathbf{w}))\mathbf{u} \right) \cdot \mathbf{v} + \mu \nabla \mathbf{u} : \nabla \mathbf{v} \right] = \int_\Omega \rho \mathbf{f}^m \cdot \mathbf{v}.$$

In particular, with $\mathbf{v} = \mathbf{u} \in V^m$, one has

$$\int_\Omega \left[\frac{\partial}{\partial t} \left(\frac{1}{2} \rho |\mathbf{u}|^2 \right) + \nabla \cdot \left(\frac{1}{2} \rho |\mathbf{u}|^2 \mathbf{w} \right) + \mu |\nabla \mathbf{u}|^2 \right] = \int_\Omega \rho \, \mathbf{f}^m \cdot \mathbf{u}.$$

Since $\mathbf{w} \in C^0([0, T]; V^m)$ and $\mathbf{u} \in C^1([0, T]; V^m)$, we see that

$$\int_\Omega \nabla \cdot \left(\frac{1}{2} \rho |\mathbf{u}|^2 \mathbf{w} \right) = \frac{1}{2} \int_{\partial\Omega} \rho |\mathbf{u}|^2 \mathbf{w} \cdot \mathbf{n} \, d\Gamma = 0,$$

whence we arrive at the differential identity

$$\frac{1}{2} \frac{d}{dt} \int_\Omega \rho |\mathbf{u}|^2 + \mu \int_\Omega |\nabla \mathbf{u}|^2 = \int_\Omega \rho \mathbf{f}^m \cdot \mathbf{u}. \tag{3.25}$$

This gives:

$$\frac{1}{2} \frac{d}{dt} \int_\Omega \rho |\mathbf{u}|^2 \le \left(\int_\Omega \rho |\mathbf{u}|^2 \right)^{1/2} \left(\int_\Omega \rho |\mathbf{f}^m|^2 \right)^{1/2}.$$

Now, we can apply Gronwall's Lemma and conclude that, for every t, the following must hold:

$$\int_\Omega \rho |\mathbf{u}|^2 \le C \left(\int_\Omega \rho_0^m |\mathbf{u}_0^m|^2 + \int_0^T \left(\int_\Omega \rho |\mathbf{f}^m|^2 \right) \right). \tag{3.26}$$

From (3.21) and the properties of ρ_0^m, $\mathbf{u}_0^m$, and $\mathbf{f}^m$, we deduce that

$\rho^{1/2}\mathbf{u}$ is bounded in $C^0([0, T]; L^2(\Omega)^3)$ (independently of m and $\mathbf{w}$).

Since $\mathbf{u}$ is of the form (3.22), ρ satisfies (3.21) and $\{\mathbf{w}^i\}$ is an orthonormal system in H, we also have

$$\sum_{j=1}^m |\phi_j(t)|^2 = \int_\Omega |\mathbf{u}|^2 \le C \quad \forall t \in [0, T]$$

for some C that can depend on m. Therefore, the functions ϕ_j are bounded in $C^0([0, T])$ and

$$\mathbf{u} \text{ is bounded in } C^0([0,T];\, V^m) \text{ (independently of } \mathbf{w}). \tag{3.27}$$

On the other hand, from the regularity of the coefficients that appear in (3.23), we see that, if $\mathbf{w}$ belongs to a bounded set in $C^0([0,T];\, V^m)$, then

$$\sum_{j=1}^{m} \rho_{ij} \frac{d\phi_j}{dt} \text{ is bounded in } C^0([0,T]) \text{ for } 1 \leq i \leq m.$$

Since the matrix $\{\rho_{ij}\}$ is symmetric and positive definite uniformly with respect to t, we find that

$$\frac{d\phi_j}{dt} \text{ is bounded in } C^0([0,T]) \text{ for } 1 \leq j \leq m,$$

which is equivalent, thanks to (3.22), to say that

$$\frac{\partial \mathbf{u}}{\partial t} \text{ is bounded in } C^0([0,T];\, V^m). \tag{3.28}$$

In view of (3.27) and (3.28), it becomes clear that, if $\mathbf{w}$ belongs to a bounded set in $C^0([0,T];\, V^m)$, then

$$\mathbf{u} \text{ is bounded in } C^1([0,T];\, V^m). \tag{3.29}$$

Substep 1.3: The Existence of Approximate Solutions
At this point, we know that, for every $\mathbf{w} \in C^0([0,T];\, V^m)$, there exists a unique solution $\rho \in C^1(\overline{Q})$ to problem (3.16) satisfying (3.21) and also a unique solution $\mathbf{u} \in C^1([0,T];\, V^m)$ to the corresponding Cauchy problem (3.17) which is bounded in $C^0([0,T];\, V^m)$ independently of $\mathbf{w}$.

Furthermore, if $\mathbf{w}$ is bounded in $C^0([0,T];\, V^m)$, $\mathbf{u}$ is bounded in $C^1([0,T];\, V^m)$. Specifically, there exists a positive constant C_0 such that

$$\|\mathbf{u}\|_{C^0([0,T];V^m)} \leq C_0 \quad \forall \mathbf{w} \in C^0([0,T];\, V^m)$$

and for every $M > 0$ there exists a positive constant $C_1(M)$ such that

$$\|\mathbf{w}\|_{C^0([0,T];V^m)} \leq M \Rightarrow \|\mathbf{u}\|_{C^1([0,T];V^m)} \leq C_1(M).$$

Let B_0 (resp. $B_1(M)$) be the closed ball centered at $\mathbf{0}$ of radius C_0 (resp. $C_1(M)$) in the space $C^0([0,T];\, V^m)$ (resp. $C^1([0,T];\, V^m)$). Let us consider the mapping $\mathbf{w} \mapsto \mathbf{u}$, that is obviously well defined from B_0 into itself.

This mapping is continuous, in view of the previous discussion in Step 1. Moreover, it maps B_0 into $B_1(C_0)$ which, in view of the Ascoli-Arzela Theorem, is a compact set of $C^0([0,T];\, V^m)$.

Consequently, Schauder's Theorem (Theorem 2.9) can be applied and we deduce that a fixed-point $\mathbf{u}^m \in B_0$ exists.

Let ρ^m be the solution to problem (3.16) corresponding to the choice $\mathbf{w} = \mathbf{u}^m$. Then $\{\mathbf{u}^m, \rho^m\}$ is a solution to (3.10)–(3.13). Furthermore, by construction, we have

$$\mathbf{u}^m \in C^1([0,T]; V^m), \quad \rho^m \in C^1(\overline{Q})$$

and

$$\frac{1}{m} + \inf_\Omega \rho_0 \le \rho^m \le \frac{1}{m} + \sup_\Omega \rho_0 \le b := 1 + \sup_\Omega \rho_0 \ \text{ in } \ Q. \tag{3.30}$$

Step 2: A Priori Estimates for the Approximate Solutions
In this step, we are going to establish some *a priori* estimates (uniform with respect to m) of the approximate solutions $\{\mathbf{u}^m, \rho^m\}$.

Substep 2.1: Strictly Positive Initial Density
Let us first indicate what can be done in the special case in which $\rho_0 \ge \alpha > 0$.

In view of (3.30), one has

$$\rho^m \ge \alpha \ge 0 \quad \text{in } Q.$$

Arguing as in (3.25) and (3.26), one gets again

$$\frac{1}{2}\frac{d}{dt}\int_\Omega \rho^m |\mathbf{u}^m|^2 + \mu \int_\Omega |\nabla \mathbf{u}^m|^2 = \int_\Omega \rho^m \mathbf{u}^m \cdot \mathbf{f}^m, \tag{3.31}$$

whence

$$\int_\Omega \rho^m |\mathbf{u}^m|^2 \le C\left(\int_\Omega \rho_0^m |\mathbf{u}_0^m|^2 + \int_0^T \left(\int_\Omega \rho^m |\mathbf{f}^m|^2\right)\right). \tag{3.32}$$

Therefore,

$$\mathbf{u}^m \text{ is uniformly bounded in } L^\infty(0,T;H). \tag{3.33}$$

On the other hand, integrating the identity (3.31) over $(0,T)$, we see that

$$\begin{aligned}
&\frac{1}{2}\left(\int_\Omega \rho^m |\mathbf{u}^m|^2\right)(T) + \mu \int_0^T \left(\int_\Omega |\nabla \mathbf{u}^m|^2\right) \\
&\le \frac{1}{2}\left(\int_\Omega \rho_0^m |\mathbf{u}_0^m|^2\right) + b \int_0^T \left(\int_\Omega |\mathbf{u}^m|^2\right)^{1/2}\left(\int_\Omega |\mathbf{f}^m|^2\right)^{1/2} \\
&\le C + b\|\mathbf{u}^m\|_{L^\infty(0,T;H)}\, \|\mathbf{f}^m\|_{L^1(0,T;L^2)} \\
&\le C
\end{aligned}$$

and consequently

$$\mathbf{u}^m \text{ is uniformly bounded in } L^2(0, T; V). \tag{3.34}$$

Since $H_0^1(\Omega) \hookrightarrow L^6(\Omega)$ with a continuous embedding, this gives

$$\mathbf{u}^m \text{ is uniformly bounded in } L^2(0, T; L^6(\Omega)^3). \tag{3.35}$$

Using (3.33), (3.35) and Proposition 2.14 for $p_0 = +\infty, q_0 = 2, p_1 = 2, q_1 = 6$, $\theta = 3/5$ and then $\theta = 3/4$, we also find that

$$\mathbf{u}^m \text{ is uniformly bounded in } L^{10/3}(Q)^3 \text{ and } L^{8/3}(0, T; L^4(\Omega)^3). \tag{3.36}$$

Let us now get estimates of $\rho^m \mathbf{u}^m \otimes \mathbf{u}^m$. One has

$$\|\mathbf{u}^m \otimes \mathbf{u}^m\|_{L^{5/3}}^{5/3} = \int_\Omega |\mathbf{u}^m \otimes \mathbf{u}^m|^{5/3} \leq \|\mathbf{u}^m\|_{L^{10/3}}^{10/3}.$$

Hence, in view of (3.36), one also has:

$$\mathbf{u}^m \otimes \mathbf{u}^m \text{ is uniformly bounded in } L^{5/3}(Q)^{3\times 3} \text{ and } L^{4/3}(0, T; L^2(\Omega)^{3\times 3}). \tag{3.37}$$

Since the ρ^m are bounded in $L^\infty(Q)$ (see (3.30)), it follows that

$$\rho^m \mathbf{u}^m \otimes \mathbf{u}^m \text{ is uniformly bounded in } L^{5/3}(Q)^{3\times 3} \text{ and } L^{4/3}(0,T;L^2(\Omega)^{3\times 3}). \tag{3.38}$$

Note that the continuity Eq. (3.14) gives

$$\frac{\partial \rho^m}{\partial t} = -\nabla \cdot (\rho^m \mathbf{u}^m). \tag{3.39}$$

Therefore, suitable bounds of $\rho^m \mathbf{u}^m$ imply bounds of $\dfrac{\partial \rho^m}{\partial t}$. More precisely, using (3.30), (3.33), and (3.35), we get:

$$\rho^m \mathbf{u}^m \text{ is uniformly bounded in } L^\infty(0, T; L^2(\Omega)^3) \text{ and } L^2(0, T; L^6(\Omega)^3) \tag{3.40}$$

and

$$\frac{\partial \rho^m}{\partial t} \text{ is uniformly bounded in } L^\infty(0, T; H^{-1}(\Omega)) \text{ and } L^2(0, T; W^{-1,6}(\Omega)). \tag{3.41}$$

We will prove now that there exists $C > 0$ (independent of m) such that, for all h with $0 < h < T$, one has

$$\|\tau_h \mathbf{u}^m - \mathbf{u}^m\|_{L^2(0,T-h;H)} \leq C\, h^{1/4}. \tag{3.42}$$

This will lead to a uniform bound of the $\mathbf{u}^m$ in a Nikolski space and, accordingly, will be viewed as an estimate of the fractional derivatives of the $\mathbf{u}^m$. Indeed, since $\mathbf{u}^m$ is uniformly bounded in $L^\infty(0, T; H)$, inequality (3.42) will prove that

$$\mathbf{u}^m \text{ is uniformly bounded in } N^{1/4,2}(0, T; H). \tag{3.43}$$

In order to prove the estimates (3.42), let us first observe that

$$\begin{aligned} \|\tau_h \mathbf{u}^m - \mathbf{u}^m\|^2_{L^2(0,T-h;H)} &= \int_0^{T-h} \|\mathbf{u}^m(t+h) - \mathbf{u}^m(t)\|^2 \, dt \\ &\leq \frac{1}{\alpha} \int_0^{T-h} \left(\int_\Omega \rho^m(t+h) |\mathbf{u}^m(t+h) - \mathbf{u}^m(t)|^2 \, d\mathbf{x} \right) dt \\ &= \frac{1}{\alpha} (I_1 - I_2), \end{aligned} \tag{3.44}$$

where

$$I_1 := \int_0^{T-h} \int_\Omega [(\rho^m \mathbf{u}^m)(t+h) - (\rho^m \mathbf{u}^m)(t)] \cdot [\mathbf{u}^m(t+h) - \mathbf{u}^m(t)]$$

and

$$I_2 := \int_0^{T-h} \int_\Omega [\rho^m(t+h) - \rho^m(t)] \mathbf{u}^m(t) \cdot [\mathbf{u}^m(t+h) - \mathbf{u}^m(t)].$$

Also, note that

$$\int_\Omega \frac{\partial \rho^m \mathbf{u}^m}{\partial t} \cdot \mathbf{v} = \frac{d}{dt} \int_\Omega \rho^m \mathbf{u}^m \cdot \mathbf{v}$$

for all $\mathbf{v} \in V^m$ and, consequently,

$$\begin{aligned} \left| \frac{d}{dt} \int_\Omega \rho^m \mathbf{u}^m \cdot \mathbf{v} \right| &= \left| \int_\Omega (\rho^m \mathbf{u}^m \otimes \mathbf{u}^m - \mu \nabla \mathbf{u}^m) : \nabla \mathbf{v} + \int_\Omega \rho^m \mathbf{f}^m \cdot \mathbf{v} \right| \\ &\leq \left(\|\rho^m \mathbf{u}^m \otimes \mathbf{u}^m - \mu \nabla \mathbf{u}^m\| + C \|\rho^m \mathbf{f}^m\| \right) \|\nabla \mathbf{v}\| \end{aligned}$$

for all $\mathbf{v} \in V^m$ (here, the constant C comes obviously from Poincaré's inequality).

Due to (3.34) and (3.38), the term $\|\rho^m \mathbf{u}^m \otimes \mathbf{u}^m - \mu \nabla \mathbf{u}^m\|$ is uniformly bounded in $L^{4/3}(0, T)$. Since ρ^m is uniformly bounded in $L^\infty(Q)$ and $\mathbf{f}^m$ is uniformly bounded in the space $L^1(0, T; L^2(\Omega)^3)$, one has that $\|\rho^m \mathbf{f}^m\|$ bounded in $L^1(0, T)$.

Hence,

$$\begin{cases} \left| \dfrac{d}{dt} \displaystyle\int_\Omega \rho^m \mathbf{u}^m \cdot \mathbf{v} \right| \leq g_m \|\nabla \mathbf{v}\| \quad \forall \mathbf{v} \in V^m \text{ in } (0, T), \\ \text{where } g_m = C \|\rho^m \mathbf{f}^m\| + \|\rho^m \mathbf{u}^m \otimes \mathbf{u}^m - \mu \nabla \mathbf{u}^m\|. \end{cases} \tag{3.45}$$

Let us show that there exists C such that $I_1 \le Ch^{1/2}$.

To this end, let $\mathbf{v} \in V^m$ be given. Due to (3.45), one has

$$\int_\Omega [(\rho^m \mathbf{u}^m)(t+h) - (\rho^m \mathbf{u}^m)(t)] \cdot \mathbf{v} = \int_t^{t+h} \left(\frac{d}{ds} \int_\Omega \rho^m \mathbf{u}^m(s) \cdot \mathbf{v} \right) ds$$
$$\le \left(\int_t^{t+h} g_m(s) \right) \|\nabla \mathbf{v}\|$$

(recall that g_m is uniformly bounded in $L^1(0, T)$).

Taking $\mathbf{v} = \mathbf{u}^m(t+h) - \mathbf{u}^m(t) \in V^m$ and integrating with respect to t over $(0, T - h)$, we get

$$I_1 \le \int_0^{T-h} \|\nabla \mathbf{u}^m(t+h) - \nabla \mathbf{u}^m(t)\| \left(\int_t^{t+h} g_m(s)\, ds \right) dt.$$

Therefore, from *Fubini's Theorem,* one has

$$I_1 \le \int_0^T g_m(s) \left(\int_{(s-h)^\star}^{s^\star} \|\nabla \mathbf{u}^m(t+h) - \nabla \mathbf{u}^m(t)\| \partial_t \right) ds,$$

where we have used the following notation:

$$\sigma^\star = \begin{cases} 0, & \text{for } \sigma \le 0, \\ \sigma, & \text{for } 0 \le \sigma \le T - h, \\ T - h, & \text{for } \sigma \ge T - h. \end{cases}$$

Using Hölder's inequality and the estimates (3.34) and (3.45), we readily see that

$$I_1 \le \int_0^T g_m(s) |s^\star - (s-h)^\star|^{1/2} \left(\int_{(s-h)^\star}^{s^\star} \|\nabla \mathbf{u}^m(t+h) - \nabla \mathbf{u}^m(t)\|^2\, dt \right)^{1/2} ds$$
$$\le Ch^{1/2} \|\nabla \mathbf{u}^m\|_{L^2(Q)} \left(\int_0^T g_m(s)\, ds \right) \le C\, h^{1/2}$$

and the estimate of I_1 holds.

Now, let us prove that $I_2 \le Ch^{1/2}$ for some $C > 0$.

To this purpose, let $t \in [0, T - h]$ be given, let us multiply (3.39) by a function $w \in W_0^{1,3/2}(\Omega)$ and let us integrate over Ω. The following is found:

$$\int_\Omega \frac{\partial \rho^m}{\partial t} w = \int_\Omega \rho^m \mathbf{u}^m \cdot \nabla w.$$

Integrating this identity over $(t, t+h)$, one has

$$\int_\Omega [\rho^m(t+h) - \rho^m(t)]\, w = \int_t^{t+h} \left(\int_\Omega (\rho^m \mathbf{u}^m)(s) \cdot \nabla w \right) ds.$$

Now, we use again Hölder's inequality and find that

$$\begin{aligned} \left| \int_\Omega [\rho^m(t+h) - \rho^m(t)] w \right| &\le C \int_t^{t+h} \|\mathbf{u}^m\|_{L^6} \|\nabla w\|_{L^{3/2}} \\ &\le C \|\nabla w\|_{L^{3/2}} \int_t^{t+h} \|\nabla \mathbf{u}^m(s)\|\, ds \\ &\le C h^{1/2} \left(\int_t^{t+h} \|\nabla \mathbf{u}^m(s)\|^2 \right)^{1/2} \|\nabla w\|_{L^{3/2}} \end{aligned}$$

and, using (3.34), we deduce that

$$\int_\Omega (\rho^m(t+h) - \rho^m(t))\, w \le C\, h^{1/2} \|\nabla w\|_{L^{3/2}} \quad \forall w \in W_0^{1,3/2}(\Omega). \tag{3.46}$$

Let us take

$$w = \mathbf{u}^m(t) \cdot [\mathbf{u}^m(t+h) - \mathbf{u}^m(t)] \in W_0^{1,3/2}(\Omega).$$

Then

$$\|\nabla w\|_{L^{3/2}} \le C \|\nabla \mathbf{u}^m\|\, \|\nabla(\mathbf{u}^m(t+h) - \mathbf{u}^m(t))\|. \tag{3.47}$$

Consequently, recalling (3.34), we find that the right-hand side in (3.47) is uniformly bounded in $L^1(0, T-h)$. Integrating the inequality (3.46) with respect to t over $(0, T-h)$, we see that I_2 is certainly bounded, as asserted.

At this point, in order to achieve the proof of (3.42), it suffices to take into account (3.44).

Remark 3.3 From (3.33) and (3.42), we see that

$$\mathbf{u}^m \text{ is uniformly bounded in } N^{1/4,2}(0, T; H)$$

and, at least for a sequence, $\mathbf{u}^m \to \mathbf{u}$ weakly-$*$ in $L^\infty(0, T; H)$. Clearly,

$$\|\tau_h \mathbf{u} - \mathbf{u}\|_{L^2(0,T;H)} \le C h^{1/4}$$

and thus $\mathbf{u} \in C^0([0, T]; V')$ (recall Theorem 2.10). Therefore, in view of Lemma 2.19, we also have

$$\mathbf{u} \in C_w^0([0, T]; H).$$

This will be used later. □

Substep 2.2: The General Case, with $\rho_0 \geq 0$
At this point, we consider the general case, where ρ_0 is not necessarily bounded from below by a positive constant, that is, we just assume that $\rho_0 \geq 0$.

Basically, in this case we will be able to deduce estimates similar to those above, but for $\rho^m \mathbf{u}^m$ instead of $\mathbf{u}^m$.

First of all, note that (3.30) yields an upper bound for ρ^m:

$$\rho^m \leq b \text{ in } Q. \tag{3.48}$$

Consequently, the ρ^m are uniformly bounded in $L^\infty(Q)$.

From the inequalities (3.32), one has

$$\left(\int_\Omega \rho^m |\mathbf{u}^m|^2 \right)^{1/2} \leq \sqrt{b} \left(C + \|\mathbf{f}^m\|_{L^1(0,T;L^2)} \right) \leq C \quad \text{in } (0, T)$$

and

$$(\rho^m)^{1/2} \mathbf{u}^m \text{ is uniformly bounded in } L^\infty(0, T; L^2(\Omega)^3), \tag{3.49}$$

which also gives:

$$\rho^m \mathbf{u}^m \text{ is uniformly bounded in } L^\infty(0, T; L^2(\Omega)^3).$$

As before, integrating the identity (3.31) over $(0, T)$, one gets

$$\begin{aligned} \mu \int_0^T \left(\int_\Omega |\nabla \mathbf{u}^m|^2 \right) &\leq \frac{1}{2} \left(\int_\Omega \rho_0^m |\mathbf{u}_0^m|^2 \right) + \int_0^T \left(\int_\Omega \rho^m \mathbf{u}^m \cdot \mathbf{f}^m \right) \\ &\leq C + \int_0^T \left(\int_\Omega |\rho^m \mathbf{u}^m|^2 \right)^{1/2} \left(\int_\Omega |\mathbf{f}^m|^2 \right)^{1/2} \\ &\leq C + \|\rho^m \mathbf{u}^m\|_{L^\infty(0,T;L^2)} \|\mathbf{f}^m\|_{L^1(0,T;L)}. \end{aligned}$$

Thus, $\nabla \mathbf{u}^m$ is uniformly bounded in $L^2(Q)^{3\times 3}$, that is, we have (3.34). As before, this implies (3.35).

From (3.35) and (3.49), we see that

$$(\rho^m)^{1/5} \mathbf{u}^m \text{ is uniformly bounded in } L^{10/3}(Q)^3$$

and

$$(\rho^m)^{1/8} \mathbf{u}^m \text{ is uniformly bounded in } L^{8/3}(0, T; L^4(\Omega)^3), \tag{3.50}$$

where the exponents $1/5$ and $1/8$ are sharp.

In fact, a careful application of Proposition 2.14 to the $(\rho^m)^a \mathbf{u}^m$ shows that

$$(\rho^m)^a \mathbf{u}^m \text{ is uniformly bounded in } L^b(0,T;L^q(\Omega)^3) \text{ if } \begin{cases} qa + q(1-2a)/6 = 1, \\ b(1-2a) = 2. \end{cases}$$

Hence, one obtains the desired properties (3.50) by choosing first $q = b = 10/3$ and $a = 1/5$ and then $q = 4$, $a = 1/8$ and $b = 8/3$.

In particular, (3.50) implies that

$$(\rho^m)^{1/2}\mathbf{u}^m \text{ is uniformly bounded in } L^{8/3}(0,T;L^4(\Omega)^3)$$

and

$$\rho^m \mathbf{u}^m \otimes \mathbf{u}^m \text{ is uniformly bounded in } L^{4/3}(0,T;L^2(\Omega)^{3\times 3}). \tag{3.51}$$

The estimates (3.41) and (3.45) for the time derivatives can be obtained as before and remain valid in this general case.

Let us show that, in this case, there exists $C > 0$ such that

$$\|\tau_h(\rho^m \mathbf{u}^m) - \rho^m \mathbf{u}^m\|_{L^2(0,T-h;W^{-1,3})} \le C\,h^{1/4}. \tag{3.52}$$

This will lead to the following:

$$\rho^m \mathbf{u}^m \text{ is uniformly bounded in } N^{1/4,2}(0,T;W^{-1,3}(\Omega)^3). \tag{3.53}$$

As before, one has

$$I_1 := \int_0^{T-h} \int_\Omega [(\rho^m \mathbf{u}^m)(t+h) - (\rho^m \mathbf{u}^m)(t)] \cdot [\mathbf{u}^m(t+h) - \mathbf{u}^m(t)] \le C\,h^{1/2}$$

and

$$I_2 := \int_0^{T-h} \int_\Omega [\rho^m(t+h) - \rho^m(t)]\mathbf{u}^m(t) \cdot [\mathbf{u}^m(t+h) - \mathbf{u}^m(t)] \le C\,h^{1/2}.$$

Therefore,

$$\int_0^{T-h} \left(\int_\Omega \rho^m(t+h)|\mathbf{u}^m(t+h) - \mathbf{u}^m(t)|^2 \right) = I_1 - I_2 \le C\,h^{1/2},$$

that is,

$$\|\tau_h \rho^m (\tau_h \mathbf{u}^m - \mathbf{u}^m)\|_{L^2(0,T-h;L^2)} \le C\,h^{1/4}. \tag{3.54}$$

Since

$$\tau_h(\rho^m \mathbf{u}^m) - \rho^m \mathbf{u}^m = (\tau_h \rho^m - \rho^m)\mathbf{u}^m + (\tau_h \rho^m)(\tau_h \mathbf{u}^m - \mathbf{u}^m),$$

in order to deduce an appropriate bound of $\tau_h(\rho^m \mathbf{u}^m) - \rho^m \mathbf{u}^m$, it only remains to establish an estimate of $(\tau_h \rho^m - \rho^m)\mathbf{u}^m$.

To do this, note that

$$(\tau_h \rho^m - \rho^m)(t) = \int_t^{t+h} \frac{\partial \rho^m}{\partial t}(s)\, ds = -\int_t^{t+h} \nabla \cdot (\rho^m \mathbf{u}^m)(s)\, ds.$$

Taking norms in $W^{-1,6}(\Omega)$ and recalling that the $\mathbf{u}^m$ are uniformly bounded in the space $L^2(0, T; L^6(\Omega)^3)$ (see (3.35)), we find that

$$\begin{aligned} \|(\tau_h \rho^m - \rho^m)(t)\|_{W^{-1,6}} &\le C \int_t^{t+h} \|\rho^m \mathbf{u}^m(s)\|_{L^6}\, ds \\ &\le C\, h^{1/2}\, \|\rho^m \mathbf{u}^m\|_{L^2(0,T;L^6)} \le C\, h^{1/2} \end{aligned}$$

for all t, whence

$$\|\tau_h \rho^m - \rho^m\|_{L^\infty(0,T-h;W^{-1,6})} \le C\, h^{1/2}.$$

On the other hand, since $\mathbf{u}^m$ is uniformly bounded in $L^2(0, T - h; V)$, we can use the continuity of the product mapping $H^1 \times W^{-1,6} \mapsto W^{-1,3}$ (Lemma 2.10, part (ii)) and get the following estimates:

$$\|(\tau_h \rho^m - \rho^m)\mathbf{u}^m\|_{L^2(0,T-h;W^{-1,3})} \le C\, h^{1/2}. \tag{3.55}$$

Clearly, the inequalities (3.54) and (3.55) imply the desired estimate (3.52).

Step 3: Extracting Convergent Subsequences, Taking Limits and Getting Conclusions

Let us summarize the uniform estimates we have obtained so far:

Property (3.48):
ρ^m is uniformly bounded in $L^\infty(Q)$

Property (3.41):
$\dfrac{\partial \rho^m}{\partial t}$ is uniformly bounded in $L^\infty(0, T; H^{-1}(\Omega)) \cap L^2(0, T; W^{-1,6}(\Omega))$

Property (3.34):
$\mathbf{u}^m$ is uniformly bounded in $L^2(0, T; V)$

Property (3.40):
$\rho^m \mathbf{u}^m$ is uniformly bounded in $L^\infty(0, T; L^2(\Omega)^3) \cap L^2(0, T; L^6(\Omega)^3)$

Property (3.53):
$\rho^m \mathbf{u}^m$ is uniformly bounded in $N^{1/4,2}(0, T; W^{-1,3}(\Omega)^3)$

Property (3.51):
$\rho^m \mathbf{u}^m \otimes \mathbf{u}^m$ is uniformly bounded in $L^{4/3}(0, T; L^2(\Omega)^{3\times 3})$.

From (3.48), (3.41) and Corollary 2.3, part *(1)*, with $X = L^\infty(\Omega)$, $B = W^{-1,\infty}(\Omega)$ and $Y = H^{-1}(\Omega)$, we conclude that

$$\rho^m \text{ belongs to a compact subset of } C^0([0,T]; W^{-1,\infty}(\Omega)).$$

From (3.40), (3.53) and Corollary 2.3, part *(2)*, this time with $X = L^6(\Omega)^3$, $B = W^{-1,\infty}(\Omega)^3$ and $Y = W^{-1,3}(\Omega)^3$, one also has:

$$\rho^m \mathbf{u}^m \text{ belongs to a compact subset of } L^2(0,T; W^{-1,\infty}(\Omega)^3).$$

Therefore, there exist subsequences of $\{\rho^m\}$ and $\{\mathbf{u}^m\}$ (again indexed by m) and functions ρ, $\mathbf{u}$, χ_1 and χ_2 satisfying the following properties:

$$\rho^m \to \rho \text{ strongly in } C^0([0,T]; W^{-1,\infty}(\Omega)) \text{ and weakly-}* \text{ in } L^\infty(Q), \tag{3.56}$$

$$\mathbf{u}^m \to \mathbf{u} \text{ weakly in } L^2(0,T; V), \tag{3.57}$$

$$\rho^m \mathbf{u}^m \to \chi_1 \begin{cases} \text{weakly in } L^2(0,T; L^6(\Omega)^3), \\ \text{weakly-}* \text{ in } L^\infty(0,T; L^2(\Omega)^3) \text{ and} \\ \text{strongly in } L^2(0,T; W^{-1,\infty}(\Omega)^3) \end{cases} \tag{3.58}$$

and

$$\rho^m \mathbf{u}^m \otimes \mathbf{u}^m \to \chi_2 \text{ weakly in } L^{4/3}(0,T; L^2(\Omega)^{3\times 3}). \tag{3.59}$$

We claim that $\chi_1 = \rho\mathbf{u}$ and $\chi_2 = \rho\mathbf{u}\otimes\mathbf{u}$. Indeed, from Lemma 2.10 *(ii)*, one has that the product mapping from $H_0^1 \times W^{-1,\infty}$ into $W^{-1,6}$ is continuous. Therefore, using (3.56) and (3.57), we find that

$$\rho^m \mathbf{u}^m \to \rho\mathbf{u} \text{ weakly in } L^2(0,T; W^{-1,6}(\Omega)^3).$$

This fact, together with (3.58), implies $\chi_1 = \rho\mathbf{u}$.

In a similar way, due to (3.57) and (3.58), one gets

$$\rho^m \mathbf{u}^m \otimes \mathbf{u}^m \to \rho\mathbf{u}\otimes\mathbf{u} \text{ weakly in } L^1(0,T; W^{-1,6}(\Omega)^{3\times 3})$$

and, consequently, $\chi_2 = \rho\mathbf{u}\otimes\mathbf{u}$.

Let us finally prove that ρ and $\mathbf{u}$ solve, together with an appropriate p, the original problem (3.3)–(3.8). Recall that

$$\begin{cases} \mathbf{u} \in L^2(0,T; V), \quad \rho \in L^\infty(Q) \cap C^0([0,T]; W^{-1,\infty}(\Omega)) \text{ and} \\ \rho\mathbf{u} \in L^\infty(0,T; L^2(\Omega)^3). \end{cases}$$

Conservation of Momentum Let $m' \geq 1$ and $\mathbf{v} \in V^{m'}$ be fixed and let us take $m \geq m'$. In view of (3.15), one has:

$$\int_\Omega \left[\frac{\partial \rho^m \mathbf{u}^m}{\partial t} \cdot \mathbf{v} + \left(\mu \nabla \mathbf{u}^m - \rho^m \mathbf{u}^m \otimes \mathbf{u}^m \right) : \nabla \mathbf{v} \right] = \int_\Omega \rho^m \mathbf{f}^m \cdot \mathbf{v}. \tag{3.60}$$

This equation holds in $C^0([0, T])$ and, consequently, also in $\mathcal{D}'(0, T)$. Our aim is to take limits as $m \to \infty$. To this purpose, we will study each term individually:

- In view of (3.58), $\rho^m \mathbf{u}^m \to \rho \mathbf{u}$ in $\mathcal{D}'(0, T; H^{-1}(\Omega)^3)$. Therefore,

$$\frac{\partial \rho^m \mathbf{u}^m}{\partial t} \to \frac{\partial \rho \mathbf{u}}{\partial t} \quad \text{in } \mathcal{D}'(0, T; H^{-1}(\Omega)^3).$$

For any $\varphi \in \mathcal{D}(0, T)$ and any $\mathbf{v} \in H_0^1(\Omega)^3$, one has

$$\begin{aligned} \langle \langle \frac{\partial \rho^m \mathbf{u}^m}{\partial t}, \varphi \rangle, \mathbf{v} \rangle_{H^{-1}, H_0^1} &= \int_\Omega \left(\int_0^T \frac{\partial \rho^m \mathbf{u}^m}{\partial t} \varphi \right) \cdot \mathbf{v} \\ &= \int_0^T \left(\int_\Omega \frac{\partial \rho^m \mathbf{u}^m}{\partial t} \cdot \mathbf{v} \right) \varphi \end{aligned} \tag{3.61}$$

Therefore,

$$\int_\Omega \frac{\partial \rho^m \mathbf{u}^m}{\partial t} \cdot \mathbf{v} \to \langle \frac{\partial \rho \mathbf{u}}{\partial t}, \mathbf{v} \rangle_{H^{-1}, H_0^1} \quad \text{in } \mathcal{D}'(0, T).$$

- From (3.56) and the strong convergence of $\mathbf{f}^m$ in $L^1(0, T; L^2(\Omega)^3)$, we also have

$$\rho^m \mathbf{f}^m \to \rho \mathbf{f} \text{ weakly in } L^1(0, T; L^2(\Omega)^3).$$

Hence,

$$\int_\Omega \rho^m \mathbf{f}^m \cdot \mathbf{v} \to \int_\Omega \rho \mathbf{f} \cdot \mathbf{v} \text{ weakly in } L^1(0, T).$$

- Using (3.59), one has

$$\int_\Omega (\rho^m \mathbf{u}^m \otimes \mathbf{u}^m) \cdot \nabla \mathbf{v} \to \int_\Omega (\rho \mathbf{u} \otimes \mathbf{u}) : \nabla \mathbf{v} \text{ weakly in } L^{4/3}(0, T).$$

In other words,

$$\langle \nabla \cdot (\rho^m \mathbf{u}^m \otimes \mathbf{u}^m), \mathbf{v} \rangle_{H^{-1}} \to \langle \nabla \cdot (\rho \mathbf{u} \otimes \mathbf{u}), \mathbf{v} \rangle_{H^{-1}, H_0^1}$$

weakly in $L^{4/3}(0, T)$.

- Finally, from (3.57), one has

$$\langle -\Delta \mathbf{u}^m, \mathbf{v}\rangle_{H^{-1},H_0^1} \to \langle -\Delta \mathbf{u}, \mathbf{v}\rangle_{H^{-1},H_0^1} \text{ weakly in } L^2(0,T).$$

Since all these convergences hold in particular in $\mathcal{D}'(0,T)$, we can take limits in (3.60) as $m \to \infty$ and obtain

$$\langle \frac{\partial \rho \mathbf{u}}{\partial t} + \nabla \cdot (\rho \mathbf{u} \otimes \mathbf{u}) - \mu \Delta \mathbf{u} + \rho \mathbf{f} \,,\, \mathbf{v}\rangle_{H^{-1}} = 0 \quad \text{in } \mathcal{D}'(0,T) \tag{3.62}$$

for all $\mathbf{v} \in V^{m'}$.

Now, by density, we see that (3.62) must also hold for all $\mathbf{v} \in V$. In particular, it holds for all $\mathbf{v} \in \mathcal{V}$.

Let us prove the existence of a pressure p such that (3.3) is satisfied (at least) in the sense of $\mathcal{D}'(Q)^3$.

Note that

$$S := \frac{\partial \rho \mathbf{u}}{\partial t} + \nabla \cdot (\rho \mathbf{u} \otimes \mathbf{u}) - \mu \Delta \mathbf{u} - \rho \mathbf{f} \in W^{-1,\infty}(0,T;H^{-1}(\Omega)^3),$$

since

$$\frac{\partial \rho \mathbf{u}}{\partial t} \in W^{-1,\infty}(0,T;L^2(\Omega)^3),\quad \nabla \cdot (\rho \mathbf{u} \otimes \mathbf{u}) \in L^{4/3}(0,T;H^{-1}(\Omega)^3),$$
$$\Delta \mathbf{u} \in L^2(0,T;H^{-1}(\Omega)^3) \text{ and } \rho \mathbf{f} \in L^1(0,T;L^2(\Omega)^3).$$

The Banach spaces $W^{-1,\infty}(0,T;H^{-1}(\Omega))$ and $H^{-1}(\Omega;W^{-1,\infty}(0,T))$ are isomorphic through the canonical mapping Ψ, defined as follows:

$$\begin{cases} \langle\langle \Psi g, v\rangle_{H^{-1},H_0^1}, h\rangle_{W^{-1,\infty}(0,T),W_0^{1,1}(0,T)} = \langle\langle g, h\rangle_{W^{-1,\infty}(0,T),W_0^{1,1}(0,T)}, v\rangle_{H^{-1},H_0^1} \\ \forall v \in H_0^1(\Omega),\quad \forall h \in W_0^{1,1}(0,T). \end{cases}$$

Consequently, $W^{-1,\infty}(0,T;H^{-1}(\Omega))^3$ is isomorphic to $H^{-1}(\Omega;W^{-1,\infty}(0,T))^3$ and we can write that $S \in H^{-1}(\Omega;W^{-1,\infty}(0,T))^3$.

It follows from Theorem 2.7 with $m = 0$, $r = 2$ and $B = W^{-1,\infty}(0,T)$ that there exists a distribution $p \in L^2(\Omega;W^{-1,\infty}(0,T))$, unique up to an additive constant (i.e., up to a real-valued distribution in $W^{-1,\infty}(0,T)$), such that $S = -\nabla p$. Taking into account that $L^2(\Omega;W^{-1,\infty}(0,T))$ is isomorphic to $W^{-1,\infty}(0,T;L^2(\Omega))$, we find that

$$\frac{\partial \rho \mathbf{u}}{\partial t} + \nabla \cdot (\rho \mathbf{u} \otimes \mathbf{u}) - \mu \Delta \mathbf{u} + \nabla p = \rho \mathbf{f} \quad \text{in } W^{-1,\infty}(0,T;H^{-1}(\Omega)^3)$$

and this proves that (3.3) is satisfied.

Conservation of Mass We can also take limits in (3.14) as $m \to +\infty$. Indeed, the convergence properties in (3.56) imply that $\rho^m \to \rho$ in $\mathcal{D}'(Q)$ and, consequently,

$$\frac{\partial \rho^m}{\partial t} \to \frac{\partial \rho}{\partial t} \ \text{ in } \mathcal{D}'(Q).$$

Moreover, in view of (3.58) and the fact that $\chi_1 = \rho \mathbf{u}$, one also has

$$\nabla \cdot (\rho^m \mathbf{u}^m) \to \nabla \cdot (\rho \mathbf{u}) \ \text{ in } \mathcal{D}'(Q).$$

Thus,

$$\frac{\partial \rho}{\partial t} + \nabla \cdot (\rho \mathbf{u}) = 0 \quad \text{in } \mathcal{D}'(Q).$$

Since $\rho \in L^\infty(Q)$, and $\rho \mathbf{u} \in L^\infty(0, T; L^2(\Omega)^3) \cap L^2(0, T; L^6(\Omega)^3)$, this equation also holds in the space

$$W^{-1,\infty}(0, T; L^\infty(\Omega)) \cap L^\infty(0, T; H^{-1}(\Omega)) \cap L^2(0, T; W^{-1,6}(\Omega)).$$

Initial Conditions From (3.56), one has $\rho^m(\cdot\,, 0) \to \rho(\cdot\,, 0)$ in $W^{-1,\infty}(\Omega)$. Thus, the initial conditions satisfied by the ρ^m and the fact that $\rho_0^m \to \rho_0$ weakly-$*$ in $L^\infty(\Omega)$ (see (3.9)) lead to the desired initial condition (3.8).

In order to prove that the initial condition (3.7) is satisfied, we fix $\mathbf{v}$ in V and we consider a sequence $\{\mathbf{v}^m\}$ such that $\mathbf{v}^m \in V^m$ for all m and

$$\mathbf{v}^m \to \mathbf{v} \quad \text{in } V.$$

Then

$$\int_\Omega \rho^m \mathbf{u}^m \cdot \mathbf{v}^m \ \text{ is uniformly bounded in } L^\infty(0, T). \tag{3.63}$$

Let us recall (3.45). One has $\|\mathbf{f}^m\| \to \|\mathbf{f}\|$ in $L^1(0, T)$. Consequently, there exists a subsequence (again indexed by m) and a function $K \in L^1(0, T)$ such that $\|\mathbf{f}^m\| \le K$ a.e. in $(0, T)$. Since the norms $\|\nabla \mathbf{v}^m\|$ are uniformly bounded, we have the estimate

$$\left| \frac{d}{dt} \int_\Omega \rho^m \mathbf{u}^m \cdot \mathbf{v}^m \right| \le C g_m \le C(bK + \psi_m), \tag{3.64}$$

where $\psi_m = \|\rho^m \mathbf{u}^m \otimes \mathbf{u}^m - \mu \nabla \mathbf{u}^m\|$ is uniformly bounded in $L^{4/3}(0, T)$.

From (3.63), (3.64) and the last part of Lemma 2.3, we see that

$$\int_\Omega \rho^m \mathbf{u}^m \cdot \mathbf{v}^m \text{ belongs to a compact subset of } C^0([0, T]).$$

On the other hand, one has

$$\int_\Omega \rho^m \mathbf{u}^m \cdot \mathbf{v}^m \to \int_\Omega \rho \mathbf{u} \cdot \mathbf{v} \quad \text{weakly-}* \text{ in } L^\infty(0, T).$$

Hence, this convergence also holds strongly in $C^0([0, T])$ and

$$\int_\Omega \rho \mathbf{u} \cdot \mathbf{v} \in C^0([0, T]).$$

In particular, at $t = 0$ one has

$$\left(\int_\Omega \rho^m \mathbf{u}^m \cdot \mathbf{v}^m\right)(0) \to \left(\int_\Omega \rho \mathbf{u} \cdot \mathbf{v}\right)(0).$$

But it is also true that

$$\left(\int_\Omega \rho^m \mathbf{u}^m \cdot \mathbf{v}^m\right)(0) = \int_\Omega \rho_0^m \mathbf{u}_0^m \cdot \mathbf{v}^m \to \int_\Omega \rho_0 \mathbf{u}_0 \cdot \mathbf{v}.$$

Consequently, the desired initial condition (3.7) holds.

Additional Properties In view of (3.56), one can take limits as $m \to \infty$ in (3.30) and obtain

$$\inf_\Omega \rho_0 \le \rho \le \sup_\Omega \rho_0 \text{ a.e. in } Q. \tag{3.65}$$

On the other hand, taking into account the regularity of ρ, $\mathbf{u}$ and $\rho\mathbf{u}$, one also has

$$\rho \mathbf{u} \in N^{1/4,2}(0, T; W^{-1,3}(\Omega)^3).$$

The proof is completely analogous to the proof of (3.53). For brevity, it will not be repeated here; see Exercise 3.3.

Another important property satisfied by the weak solution is the energy inequality.

To see this, let us recall the identity (3.25), satisfied by the semi-Galerkin approximations:

$$\frac{1}{2}\frac{d}{dt}\int_\Omega \rho^m |\mathbf{u}^m|^2 + \mu \int_\Omega |\nabla \mathbf{u}^m|^2 = \int_\Omega \rho^m \mathbf{f}^m \cdot \mathbf{u}^m.$$

After integration with respect to time, taking into account the initial conditions satisfied by ρ^m and $\mathbf{u}^m$, we find that

$$\frac{1}{2}\left(\int_\Omega \rho^m|\mathbf{u}^m|^2\right)(t)+\mu\iint_{\Omega\times(0,t)}|\nabla\mathbf{u}^m|^2=\frac{1}{2}\int_\Omega \rho_0^m|\mathbf{u}_0^m|^2+\iint_{\Omega\times(0,t)}\rho^m\mathbf{f}^m\cdot\mathbf{u}^m$$

for all $t \in [0, T]$.

Now, in view of the convergence properties deduced above, taking limits along a well chosen subsequence, we see that

$$\frac{1}{2}\left(\int_\Omega \rho|\mathbf{u}|^2\right)(t)+\mu\iint_{\Omega\times(0,t)}|\nabla\mathbf{u}|^2 \le \frac{1}{2}\int_\Omega \rho_0|\mathbf{u}_0|^2+\iint_{\Omega\times(0,t)}\rho\mathbf{f}\cdot\mathbf{u} \tag{3.66}$$

for all t.

The change from an identity to an inequality comes from the fact that we can ensure that

$$\liminf_{m\to+\infty}\iint_{\Omega\times(0,t)}|\nabla\mathbf{u}^m|^2 \ge \iint_{\Omega\times(0,t)}|\nabla\mathbf{u}|^2$$

but no more, as a consequence of the weak (and not necessarily strong) convergence of $\mathbf{u}^m$ in $L^2(0, T; V)$.

It is usual to say that (3.66) is the *energy inequality* satisfied by ρ and $\mathbf{u}$.

In physical terms, the first and second terms on the left-hand side of (3.66) must be viewed as the instantaneous *kinetic energy* and the total amount of dissipated (i.e. lost) *viscous energy* of the fluid during the time interval $[0, t]$.

The first term on the right-hand side is the kinetic energy at time 0.

Finally, the second term on the right-hand side must be regarded as the amount of work performed by the external forces along the time interval $[0, t]$.

Under the assumptions of Theorem 3.1, the existence of a global weak solution to (3.3)–(3.8) satisfying the energy identity is unknown. In fact, it is also unknown in the case of the classical (constant density) three-dimensional Navier-Stokes equations.

3.2 Some Additional Results

In this section, we recall some other results, similar and/or complementary to Theorem 3.1. In several cases, the proofs are not too different from the previous one.

We will also mention several open problems.

- **Navier-Stokes-Fourier fluids with variable density.**

As already seen, if heat effects are important, we must complete the system with an equation for the temperature. In particular, the following PDEs appear in a real-world situation:

$$
\begin{cases}
\rho_t + \nabla \cdot (\rho \mathbf{u}) = 0, \\
(\rho \mathbf{u})_t + \nabla \cdot (\rho \mathbf{u} \otimes \mathbf{u}) - \mu \Delta \mathbf{u} + \nabla p = \rho \theta \mathbf{k} + \rho \mathbf{f}, \\
\nabla \cdot \mathbf{u} = 0, \\
(\rho \theta)_t + \nabla \cdot (\rho \theta \mathbf{u}) - \kappa \Delta \theta = \mu D\mathbf{u} : \nabla \mathbf{u} + \rho g,
\end{cases} \tag{3.67}
$$

where $\mathbf{k}$ is the gravity acceleration (a constant vector), $\kappa > 0$ and $\mathbf{f}$ and g are appropriate functions.

Of course, boundary and initial conditions must be added to (3.67). For example, we can impose homogeneous Dirichlet conditions for $\mathbf{u}$ and θ on Σ and initial conditions for $\mathbf{u}$, ρ and θ at $t = 0$.

The existence of an associated weak solution is unknown. Note that the right-hand side of the energy equation (that is, the PDE for θ) contains a term $2\mu D\mathbf{u} : \nabla \mathbf{u}$ that is at most expected to be in $L^1(Q)$ and this introduces a nontrivial difficulty in the proof; see to this respect [4–7, 9] for some results.

If we simplify the problem by neglecting heat effects due to viscous forces, the system becomes simpler and the following result holds:

Theorem 3.2 *Let $T > 0$ be given. Assume that $\mathbf{u}_0 \in H$, $\rho_0 \in L^\infty(\Omega)$ with $\rho_0 \geq 0$, $\theta_0 \in L^2(\Omega)$, $\mathbf{f} \in L^1(0, T; L^2(\Omega)^3)$ and $g \in L^1(0, T; L^2(\Omega))$ Then, there exist $\mathbf{u} \in L^2(0, T; V)$, $\rho \in L^\infty(Q)$ and $p \in W^{-1,\infty}(0, T; L^2(\Omega))$ and $\theta \in L^2(0, T; H_0^1(\Omega))$ such that*

$$
\rho \mathbf{u} \in L^\infty(0, T; L^2(\Omega)^3) \cap N^{1/4,2}(0, T; W^{-1,3}(\Omega)^3), \tag{3.68}
$$

$$
\inf_\Omega \rho_0 \leq \rho(x, t) \leq \sup_\Omega \rho_0 \quad \text{a.e. in } Q, \tag{3.69}
$$

$$
\rho \theta \in L^\infty(0, T; L^2(\Omega)) \cap N^{1/4,2}(0, T; W^{-1,3}(\Omega)), \tag{3.70}
$$

the equations

$$
\frac{\partial \rho \mathbf{u}}{\partial t} + \nabla \cdot (\rho \mathbf{u} \otimes \mathbf{u}) + \nabla p = \mu \Delta \mathbf{u} + \rho(\theta \mathbf{e}^N + \mathbf{f}), \tag{3.71}
$$

$$
\nabla \cdot \mathbf{u} = 0, \tag{3.72}
$$

$$
\frac{\partial \rho}{\partial t} + \nabla \cdot (\rho \mathbf{u}) = 0 \tag{3.73}
$$

and

$$
\frac{\partial \rho \theta}{\partial t} + \nabla \cdot (\rho \mathbf{u} \theta) = k \Delta \theta + \rho g \tag{3.74}
$$

are satisfied in Q (in the distributional sense), the boundary conditions

$$
\mathbf{u} = \mathbf{0} \text{ and } \theta = 0 \tag{3.75}
$$

is satisfied on $\Sigma = \partial\Omega \times (0, T)$ *and the following initial conditions hold:*

$$\left(\int_\Omega \rho\mathbf{u}\cdot\mathbf{v}\right)\Big|_{t=0} = \int_\Omega \rho_0\mathbf{u}_0\cdot\mathbf{v} \quad \forall\mathbf{v}\in V, \tag{3.76}$$

$$\theta|_{t=0} = \theta_0 \ \textit{in} \ \Omega, \tag{3.77}$$

$$\rho|_{t=0} = \rho_0 \ \textit{in} \ \Omega. \tag{3.78}$$

The proof can be obtained by adapting the arguments of the proof of Theorem 3.1. More precisely, we can again introduce appropriate semi-Galerkin approximations $\{\mathbf{u}^m, \rho^m, p^m, \theta^m\}$ associated to a basis of $V \times H^1_0(\Omega)$, uniform estimates for the $\mathbf{u}^m$, ρ^m and θ^m and their time derivatives can be deduced, converging subsequences can be extracted, etc.

- **Unbounded domains.**

Let us assume that $\Omega \subset \mathbb{R}^3$ is an unbounded connected open set with Lipschitz-continuous boundary.

Recall that an important ingredient of the proof of Theorem 3.1 is the compactness of the embedding of (for instance) V into H. Indeed, this was crucial to pass to the limit in the nonlinear terms and confirm that the limit of a (sub)sequence of approximate solutions solves the original PDEs.

Unfortunately, compactness fails when Ω is not bounded. However, an existence result similar to Theorem 3.1 in Q can be obtained by applying the following strategy:

1. First, we approximate Ω by regular bounded regions Ω_R, with $\Omega_R \nearrow \Omega$ as $R \to +\infty$.
2. Then, as in Sect. 3.1, we prove the existence of a solution in $\Omega_R \times (0, T)$ for each $R > 0$.
3. In a third step, we deduce uniform estimates for these solutions.
4. Finally, a convergent subsequence is extracted and it is checked that the corresponding limit solves the original problem.

Let us present a more precise result.

To fix ideas, it will be assumed that $\{\Omega_R\}$ is a family of bounded open sets with Lipschitz-continuous boundaries such that $\Omega_R \subset \Omega$ for all $R > 0$, $\Omega_R \subset \Omega_S$ for $R < S$ and $\cup_{R>0}\Omega_R = \Omega$.

We will need the spaces $K(\Omega)$, $K_0(\Omega)$ and $W(\Omega)$, introduced near the end of Sect. 2.4.

One has the following result (see for instance [17]):

Theorem 3.3 *Assume that*

$$\mathbf{u}_0 \in H, \quad \rho_0 \in L^\infty(\Omega), \quad \rho_0 \geq 0$$

and

$$\mathbf{f} \in L^1(0, T; L^2(\Omega)^3) \cap L^1(0, T; L^{6/5}(\Omega)^3).$$

Then, there exists at least one triplet $\{\mathbf{u}, \rho, p\}$, *with*

$$\mathbf{u} \in L^2(0, T; W(\Omega)), \tag{3.79}$$

$$\rho \in L^\infty(Q) \cap C^0([0, T]; W_{\mathrm{loc}}^{-1,\infty}(\Omega)), \tag{3.80}$$

$$p \in W^{-1,\infty}(0, T; L^2_{\mathrm{loc}}(\Omega)), \quad \nabla p \in W^{-1,\infty}(0, T; H^{-1}(\Omega)^3), \tag{3.81}$$

$$\rho\mathbf{u} \in L^\infty(0, T; L^2(\Omega)^3), \quad \int_\Omega \rho\mathbf{u} \cdot \mathbf{v} \in C^0([0, T]) \quad \forall \mathbf{v} \in W(\Omega), \tag{3.82}$$

that satisfies (3.3)–(3.5) *in the distributional sense, the initial condition (3.7) in the sense*

$$\left(\int_\Omega \rho\mathbf{u} \cdot \mathbf{v}\, d\mathbf{x} \right)\Bigg|_{t=0} = \int_\Omega \rho_0\mathbf{u}_0 \cdot \mathbf{v} \quad \forall \mathbf{v} \in W(\Omega)$$

and the initial condition (3.8) in the usual sense in $W_{\mathrm{loc}}^{-1,\infty}(\Omega)$.

- **Two-dimensional flows.**

It makes sense to consider variable density Newtonian flows in two dimensions. They are modeled by (3.3)–(3.5), where $\mathbf{u}$ and $\mathbf{f}$ have now only two components and $(\mathbf{x}, t)$ must belong to a set $Q = \Omega \times (0, T)$, where $\Omega \subset \mathbb{R}^2$ is a non-empty connected open set.

Obviously, conclusions like those in Theorem 3.1 hold again in this case. In fact, the regularity of $\mathbf{u}$ is slightly improved: for instance, if Ω is bounded and $\mathbf{f} \in L^2(Q)^2$, additionally to (3.1) one has

$$(\rho\mathbf{u})_t \in L^2(0, T; V') \tag{3.83}$$

and, consequently, regarded as a V'-valued function, $t \mapsto (\rho\mathbf{u})(\cdot, t)$ is absolutely continuous. Accordingly, the initial condition (3.7) can be replaced by the slightly stronger identity

$$\rho\mathbf{u}|_{t=0} = \rho_0\mathbf{u}_0 \text{ in } V'.$$

However, this is not sufficient to ensure uniqueness. We refer the reader to the last paragraph in this section, where the best known uniqueness results at present are indicated.

Without leaving this two-dimensional model, it makes sense to ask what happens when we start from a not natural situation, for instance, with a velocity field close to zero and an initial density that is increasing in the vertical direction, contrarily to the action of gravity (downwards directed).

It is absolutely reasonable to expect that, at large times, there is a final mass density with reverse properties. In the meanwhile, we must find some kind of restless motion: the so called Rayleigh-Taylor instabilities, see for instance [12]. The process is illustrated in Fig. 3.1.

For three-dimensional flows, the expected situation is the same. We leave to the reader the task to provide a rigorous proof of this asymptotic behavior.

- **Other boundary conditions.**

Other boundary conditions can be used to complement the PDEs (3.3). Some of them were introduced in Chap. 1.

For example, the arguments in the proof of Theorem 3.1 can be adapted to the case in which the fluid slips on the boundary and the tangential components of the normal stress are proportional to the tangential components of the velocity field, that is,

$$\mathbf{u}\cdot\mathbf{n}=0, \quad (\sigma\cdot\mathbf{n})_\tau+\kappa(\mathbf{u}-\mathbf{a})_\tau=0 \text{ on } \Sigma, \tag{3.84}$$

with $\kappa \geq 0$ and $\mathbf{a} \in \mathbb{R}^3$.

Let us introduce the space

$$\tilde{V}=\{\mathbf{v}\in H^1(\Omega)^3 : \nabla\cdot\mathbf{v}=0 \text{ in } \Omega,\ \mathbf{v}\cdot\mathbf{n}=0 \text{ on } \partial\Omega\}.$$

The following result holds:

Theorem 3.4 *Let $T>0$ be given. Assume that $\mathbf{u}_0 \in H$, $\rho_0 \in L^\infty(\Omega)$ with $\rho_0 \geq 0$ and $\mathbf{f} \in L^1(0,T;L^2(\Omega)^3)$. Then, there exist*

$$\mathbf{u}\in L^2(0,T;\tilde{V}),\ \rho\in L^\infty(Q)\cap C^0([0,T];W^{-1,\infty}(\Omega)) \text{ and } p\in W^{-1,\infty}(0,T;L^2(\Omega))$$

such that

$$\rho\mathbf{u}\in L^\infty(0,T;L^2(\Omega)^3)\cap N^{1/4,2}(0,T;W^{-1,3}(\Omega)^3),$$

$$\inf_\Omega \rho_0 \leq \rho(x,t) \leq \sup_\Omega \rho_0 \quad \text{a.e. in } Q,$$

Equations (3.3) *are satisfied in Q, the boundary conditions* (3.84) *are satisfied on Σ in the sense that*

$$\mathbf{u}(\cdot,t)\in\tilde{V} \text{ a.e. in } (0,T)$$

and

$$\begin{cases} \displaystyle\int_\Omega \Big[\rho(\frac{\partial\mathbf{u}}{\partial t}+(\mathbf{u}\cdot\nabla)\mathbf{u})\cdot\mathbf{v}+\mu\nabla\mathbf{u}:\nabla\mathbf{v}\Big]+\kappa\int_{\partial\Omega}(\mathbf{u}-\mathbf{a})\cdot\mathbf{v}\,d\Gamma=\int_\Omega \rho\mathbf{f}\cdot\mathbf{v} \\ \forall\mathbf{v}\in\tilde{V} \end{cases}$$

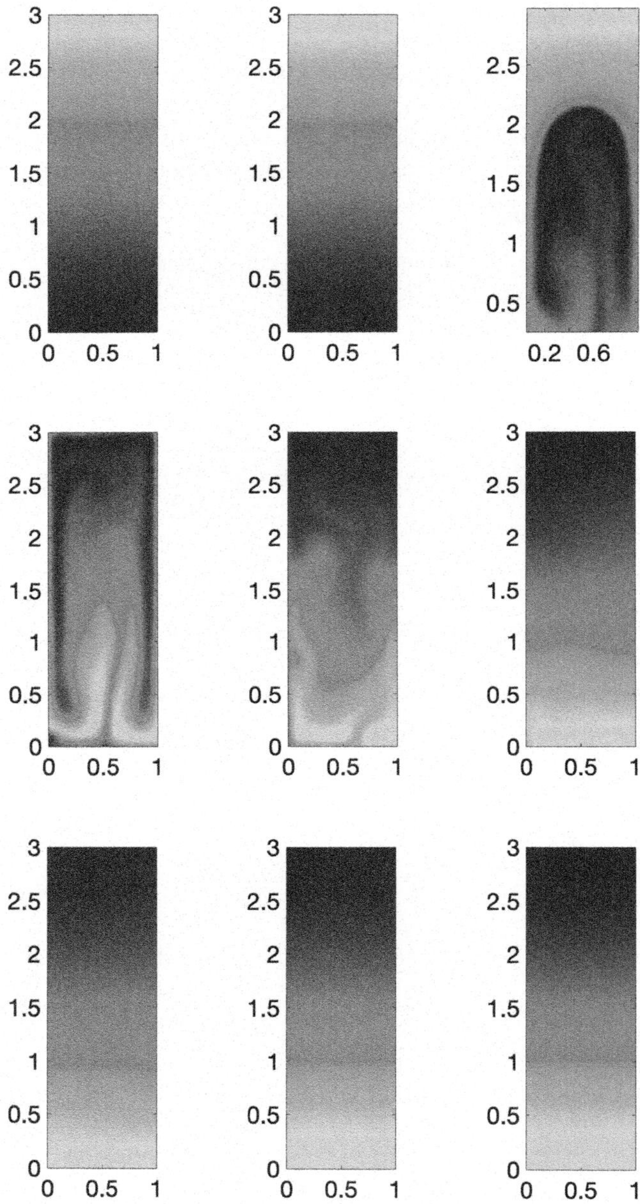

Fig. 3.1 Evolution of the density for times in (0, 60). From left to right and top to bottom: $\rho(\cdot, t)$ at $t = 0$ (initial state), 1.80, 2.30, 3.45, 4.60, 6.90, 9.20, 36.80 and 55.20

and the following initial conditions are fulfilled:

$$\left(\int_\Omega \rho\mathbf{u}\cdot\mathbf{v}\right)\bigg|_{t=0} = \int_\Omega \rho_0\mathbf{u}_0\cdot\mathbf{v} \quad \forall\mathbf{v}\in\tilde{V}, \tag{3.85}$$

$$\rho|_{t=0} = \rho_0. \tag{3.86}$$

A similar result can be obtained if we impose different boundary conditions on different parts of Σ; see to this respect Exercises 3.6 and 3.7.

- **The existence of a weak solution when** $\mu = \mu(\rho)$.

As shown in Chap. 1, it is meaningful to consider fluids with density dependent viscosity for which the motion equation is

$$\frac{\partial\rho\mathbf{u}}{\partial t} + \nabla\cdot(\rho\mathbf{u}\otimes\mathbf{u}) - \nabla\cdot(2\mu(\rho)D(\mathbf{u})) + \nabla p = \mathbf{f},$$

where $\mu : \mathbb{R}_+ \mapsto \mathbb{R}_+$ is a function satisfying

$$\mu\in C^0(\mathbb{R}), \quad \mu(r)\geq\mu_0>0.$$

In this case, the existence of a weak solution, that is, a result similar to Theorem 3.1, can also be established. See a proof in [17].

- **The convergence towards a solution to the Navier-Stokes equations.**

The situation is the following.

The initial velocity field $\mathbf{u}_0$ is given in H is given and $\{\rho_0^\varepsilon\}$ is a family of initial densities in $L^\infty(\Omega)$ with

$$0<\alpha\leq\rho_0^\varepsilon\leq\beta, \quad \rho_0^\varepsilon\to\overline{\rho} \text{ weakly-}* \text{ in } L^\infty(\Omega),$$

where $\overline{\rho}$ is a positive constant.

Let $\{\mathbf{u}^\varepsilon, \rho^\varepsilon, p^\varepsilon\}$ be an associated solution to (3.3)–(3.8) for every $\varepsilon>0$, l Then, at least for a subsequence, one has $\mathbf{u}^\varepsilon\to\mathbf{u}$ and $\rho^\varepsilon\to\rho$ (in an appropriate sense), where $\mathbf{u}$ solves, together with some p, the Navier-Stokes system with constant density $\overline{\rho}$.

The proof of this claim is not difficult. Indeed, the uniform estimates in the proof of Theorem 3.1 are also satisfied in this case uniformly with respect to ε. We deduce that there exists a subsequence (again indexed by ε) that converges in the sense of (3.56), (3.57), etc. where the limits $\mathbf{u}$ and ρ solve, together with some p, the variable density Navier-Stokes problem corresponding to the data $\mathbf{u}_0$, $\mathbf{f}$ and $\rho_0\equiv\overline{\rho}$.

Since the limit density ρ must satisfy (3.65), the assertion holds.

- **Stationary solutions.**

As noticed in [17], a "good" framework for the analysis of the stationary variable density Navier-Stokes equations is unknown.

This is because, in the stationary case, the continuity equation and the incompressibility condition are too similar and, in some sense, we are led to an underdetermined system. Indeed, it is easy to see that the stationary system

$$\begin{cases} \nabla \cdot (\rho \mathbf{u} \otimes \mathbf{u}) - \mu \Delta \mathbf{u} + \nabla p = \rho \mathbf{f}, \quad \nabla \cdot \mathbf{u} = 0, \quad \mathbf{x} \in \Omega, \\ \nabla \cdot (\rho \mathbf{u}) = 0, \quad \mathbf{x} \in \Omega, \\ \mathbf{u} = 0, \quad \mathbf{x} \in \partial\Omega, \end{cases} \tag{3.87}$$

where $\mathbf{f} \in L^2(\Omega)^3$, possesses many solutions. Thus, for any constant positive density, there exists at least one solution $\{\mathbf{u}, p\}$, with $\mathbf{u} \in D(A)$ and $p \in H^1(\Omega)$.

In fact, something still more interesting happens: let us assume that (for instance) $N = 2$, $\Omega. = B(0; 1)$ and $\mathbf{f}(\mathbf{x}) \equiv (x_2 g(|\mathbf{x}|), -x_1 g(|\mathbf{x}|))$ for some $g \in L^2(0, 1)$; then, for any $\overline{\rho} \in L^\infty(0, 1)$, the triplet $\{\mathbf{u}, \rho, p\}$, where

$$\begin{cases} \mathbf{u}(\mathbf{x}) = (x_2 \psi(|\mathbf{x}|), -x_1 \psi(|\mathbf{x}|)) \quad \forall \mathbf{x} \in B(0; 1), \\ \psi(r) = c_0 - \dfrac{1}{2} \displaystyle\int_0^r s \left(1 - \left(\frac{s}{r}\right)^2\right) \overline{\rho}(s)\, g(s)\, ds \quad \forall r \in [0, 1], \quad \psi(1) = 0, \\ \rho(\mathbf{x}) = \overline{\rho}(|\mathbf{x}|) \text{ and } p(\mathbf{x}) = |\mathbf{x}|\, \overline{\rho}(|\mathbf{x}|)\, \psi(|\mathbf{x}|) \quad \forall \mathbf{x} \in B(0; 1), \end{cases}$$

is a solution to (3.87).

Accordingly, in view of the known results for the classical Navier-Stokes equations, a natural question is the following: which condition(s) must be added to (3.87) in order to get (a) existence for any μ and (b) existence and uniqueness for large μ?

- **Parabolic regularization.**

In this paragraph, we will recall an alternative method of solution of (3.3)–(3.8).

The nonhomogeneous Navier-Stokes equations can be solved by first considering and solving, for each $\varepsilon > 0$, the regularized system

$$\begin{cases} \dfrac{\partial \rho \mathbf{u}}{\partial t} + \nabla \cdot (\rho \mathbf{u} \otimes \mathbf{u}) - \mu \Delta \mathbf{u} + \nabla p = \rho \mathbf{f}, \quad \nabla \cdot \mathbf{u} = 0, \quad (\mathbf{x}, t) \in Q, \\ \dfrac{\partial \rho}{\partial t} + \nabla \cdot (\rho \mathbf{u}) - \varepsilon \Delta \rho = 0, \quad (\mathbf{x}, t) \in Q, \\ \mathbf{u} = 0, \quad \dfrac{\partial \rho}{\partial n} = 0, \quad (\mathbf{x}, t) \in \Sigma, \\ \rho|_{t=0} = \rho_0, \quad (\rho \mathbf{u})|_{t=0} = \rho_0 \mathbf{u}_0, \quad \mathbf{x} \in \Omega, \end{cases} \tag{3.88}$$

and then taking limits as $\varepsilon \to 0^+$. Indeed, it is not difficult to prove that, for each $\varepsilon > 0$, there exists at least one solution $\{\mathbf{u}^\varepsilon, \rho^\varepsilon, p^\varepsilon\}$ to (3.88), with appropriate uniform estimates.

Consequently, at least for a subsequence, one must have convergence in adequate spaces and these convergence properties suffice to pass to the limit in system (3.88).

It would be pertinent to identify the properties of the solutions that can be obtained this way, as the limit of a regularized parabolic system like (3.88).

By analogy with other (simpler) systems of conservation laws, it is natural to expect that these properties provide some kind of uniqueness; see for instance [2, 16].

This is a very interesting open problem.

- **The uniqueness problem.**

The uniqueness of a weak solution is a major open problem. It is also open for the standard constant density Navier-Stokes equations.[3]

So far, the best uniqueness results are due to Danchin and Mucha [8]. They are as follows:

1. Assume that $\partial\Omega$ is of class C^2 and let $\{\mathbf{u}^1, \rho^1, p^1\}$ and $\{\mathbf{u}^2, \rho^2, p^2\}$ be two weak solutions to (3.3)–(3.8). Assume that $\rho_0 \geq \alpha > 0$ a.e., the $\mathbf{u}^i$ belong to $L^2(0, T; W^{2,3}(\Omega)^3) \cap L^2(0, T; W^{1,\infty}(\Omega)^3)$, $\mathbf{u}^i_t \in L^2(0, T; L^3(\Omega)^3)$ and $\nabla p^i \in L^2(0, T; L^3(\Omega)^3)$ for $i = 1, 2$. Then $\{\mathbf{u}^1, \rho^1, \nabla p^1\} = \{\mathbf{u}^2, \rho^2, \nabla p^2\}$.
2. In the two-dimensional case, the same can be said under the following weaker regularity hypotheses on the $\{\mathbf{u}^i, \rho^i, p^i\}$: there exists $q > 2$ such that $\mathbf{u}^i \in L^2(0, T; W^{2,q}(\Omega)^3)$, $\mathbf{u}^i_t \in L^2(0, T; L^q(\Omega)^3)$ and $\nabla p^i \in L^2(0, T; L^q(\Omega)^3)$ for $i = 1, 2$.

The proofs rely on an analysis of Eqs. (3.3) and (3.4) rewritten in Lagrangian coordinates. This viewpoint is connected to the numerical approximations in time and space considered in Chap. 4.

This approach only requires upper and lower L^∞ bounds for the density. Accordingly, the results hold, for example, for piecewise constant ρ_0, which is a physically relevant case.

Other results and comments concerning uniqueness are given in Chap. 5 (see Sect. 5.2).

3.3 Exercises of Chapter 3

Exercise 3.1* Prove the assertions in Remark 3.2.

Exercise 3.2* Prove that the unique solution to (3.16) with $\rho_0^m = \eta_0$ is given by

$$\eta(\mathbf{y}, t) = \eta_0(S_t^{-1}(\mathbf{y})) \quad \forall \mathbf{y} \in \overline{\Omega}, \ \ \forall t \in [0, T].$$

Exercise 3.3* Let $\Omega \subset \mathbb{R}^2$ be a non-empty bounded connected open set with Lipschitz-continuous boundary and assume that $T > 0$. State and prove a result

[3] See, however, [3], where the authors show some evidence of non-uniqueness in a weakened class of solutions for the spatially periodic case.

similar to Theorem 3.1 in $Q := \Omega \times (0, T)$. What can be said now about $\mathbf{u}$, $\rho\mathbf{u}$ and $\rho\mathbf{u} \otimes \mathbf{u}$?

Exercise 3.4* Let $\{\mathbf{u}, \rho, p\}$ be a solution obtained as in Exercise 3.3. Assume that $\rho_0 \geq \alpha > 0$ in Ω. Prove that

$$\mathbf{f} \in L^2(Q)^2 \Rightarrow (\rho\mathbf{u})_t \in L^2(0, T; V').$$

HINT: Compute $\langle(\rho\mathbf{u})_t, \mathbf{v}\rangle_{V',V}$ and use that $\rho^{1/4}\mathbf{u} \in L^4(Q)^2$.

Exercise 3.5* Prove Theorem 3.4.

Exercise 3.6* Prove the analog of Theorem 3.4 with the boundary condition (3.84) replaced by the Fourier-like condition

$$\sigma \cdot \mathbf{n} + \kappa(\mathbf{u} - \mathbf{a}) = 0 \text{ on } \Sigma,$$

where $\kappa \geq 0$ and $\mathbf{a} \in \mathbb{R}^3$.

Exercise 3.7* Prove the analog of Theorem 3.4 with the boundary condition (3.84) replaced by

$$\mathbf{u} = \mathbf{a} \text{ on } \Sigma_0 \text{ and } \sigma \cdot \mathbf{n} = 0 \text{ on } \Sigma_1,$$

where $\Sigma_i = \Gamma_i \times (0, T)$ for $i = 0, 1$, $\{\Gamma_0, \Gamma_1\}$ is a partition of $\partial\Omega$ and Γ_0 is non-empty and open in $\partial\Omega$.

Exercise 3.8* Prove Theorem 3.3.

Exercise 3.9** Consider the following PDE system, that can be viewed as a one-dimensional version of the variable density Navier-Stokes equations:

$$\begin{cases} \rho_t + u\rho_x = 0, \\ \rho(u_t + uu_x) - \mu u_{xx} = \rho f. \end{cases}$$

This is the variable density viscous Burgers system. Complete these PDEs with appropriate boundary and initial conditions, define the concept of weak solution and state and prove a related existence result. Is the weak solution unique?

HINT: Try to adapt the arguments in the proof of Theorem 3.1.

Exercise 3.10** Same statement for the system

$$\begin{cases} \rho_t + u\rho_x = 0, \\ \rho(u_t + uu_x) - \mu u_{xx} = \rho\theta + \rho f. \\ \rho(\theta_t + u\theta_x) - \kappa\theta_{xx} = \rho g. \end{cases}$$

This corresponds to a one-dimensional variable density viscous Burgers fluid where heat effects are important.

Exercise 3.11** Let $\mu = \mu(\mathbf{x}, t)$ be given, with

$$\mu \in L^\infty(Q), \quad \mu \geq \mu_0 > 0 \text{ a.e. in } Q.$$

The following system models the behavior of a Newtonian fluid that is nonhomogeneous in density and viscosity:

$$\begin{cases} \rho_t + \nabla \cdot (\rho \mathbf{u}) = 0, \\ (\rho \mathbf{u})_t + \nabla \cdot (\rho \mathbf{u} \otimes \mathbf{u}) - \nabla \cdot (2\mu(\mathbf{x}, t) D\mathbf{u}) + \nabla p = \rho \mathbf{f}, \\ \nabla \cdot \mathbf{u} = 0. \end{cases}$$

Complete this system with boundary and initial conditions for $\mathbf{u}$ and ρ and prove an existence result similar to Theorem 3.1.

Exercise 3.12** Let $M = M(\mathbf{x}, t)$ be given, with $M \in L^\infty(Q)^{N \times N}$,

$$M(\mathbf{x}, t) S \cdot S \geq \mu_0 |S|^2 \ \forall S \in \mathbb{R}^{N \times N} \text{ with } S^T = S, \text{ a.e. in } Q.$$

The following system models the behavior of a variable density anisotropic Newtonian fluid:

$$\begin{cases} \rho_t + \nabla \cdot (\rho \mathbf{u}) = 0, \\ (\rho \mathbf{u})_t + \nabla \cdot (\rho \mathbf{u} \otimes \mathbf{u}) - \nabla \cdot (2M(\mathbf{x}, t) D\mathbf{u}) + \nabla p = \rho \mathbf{f}, \\ \nabla \cdot \mathbf{u} = 0. \end{cases}$$

Again, complete this system with boundary and initial conditions for $\mathbf{u}$ and ρ and prove an existence result similar to Theorem 3.1.

Exercise 3.13* Prove Theorem 3.2.

HINT: Introduce a family of semi-Galerkin approximations $\{\mathbf{u}_m, \rho_m, \theta_m\}$ where the $\mathbf{u}_m$ and θ_m take values in finite-dimensional spaces and try to adapt the arguments in the proof of Theorem 3.1.

Exercise 3.14* Let the ρ_0^ε be given in $L^\infty(\Omega)$, with

$$0 < \alpha \leq \rho_0^\varepsilon \leq \beta, \quad \rho_0^\varepsilon \to \overline{\rho} \text{ weakly-} * \text{ in } L^\infty(\Omega) \text{ as } \varepsilon \to 0,$$

where $\overline{\rho}$ is a positive constant. For each $\varepsilon > 0$, let $\{\mathbf{u}^\varepsilon, \rho^\varepsilon, p^\varepsilon\}$ be a solution to (3.3)–(3.8). Prove that, at least for a subsequence, one has $\mathbf{u}^\varepsilon \to \mathbf{u}$ and $\rho^\varepsilon \to \rho$ in appropriate senses, where $\mathbf{u}$ solves, together with some p, the Navier-Stokes system with constant density $\overline{\rho}$.

References

1. Antontsev, S.N., Kazhikhov, A.V., Monakhov, V.N.: Boundary Value Problems in Mechanics of Nonhomogeneous Fluids, volume 22 of Studies in Mathematics and its Applications. North-Holland Publishing Co., Amsterdam (1990). Translated from the Russian
2. Bénilan, P., Kružkov, S.: Conservation laws with continuous flux functions. NoDEA Nonlinear Differential Equations Appl. **3**(4), 395–419 (1996)
3. Buckmaster, T., Vicol, V.: Nonuniqueness of weak solutions to the Navier-Stokes equation. Ann. Math. (2) **189**(1), 101–144 (2019)
4. Bulicek, M., Feireisl, E., Malek, J.: A Navier-Stokes-Fourier system for incompressible fluids with temperature dependent material coefficients. Nonlinear Anal. Real World Appl. **10**(2), 992–1015 (2009)
5. Bulicek, M., Lewandowski, R., Malek, J.: On evolutionary Navier-Stokes-Fourier type systems in three spatial dimensions. Comment. Math. Univ. Carolin. **52**(1), 89–114 (2011)
6. Chaudhuri, N., Feireisl, E.: Navier-Stokes-Fourier system with Dirichlet boundary conditions. Appl. Anal. **101**(12), 4076–4094 (2022)
7. Chiodaroli, E., Feireisl, E.: On the long-time behavior of solutions to the Navier–Stokes–Fourier system on unbounded domains. J. Lond. Math. Soc. (2) **111**(1), Paper No. e70067 (2025)
8. Danchin, R., Mucha, P.B.: Incompressible flows with piecewise constant density. Arch. Ration. Mech. Anal. **207**(3), 991–1023 (2013)
9. Feireisl, E.: On strict positivity of the temperature in the Navier-Stokes-Fourier system. Pure Appl. Funct. Anal. **10**(1), 45–57 (2025)
10. Fernández-Cara, E., Guillén-González, F.: The existence of nonhomogeneous, viscous and incompressible flow in unbounded domains. Comm. Partial Differential Equ. **17**(7–8), 1253–1265 (1992)
11. Fernández-Real, X., Ros-Oton, X.: Regularity Theory for Elliptic PDE, volume 28 of Zurich Lectures in Advanced Mathematics. EMS Press, Berlin (2022)
12. Gaissinski, I., Rovenski, V.: Modeling in Fluid Mechanics. CRC Press, Boca Raton, FL (2018). Instabilities and turbulence, With a foreword by Irina Albinsky
13. Kazhikov, A.V.: Solvability of the initial-boundary value problem for the equations of motion of an inhomogeneous viscous incompressible fluid. Dokl. Akad. Nauk SSSR **216**, 1008–1010 (1974)
14. Kazhikhov, A.V., Antontsev, S.N., Monakhov, V.N.: Boundary Value Problems in Mechanics of Nonhomogeneous Fluids, volume 22 of Studies in Mathematics and its Applications. North-Holland Publishing, Amsterdam (1990). Translated from the Russian
15. Kim, J.U.: Weak solutions of an initial-boundary value problems for an incompressible viscous fluid with nonnegative density. SIAM J. Math. Anal. **18**(1), 89–96 (1987)
16. Kružkov, S.N.: First order quasilinear equations with several independent variables. Mat. Sb. (N.S.) **81**(123), 228–255 (1970)
17. Lions, P.-L.: Mathematical Topics in Fluid Mechanics. Vol I: Incompressible Models, volume 3 of Oxford Lecture Series in Mathematics and its Applications. The Clarendon Press/Oxford University Press, New York (1996)
18. Simon, J.: Existencia de solución del problema de Navier-Stokes con densidad variable. (Spanish) [existence of solution for the variable density Navier-Stokes problem]. Lectures at the University of Sevilla, Spain (1989)
19. Simon, J.: Nonhomogeneous viscous incompressible fluids: existence of velocity, density and pressure. SIAM J. Math. Anal. **21**(5), 1093–1117 (1990)

Chapter 4
Finite-Dimensional Approximation

This chapter is devoted to present some techniques that allow to get quantitative information on the solution $\{\mathbf{u}, \rho, p\}$ to (3.3)–(3.8). This will provide a practical complement to the theoretical results in Chap. 3.

Of course, we will only be able to compute finite-dimensional (numerical) approximations, in view of the complexity of the system. Indeed, explicit resolution is impracticable and, for a suitable description of the solution, numerical methods are therefore indispensable.

It will be seen that, with a reasonable effort, acceptable results can be obtained in many situations.

4.1 General Ideas

What we are going to show in these paragraphs corresponds to a few of a large set of possibilities. For a more complete information, the reader is referred, for instance, to [4, 14, 17].

To fix ideas and simplify the presentation, let us assume that $N = 3$, Ω is a polyhedron and $\rho_0 \geq \alpha > 0$.

Note that the resolution of the system must overcome several nontrivial difficulties: time evolution effects, nonlinearity, non-scalar variables linked through incompressibility, etc.

In view of the proof of Theorem 3.1, a first "naive" idea may be to compute the solutions to semi-Galerkin approximated problems. For example, this can be done by previously computing a finite family of (approximations to the) eigenfunctions and eigenvalues of the Stokes system in Ω.

P. Braz e Silva et al., *Analysis and Control of the Variable Density Incompressible Navier-Stokes Equations*, MS&A 22, https://doi.org/10.1007/978-3-032-14510-9_4

This is the origin of spectral methods, see [18] for an introduction. This way, we are led to the nonlinear ODE systems (3.10) and (3.11) that, completed with the initial conditions (3.12) and (3.13) for ρ^m and $\mathbf{u}^m$, can be solved numerically for instance with the help of a suitable Runge-Kutta method.

However, the computation of the eigenfunctions is usually very expensive, specially if we need many of them and we will consequently prefer another strategy.[1]

In what follows, we will approximate the problem at two levels: first, by performing time discretization, making time derivatives disappear and reducing the original problem to a finite family of stationary nonlinear problems; then, by introducing spatial approximations of these stationary systems, which will finally lead to a family of finite-dimensional nonlinear algebraic systems.

In order to explain clearly the meaning and outcome of each of these steps, let us first indicate the process in the much simpler case of the Cauchy-Dirichlet problem for the heat PDE:

$$\begin{cases} \dfrac{\partial \theta}{\partial t} - \Delta\theta = f(\mathbf{x}, t), & (\mathbf{x}, t) \in Q, \\ \theta = 0, \quad (\mathbf{x}, t) \in \Sigma, \\ \theta|_{t=0} = \theta_0, \quad \mathbf{x} \in \Omega, \end{cases} \tag{4.1}$$

where $f \in L^2(Q)$ and $\theta_0 \in L^2(\Omega)$.

Let $\theta \in L^2(0, T; H_0^1(\Omega)) \cap C^0([0, T]; L^2(\Omega))$ be the unique solution to (4.1).

Then, for use in the first step, we introduce a partition $t^0 = 0 < t^1 < \cdots < t^M = T$ of $[0, T]$ and we look for approximations θ^n of the functions $\theta(\cdot\,, t^n)$. For simplicity, we will assume that the partition is uniform, that is, $t^n = n\tau$ for all n, where $\tau = T/M$.

The θ^n are given as follows:

(a) $\theta^0 = \theta_0$.
(b) Then, for $n = 0, 1, \ldots, M - 1$, θ^{n+1} is the unique solution to the stationary system

$$\begin{cases} \dfrac{1}{\tau}(\theta - \theta^n) - \Delta\theta = f^{n+1}(\mathbf{x}), & \mathbf{x} \in \Omega, \\ \theta = 0, \quad \mathbf{x} \in \partial\Omega, \end{cases} \tag{4.2}$$

[1] Note that this strategy can also work with other more amenable basis functions.

where we have denoted by f^{n+1} the following approximation of f in the time interval (t^n, t^{n+1}):

$$f^{n+1}(\mathbf{x}) := \frac{1}{\tau} \int_{t^n}^{t^{n+1}} f(\mathbf{x}, t)\, dt. \tag{4.3}$$

It is clear that the linear systems (4.2) are uniquely solvable. This makes reasonable to think that a first approximation of θ is given by the continuous piecewise affine function $\theta_\tau : [0, T] \mapsto H_0^1(\Omega)$ that satisfies

$$\theta_\tau|_{t=t^n} = \theta^n \ \text{ for } \ n = 0, 1, \ldots, M.$$

Remark 4.1 In (4.2), the time derivative has been replaced by a difference quotient. This is a realization of the *implicit Euler* discretization scheme, first-order in time. Many other schemes can be used, with higher complexity. Among them, let us mention the so called *implicit Gear* scheme, that is of second order and is defined as follows:

(a) $\theta^0 = \theta_0$ and θ^1 as in (4.2) for $n = 0$.
(b) Then, for $n = 1, \ldots, M-1$, θ^{n+1} is the solution to the stationary system

$$\begin{cases} \dfrac{1}{2\tau}(3\theta - 4\theta^n + \theta^{n-1}) - \Delta\theta = f^{n+1}(\mathbf{x}), & \mathbf{x} \in \Omega, \\ \theta = 0, \quad \mathbf{x} \in \partial\Omega, \end{cases} \tag{4.4}$$

where f^{n+1} is given by (4.3).

□

Now, in a second step, what we have to do is to solve numerically a (maybe large but finite) family of linear elliptic problems (4.2). To this purpose, there exist many techniques. Let us choose the standard *finite element* method in its simplest version, see for instance [5].

Thus, let us introduce a family of triangulations $\{\mathcal{T}_h\}_{h\in(0,1)}$ of the polyhedron $\overline{\Omega}$. By definition, this means that each $\mathcal{T}_h$ is a finite family of tetrahedra $K \subset \overline{\Omega}$ with the following properties:

- $\bigcup_{K\in\mathcal{T}_h} K = \overline{\Omega}$.
- If $K, K' \in \mathcal{T}$, then either $K = K'$ or one and only one of these possibilities hold:
  - K and K' are disjoint,
  - $K \cap K'$ is a vertex,
  - $K \cap K'$ is an edge or a face common to K and K',

Recall that, for any compact set $B \subset \mathbb{R}^3$, the associated "thickness" $\rho(B)$ is the diameter of the largest ball contained in B.

It will be assumed that $\{\mathcal{T}_h\}_{h\in(0,1)}$ is *regular* in the following sense:

1. There exists $C_1 > 0$ such that the diameters $\delta(K)$ of the $K \in \mathcal{T}_h$ satisfy

$$\max_{K\in\mathcal{T}_h} \delta(K) \leq C_1 h \quad \forall h \in (0, 1]. \tag{4.5}$$

2. There exists $C_2 > 0$ such that

$$\frac{\delta(K)}{\rho(K)} \leq C_2 \quad \forall K \in \mathcal{T}_h, \quad \forall h \in (0, 1]. \tag{4.6}$$

The requirements (4.5) and (4.6) mean that the simplices K become uniformly small and cannot degenerate as $h \to 0$; see [5].

Let a triangulation $\mathcal{T}_h$ be fixed. For every $K \in \mathcal{T}_h$ and every integer $\ell \geq 1$, we will denote by $\mathbb{P}_\ell(K)$ the linear space of polynomial functions on K of degree $\leq \ell$. Then, it is no difficult to check that the spaces

$$W_h^\ell := \{v \in C^0(\overline{\Omega}) : v|_K \in \mathbb{P}_\ell(K) \ \ \forall K \in \mathcal{T}_h\} \tag{4.7}$$

are finite-dimensional subspaces of $H^1(\Omega)$.

In particular, the dimension of W_h^1 (resp. W_h^2) is equal to the number of vertices of $\mathcal{T}_h$ (resp. the number of vertices and edges of $\mathcal{T}_h$).

In fact, if we denote by $\mathbf{a}^1, \dots, \mathbf{a}^{I_h}$ the vertices of $\mathcal{T}_h$, the linear mapping

$$v \mapsto (v(\mathbf{a}^1), \dots, v(\mathbf{a}^{I_h}))$$

is an isomorphism from W_h^1 into $\mathbb{R}^{I_h}$. Similarly, if the midpoints of the edges are denoted by $\mathbf{b}^1, \dots, \mathbf{b}^{J_h}$, the mapping

$$v \mapsto (v(\mathbf{a}^1), \dots, v(\mathbf{a}^{I_h}), v(\mathbf{b}^1), \dots, v(\mathbf{b}^{J_h}))$$

is an isomorphism from W_h^2 into $\mathbb{R}^{I_h+J_h}$.

Now, let us introduce the spaces Y_h^ℓ, with

$$Y_h^\ell := \{v \in W_h^\ell : v = 0 \ \text{ on } \ \partial\Omega\}. \tag{4.8}$$

Then the Y_h^ℓ are finite-dimensional subspaces of $H_0^1(\Omega)$.

Let I_{0h} and J_{0h} respectively denote the numbers of vertices and midpoints of edges interior to Ω. Then one has $\dim Y_h^1 = I_{0h}$ and $\dim Y_h^2 = I_{0h} + J_{0h}$ and, as before, we can establish canonical isomorphisms from Y_h^1 (resp. Y_h^2) onto $\mathbb{R}^{I_{0h}}$ (resp. $\mathbb{R}^{I_{0h}+J_{0h}}$).

With the spaces Y_h^1 in mind, we can easily introduce spatial approximations of the problems (4.2). It suffices to recall the weak formulations

$$\begin{cases} \displaystyle\int_\Omega (\theta v + \tau \nabla\theta \cdot \nabla v)\, d\mathbf{x} = \int_\Omega \left(\theta^n + \tau f^{n+1}(\mathbf{x})\right) v\, d\mathbf{x} \\ \forall v \in H_0^1(\Omega), \quad \theta \in H_0^1(\Omega) \end{cases} \tag{4.9}$$

and replace $H_0^1(\Omega)$ by Y_h^1:

$$\begin{cases} \displaystyle\int_\Omega (\theta_h v + \tau \nabla\theta_h \cdot \nabla v)\, d\mathbf{x} = \int_\Omega \left(\theta_h^n + \tau f^{n+1}(\mathbf{x})\right) v\, d\mathbf{x} \\ \forall v \in Y_h^1, \quad \theta_h \in Y_h^1. \end{cases} \tag{4.10}$$

Of course, (4.10) possesses exactly one solution. On the other hand, it is clear that similar approximations can be obtained using Y_h^2 or even Y_h^ℓ for larger ℓ.

It is usual to say that W_h^ℓ and Y_h^ℓ are spaces of $\mathbb{P}_\ell$-Lagrange finite elements; in particular, (4.10) is a $\mathbb{P}_1$-Lagrange finite element approximation of (4.9).

For each $n = 0, 1, \ldots, M-1$, let us denote by θ_h^{n+1} the unique solution to (4.10). Then, by considering first the elliptic problems (4.2) and then the algebraic systems (4.10), we see that a space-time approximation of θ is given now by the continuous piecewise affine function $\theta_{\tau,h} : [0,T] \mapsto H_0^1(\Omega)$ satisfying

$$\theta_{\tau,h}|_{t=t^n} = \theta_h^n \quad \text{for } n = 0, 1, \ldots, M.$$

This double approximation process has been used for many linear and nonlinear problems, including Navier-Stokes systems.

As already said, this is not the unique way to get numerical approximations of (4.1). For instance, one can proceed in the reverse order: first, we approximate in space, then in time.

Thus, we can rewrite the original problem in the form

$$\begin{cases} \displaystyle\frac{d}{dt}\int_\Omega \theta\, v\, d\mathbf{x} + \int_\Omega \nabla\theta \cdot \nabla v\, d\mathbf{x} = \int_\Omega f(\mathbf{x},t)\, v\, d\mathbf{x} \ \text{ in } (0,T), \\ \forall v \in H_0^1(\Omega), \quad \theta \in L^2(0,T; H_0^1(\Omega)) \cap C^0([0,T]; L^2(\Omega)), \\ \theta|_{t=0} = \theta_0, \end{cases}$$

then introduce the (large) ODE system

$$\begin{cases} \displaystyle\frac{d}{dt}\int_\Omega \theta\, v\, d\mathbf{x} + \int_\Omega \nabla\theta \cdot \nabla v\, d\mathbf{x} = \int_\Omega f(\mathbf{x},t)\, v\, d\mathbf{x} \ \text{ in } (0,T), \\ \forall v \in Y_h^1, \quad \theta \in H^1(0,T; Y_h^1), \\ \theta|_{t=0} = \theta_{0h} \end{cases}$$

(where θ_{0h} is an approximation in V_h^1 of θ_0) and then, with the help of an adequate Runge-Kutta or a similar method, numerical solutions can be computed.

4.2 Time and Space Approximation of (3.3)–(3.8)

The problem to solve is (3.3)–(3.8). As a working hypothesis, we will assume that there exists one solution $\{\mathbf{u}, \rho, p\}$, with

$$\mathbf{u} \in C^0([0,T]; H), \quad \rho \in C^0([0,T]; L^2(\Omega)), \quad p \in C^0([0,T]; H^{-1}(\Omega)).$$

4.2.1 Time Discretization Along Characteristics

As before, let a (large) integer M be given and set $\tau = T/M$ and $t^n = n\tau$ for all $n = 0, 1, \dots, M$.[2]

Also as before, our goal in this step is to find "approximations" $\mathbf{u}^n$, ρ^n and p^n respectively of $\mathbf{u}(\cdot\,, t^n)$, $\rho(\cdot\,, t^n)$ and $p(\cdot\,, t^n)$ at times t^n.

We can do the following:

1. Take $\mathbf{u}^0 = \mathbf{u}_0$ and $\rho^0 = \rho_0$ (the initial data).
2. Then, for any given $n \geq 0$, $\mathbf{u}^n$ and ρ^n, find $\mathbf{u}^{n+1}$, ρ^{n+1} and p^{n+1} satisfying

$$\begin{cases} \rho^{n+1} = \rho^n \circ \mathbf{X}^n \ \text{ in } \Omega, \\ \dfrac{1}{\tau}\rho^{n+1}(\mathbf{u}^{n+1} - \mathbf{u}^n \circ \mathbf{X}^n) - \mu\Delta\mathbf{u}^{n+1} + \nabla p^{n+1} = \rho^{n+1}\mathbf{f}^{n+1} \ \text{ in } \Omega, \\ \nabla\cdot\mathbf{u}^{n+1} = 0 \ \text{ in } \Omega, \\ \mathbf{u}^{n+1} = 0 \ \text{ on } \partial\Omega. \end{cases} \tag{4.11}$$

Here, we have introduced $\mathbf{X}^n : \overline{\Omega} \mapsto \overline{\Omega}$, with

$$\mathbf{X}^n(\mathbf{x}) := \mathbf{X}^{(n)}(\mathbf{x}, t^{n+1}; t^n) \quad \forall \mathbf{x} \in \Omega,$$

where $\mathbf{X}^{(n)}$ is the inverse Lagrangian coordinate associated to $\mathbf{u}^n$.

That is, $\mathbf{X}^{(n)}(\mathbf{x}, t; s)$ is the value at s of the solution to the ODE problem

$$\begin{cases} \dfrac{d\mathbf{z}}{ds} = \mathbf{u}^n(\mathbf{z}), \\ \mathbf{z}(t) = \mathbf{x}. \end{cases}$$

Clearly, $\mathbf{X}^n(\mathbf{x})$ can be viewed as an approximation to the position at time t^n of the fluid particle that is located at $\mathbf{x}$ at time t^{n+1}.

[2] However, what follows can also be carried out at some not necessarily equispaced t^n under the conditions $t^0 = 0$ and $t^M = T$. See to this purpose Exercise 4.9.

The first equality in (4.11) is justified by the fact that the solution to (3.5), (3.8) is given by

$$\rho(\mathbf{x}, t) = \rho_0(\mathbf{X}(\mathbf{x}; t, 0)) \text{ in } Q$$

and consequently satisfies

$$\rho(\mathbf{x}, t^{n+1}) = \rho(\mathbf{X}(\mathbf{x}; t^{n+1}, t^n), t^n) \text{ in } \Omega$$

for all n. On the other hand, note that in the second equality in (4.11) the first term is the product of ρ^{n+1} times a finite difference approximation of

$$\frac{d}{ds}\mathbf{u}(\mathbf{X}(\mathbf{x}, t; s), s)\bigg|_{s=t} = \left(\frac{\partial \mathbf{u}}{\partial t} + (\mathbf{u} \cdot \nabla)\mathbf{u}\right)(\mathbf{x}, t)$$

at $t = t^{n+1}$. Accordingly, it can be viewed as a time discretization of $\rho\dfrac{\partial \mathbf{u}}{\partial t}$ at this time.

It is usual to say that (4.11) has been obtained after *time approximation along characteristics.*

Remark 4.2 As in Remark 4.1, we can use more sophisticated and accurate approximations of the time derivatives. Thus, an alternative to the previous discretization schemes is the following, where we follow a modified Gear strategy:

1. Take $\mathbf{u}^0 = \mathbf{u}_0$, $\rho^0 = \rho_0$ and $\{\mathbf{u}^1, \rho^1, p^1\}$ as in (4.11) for $n = 0$.
2. Then, given $n \geq 1$, $\mathbf{u}^n$ and ρ^n, find $\mathbf{u}^{n+1}$, ρ^{n+1} and p^{n+1} satisfying

$$\begin{cases} \rho^{n+1} = \dfrac{1}{3}\left(4\rho^n \circ \mathbf{X}^n - \rho^{n-1} \circ \mathbf{X}^{n-1}\right) \text{ in } \Omega, \\ \dfrac{1}{2\tau}\rho^{n+1}(3\mathbf{u}^{n+1} - 4\mathbf{u}^n \circ \mathbf{X}^n + \mathbf{u}^{n-1} \circ \mathbf{X}^{n-1}) - \mu\Delta\mathbf{u}^{n+1} + \nabla p^{n+1} = \rho^{n+1}\mathbf{f}^{n+1} \text{ in } \Omega, \\ \nabla \cdot \mathbf{u}^{n+1} = 0 \text{ in } \Omega, \\ \mathbf{u}^{n+1} = 0 \text{ on } \partial\Omega. \end{cases}$$

We can also go further and formulate other schemes relying on higher order time approximations. □

In practice, in order to compute $\mathbf{X}^n(\mathbf{x}) = \mathbf{X}^{(n)}(\mathbf{x}, t^{n+1}; t^n)$ at a point $\mathbf{x}$, we have to find the location at past time t^n of a particle transported by $\mathbf{u}^n$ that is now (at time t^{n+1}) at $\mathbf{x}$.

Later, it will be seen that, in practice, we have to do this at all nodes (the $\mathbf{a}^i$, the $\mathbf{b}^j$ and eventually other additional points).

Remark 4.3 Note that

$$\mathbf{X}^n(\mathbf{x}) = \mathbf{x} - \int_{t^n}^{t^{n+1}} \mathbf{u}^n(\mathbf{X}^{(n)}(\mathbf{x}, t^{n+1}; s)) \quad \forall \mathbf{x} \in \Omega.$$

Consequently, we get a simplified variant of (4.11) by replacing $\mathbf{X}^n$ by $\tilde{\mathbf{X}}^n$, with

$$\tilde{\mathbf{X}}^n(\mathbf{x}) := \mathbf{x} - \tau \mathbf{u}^n(\mathbf{x}) \quad \forall \mathbf{x} \in \overline{\Omega}.$$

Of course, a similar variant of second order can be obtained arguing as in Remark 4.2. □

After computing the $\{\mathbf{u}^n, \rho^n, p^n\}$ and arguing as in Sect. 4.2.1, we can construct "approximations in time" of the solution. Indeed, it suffices to consider the continuous piecewise affine functions $\mathbf{u}_\tau$, ρ_τ and p_τ, with

$$\{\mathbf{u}_\tau, \rho_\tau, p_\tau\}|_{t=t^n} = \{\mathbf{u}^n, \rho^n, p^n\} \text{ for } n = 0, 1, \dots, M.$$

To our knowledge, it is unknown whether uniform "energy" estimates, similar to those in the proof of Theorem 3.1, can be established for the $\mathbf{u}_\tau$. This would lead to satisfactory convergence results, as those in Exercises 4.2 and 4.7.

Note that, for the classical (constant density) Navier-Stokes PDEs, time discretization along characteristics (and even modified characteristics) lead to stable and convergent approximations, see [3, 19].

4.2.2 *Spatial Approximation by a Mixed Finite Element Method*

At this second level, what we have to solve are linear stationary quasi-Stokes systems like those in (4.11). Obviously, they are still infinite-dimensional and, to get some quantitative information, a (spatial) numerical approximation is needed.

For simplicity, we will introduce and describe a method that furnishes $\mathbb{P}_1$-Lagrange approximations of the density and mixed $(\mathbb{P}_2, \mathbb{P}_1)$-Lagrange approximations of the velocity-pressure couple.

Of course, there exist other possibilities, see Sect. 4.3.

Note that the second to fourth identities in (4.11) possess the following mixed weak reformulation:

$$\begin{cases} \displaystyle\int_\Omega \left(\rho^{n+1}\mathbf{u}^{n+1} \cdot \mathbf{v} + \tau\mu\nabla\mathbf{u} : \nabla\mathbf{v}\right) - \tau\int_\Omega p^{n+1}(\nabla\cdot\mathbf{v}) \\ \qquad = \displaystyle\int_\Omega \rho^{n+1}\left(\mathbf{u}^n \circ \mathbf{X}^n + \tau\mathbf{f}^{n+1}\right)\cdot\mathbf{v} \quad \forall \mathbf{v} \in H_0^1(\Omega)^3, \\ \displaystyle\int_\Omega q\,(\nabla\cdot\mathbf{u}^{n+1}) = 0 \quad \forall q \in L^2(\Omega), \\ \mathbf{u}^{n+1} \in H_0^1(\Omega)^3, \quad p^{n+1} \in L^2(\Omega). \end{cases} \tag{4.12}$$

Let us fix a triangulation $\mathcal{T}_h$ of $\overline{\Omega}$ and let us introduce the following associated subspace of $H_0^1(\Omega)^3$:

$$S_h := Y_h^2 \times Y_h^2 \times Y_h^2 = \{\mathbf{v} \in C^0(\overline{\Omega})^3 : \mathbf{v}|_K \in \mathbb{P}_2(K)^3 \ \ \forall K \in \mathcal{T}_h, \ \mathbf{v} = 0 \text{ on } \partial\Omega\};$$

recall (4.7) and (4.8) for the definition of Y_h^2.

As in Sect. 4.1, we will denote by $\mathbf{a}^i$ with $1 \le i \le I_h$ (resp. $\mathbf{b}^j$ with $1 \le j \le J_h$) the vertices (resp. the midpoints of the edges) in $\mathcal{T}_h$.

Clearly, it is in practice interesting to work with triangulations with many tetrahedra, that is, I_h and J_h must be large. To fix ideas, it will be assumed that $\mathbf{a}^i \in \Omega$ (resp. $\mathbf{b}^j \in \Omega$) if and only if $1 \le i \le I_{0h}$ (resp. $1 \le j \le J_{0h}$).

In this second step, we do as follows:

1. Choose finite-dimensional approximations in S_h and W_h^1 of the initial data, that is,

$$\mathbf{u}_h^0 \approx \mathbf{u}_0 \ \text{ and } \ \rho_h^0 \approx \rho_0, \ \text{ with } \ (\mathbf{u}_h^0, \rho_h^0) \in S_h \times W_h^1. \tag{4.13}$$

2. Then, given $n \ge 0$, $\mathbf{u}_h^n$ and ρ_h^n, compute $\rho_h^{n+1} \in W_h^1$ and $(\mathbf{u}^{n+1}, p^{n+1}) \in S_h \times W_h^1$ satisfying

$$\rho_h^{n+1}(\mathbf{a}^i) = \rho_h^n(\mathbf{X}_h^n(\mathbf{a}^i)) \ \ \forall i = 1, \dots, I_h, \tag{4.14}$$

$$\begin{cases} \displaystyle\int_\Omega (\rho_h^{n+1}\mathbf{u}_h^{n+1}\cdot\mathbf{v} + \tau\mu\nabla\mathbf{u}_h^{n+1} : \nabla\mathbf{v}) - \tau\int_\Omega p_h^{n+1}(\nabla\cdot\mathbf{v}) = \int_\Omega \mathbf{g}_h^n\cdot\mathbf{v} \quad \forall\mathbf{v}\in S_h, \\ \displaystyle\int_\Omega q\,(\nabla\cdot\mathbf{u}_h^{n+1}) = 0 \quad \forall q \in W_h^1. \end{cases} \tag{4.15}$$

In (4.15), we have used the notation

$$\mathbf{g}^n := \rho_h^{n+1}(\tilde{\mathbf{u}}_h^n + \tau\mathbf{f}^{n+1}), \tag{4.16}$$

where the function $\tilde{\mathbf{u}}_h^n$ belongs to S_h and satisfies

$$\tilde{\mathbf{u}}_h^n(\mathbf{a}^i) = \mathbf{u}_h^n(\mathbf{X}_h^n(\mathbf{a}^i)) \ \ \forall i \le I_{0h} \ \text{ and } \ \tilde{\mathbf{u}}_h^n(\mathbf{b}^j) = \mathbf{u}_h^n(\mathbf{X}_h^n(\mathbf{b}^j)) \ \ \forall j \le J_{0h}. \tag{4.17}$$

In (4.14) and (4.17), we have used $\mathbf{X}_h^n$, with

$$\mathbf{X}_h^n(\mathbf{x}) := \mathbf{X}_h^{(n)}(\mathbf{x}, t^{n+1}; t^n),$$

where $\mathbf{X}_h^{(n)}$ is defined as $\mathbf{X}^{(n)}$, but replacing $\mathbf{u}^n$ by $\mathbf{u}_h^n$. That is, $\mathbf{X}_h^{(n)}(\mathbf{x}, t; s)$ is the value at s of the solution to

$$\begin{cases} \dfrac{d\mathbf{z}}{ds} = \mathbf{u}_h^n(\mathbf{z}), \\ \mathbf{z}|_{s=t} = \mathbf{x}. \end{cases}$$

Recall that any function in W_h^1 is uniquely determined by its values at the $\mathbf{a}^i$ with $1 \le i \le I_h$. Also, any function in S_h is uniquely defined by its values at the $\mathbf{a}^i$ and $\mathbf{b}^j$ with $1 \le i \le I_{0h}$ and $1 \le j \le J_{0h}$. Consequently, (4.14)–(4.17) is a consistent approximation of (4.11).

As in Remark 4.2, we can use second-order approximations of the time derivatives and get a Gear-like scheme for the computation of $\{\mathbf{u}_h^n, \rho_h^n, p_h^n\}$.

Note that $\mathbf{X}_h^n(\mathbf{x})$ may be viewed as an approximation of $\mathbf{X}^{(n)}(\mathbf{x}, t^{n+1}; t^n)$; in other words, $\mathbf{X}_h^n(\mathbf{x})$ is an approximation of the position at time t^n of the particle that is located at $\mathbf{x}$ at time t^{n+1} and has been transported by $\mathbf{u}_h^n$.

On the other hand, it is again possible to simplify the method by replacing $\mathbf{X}_h^n$ by $\tilde{\mathbf{X}}_h^n$, with

$$\tilde{\mathbf{X}}_h^n(\mathbf{x}) := \mathbf{x} - \tau \mathbf{u}_h^n(\mathbf{x}) \quad \forall \mathbf{x} \in \overline{\Omega}.$$

With the help of the computed $\{\mathbf{u}_h^n, \rho_h^n, p_h^n\}$, we can construct continuous piecewise affine "space-time approximations" $\mathbf{u}_{\tau,h}$, $\rho_{\tau,h}$ and $p_{\tau,h}$ of the solution, by setting

$$\{\mathbf{u}_{\tau,h}, \rho_{\tau,h}, p_{\tau,h}\}|_{t=t^n} = \{\mathbf{u}_h^n, \rho_h^n, p_h^n\} \text{ for } n = 0, 1, \dots, M.$$

As before, uniform estimates and convergence results are unfortunately unknown at present.

4.2.3 Computational Aspects

In order to produce numerical approximations of the solution, we have used the method described by (4.11)–(4.17), in combination with the `Freefem` package.

The numerical tasks at time t^n have been:

- To compute $\mathbf{X}_h^n(\mathbf{a}^i)$ for all i and determine ρ_h^{n+1} though (4.14).
- Then, to compute $\mathbf{X}_h^n(\mathbf{b}^j)$ for all j, determine $\mathbf{g}^n$ through (4.16) and solve the Stokes-like problem (4.15).

For the first task, a search algorithm must be executed at each vertex; this can be successfully achieved by `Freefem`, see below.

For the second one, there are a lot of possibilities: direct resolution, Uzawa or Arrow-Hurwicz iterates, augmented Lagrangian techniques, etc. For simplicity, we have just used a direct resolution method together with penalization.

More precisely, we first approximate (4.15) by

$$\begin{cases} \int_\Omega (\rho_h^{n+1}\mathbf{u}^{n+1}\cdot\mathbf{v}+\tau\mu\nabla\mathbf{u}_h^{n+1}:\nabla\mathbf{v})-\tau\int_\Omega(\nabla\cdot\mathbf{v})\,p_h^{n+1}=\int_\Omega \mathbf{g}_h^n\cdot\mathbf{v} \quad \forall\mathbf{v}\in S_h, \\ \int_\Omega(\nabla\cdot\mathbf{u}_h^{n+1})\,q-\varepsilon_0\int_\Omega p_h^{n+1}q=0 \quad \forall q\in W_h^1, \end{cases} \tag{4.18}$$

where $\varepsilon_0 > 0$ is a small parameter and we then compute the corresponding solution by applying the *Unsymmetric MultiFrontal* method `UMFPACK`, see [6] for details.

A three-dimensional experiment is depicted in Figs. 4.1, 4.2, and 4.3. It concerns the fluid in a corner.

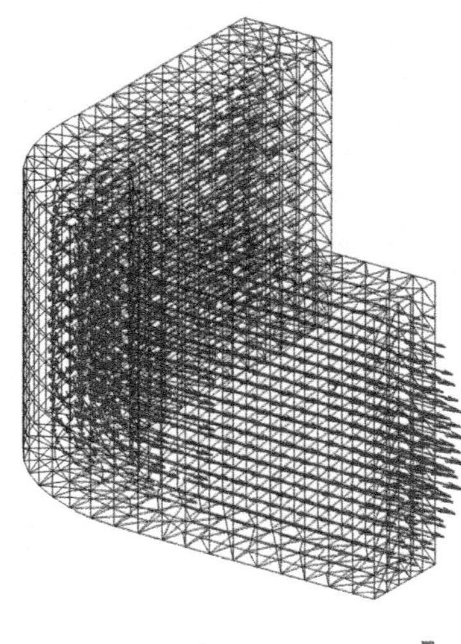

Fig. 4.1 A 3D experiment: the domain, the mesh and the final velocity field (the domain is of the form $D \times (0, 3)$; number of elements: 19,620; number of vertices: 4256). The flow is from back to front, entering the domain with a parabolic profile and leaving freely

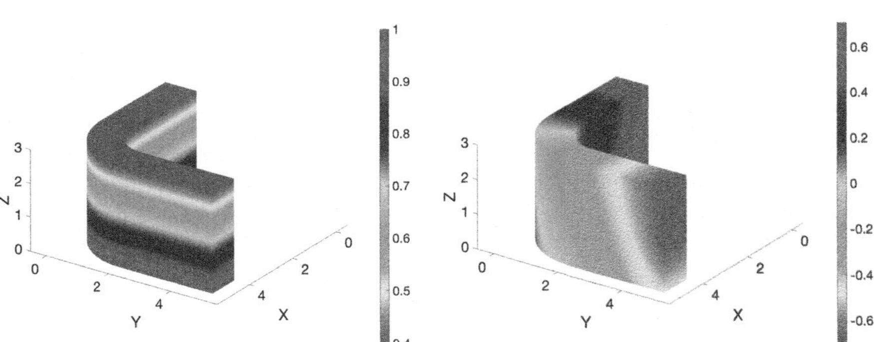

Fig. 4.2 A 3D experiment: the final density (left) and the final pressure (right)

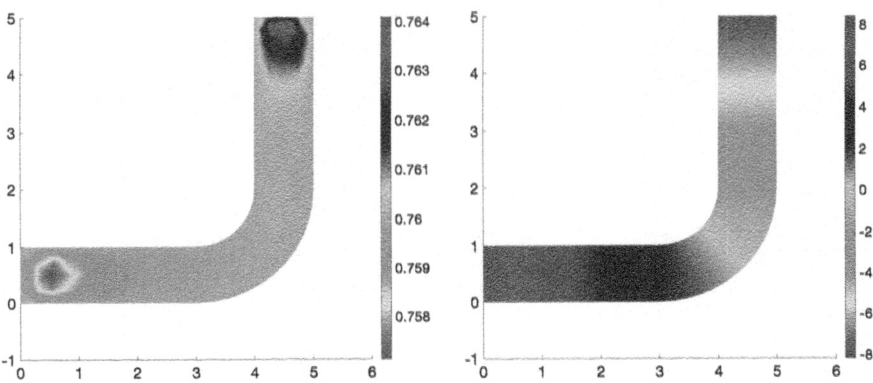

Fig. 4.3 A 3D experiment: horizontal cuts at $x_3 = 1.5$ of the final density (left) and the final pressure (right)

See other results in Figs. 1.1–1.7 in Chap. 1 and Fig. 3.1 in Chap. 3. The numerical codes used for these experiments (and also other tests) have been left accessible to the reader in https://hdvirtual.us.es/discovirt/index.php/s/o6qE66kYktdFECS.

For more details concerning (4.15) and (4.16) and some variants, see for instance [2, 8, 16].

As usual, the code appropriate for these experiments must be composed of three blocks that correspond to pre-processing, processing and post-processing. Let us give more details:

(1) The pre-processing block, where the following is done:

- We determine the domain, the time interval, the constants, the right-hand side, and boundary and initial data.
- We fix the approximation parameters: mesh generation and refinement parameters, time discretization parameters, penalization parameter, auxiliary variables, etc.
- We generate the mesh by indicating the boundary of the domain, the number of points on the boundary pieces and appropriate labels to identify the role played by each of them.
  This is relatively easy with the help of `Freefem` tools. Indeed, using the previous information, `Freefem` is able to build a family $\mathcal{T}_h$ of tetrahedra (as close to equilateral tetrahedra as possible) that serves as a (first) triangulation of Ω.
- We construct the elementary mass and rigidity matrices concerning (4.18) and determine and factorize the matrices of coefficients to be used (that depend on ρ_h^{n+1}).

Note that, if we were to solve (4.18) without taking into account that $\mathbf{u}^{n+1}$ must vanish on the boundary, we would have to consider a large algebraic linear system where the matrix coefficients are of the form

$$\int_{\Omega} \left(\rho^{n+1} v w + \tau \mu \nabla v \cdot \nabla w\right) - \tau \int_{\Omega} y \, (\nabla \cdot (w \, \mathbf{e}))$$

or

$$\int_{\Omega} (\nabla \cdot (v \, \mathbf{e})) \, z - \varepsilon_0 \int_{\Omega} yz,$$

where the v and w are arbitrary $\mathbb{P}_2$ basis functions, the y and z are arbitrary $\mathbb{P}_1$ basis functions and $\mathbf{e}$ is a vector of the canonical basis in $\mathbb{R}^3$.

Therefore, in order to prepare the solution of (4.18), it is appropriate to compute for every $K \in \mathcal{T}_h$ the quantities

$$\int_K vw, \quad \int_K \nabla v \cdot \nabla w, \quad \int_K \partial_i v \, z \;\text{ and }\; \int_K yz.$$

Remark that, for each K, only a relatively small amount of these integrals are different from zero. Furthermore, in all cases, the corresponding integral over Ω can be written as a sum with a reduced number of integrals over the $K \in \mathcal{T}_h$. After assembling these elementary integrals, the matrix coefficients are computed and ready for use.

Let us forget for the moment the boundary conditions that must be imposed to $\mathbf{u}_h^{n+1}$ and let the system (4.18) be rewritten in the form

$$Qy = b, \quad y \in \mathbb{R}^{3I_h + J_h - 1}.$$

Then, it is easy to see that Q is a sparse matrix of the form

$$Q = \begin{bmatrix} Q_{11} & Q_{12} \\ -Q_{12}^T & \varepsilon_0 \, \text{Id.} \end{bmatrix},$$

where Q_{11} is a square definite positive matrix of dimension $3I_h \times 3I_h$.

Consequently, it seems appropriate to solve (4.18) with the help of the UMFPACK method.

- We take into account the boundary conditions to be imposed.

This is done by assigning specific labels to the edges and faces on the boundary. that will serve later to identify the nodes where the values of the variables are prescribed.

In order to block or freeze the value of (say) the unknown $u_1^{n+1}(\mathbf{a}^i)$ to zero, the `Freefem` "trick" consists of identifying the line in the system matrix where this unknown is located in the main diagonal and then multiply the diagonal coefficient by a large quantity (10^{15} or larger) and replace the associated right-hand term by zero. Obviously, the same can be done for the other unknowns $u_k(\mathbf{a}^i)$ and $u_\ell(\mathbf{b}^j)$. This is a fast and accurate way to incorporate and impose Dirichlet boundary conditions.

- We initialize the working variables.

This refers to (4.13). In practice (at least for sufficiently regular $\mathbf{u}_0$ and ρ_0), what we do is to fix the values of $\mathbf{u}_h^0$ at the points $\mathbf{a}^i$ and $\mathbf{b}^j$ and the values of ρ_h^0 at the $\mathbf{a}^i$.

(2) **The processing block:**

This is the practical implementation of (4.14)–(4.17). We perform the computations in a loop in time. As already said, two tasks have to be achieved at each step.

- The first task is given by (4.14).

Note that everything is reduced to the computation of the $\mathbf{X}_h^n(\mathbf{a}^i)$ and $\mathbf{X}_h^n(\mathbf{b}^j)$, that is, the approximations of the positions at time t^n of the particles that are located at the nodes at time t^{n+1} and have been convected by $\mathbf{u}_h^n$. This amounts to solve numerically a family of backwards ODE Cauchy problems (from t^{n+1} to t^n).

The task is considerably easier if we apply the simplified or modified characteristics method described above where the $\mathbf{X}_h^n$ are replaced by the $\tilde{\mathbf{X}}_h^n$.

- The second task is to solve (4.15) or, in fact, (4.18). As already mentioned, although there are several possibilities, we have just applied a factorization method. This amounts to rewrite the matrix of coefficients as the product of two triangular matrices and then compute the solution by solving two triangular systems.
- In order to increase efficiency and make the computations more accurate, it is very convenient to adapt the mesh (at least) periodically, after a prescribed amount of iterates.

The mesh adaptation process is also performed with the help of `Freefem` tools. For details on the algorithm, see [15].

(3) **The post-processing block:**

- Here, the first goal is to store the results and, eventually, useful observations that the numerical outcomes can produce.

Thus, the computed velocity, density or pressure may work as relevant sources of information for subsequent questions and problems.

- A second post-process goal is to visualize what we have been able to do, its reliability and the most significant effects on the fluid behavior.

4.3 A Brief Summary of Other Methods

The main advantage of the method in Sects. 4.2.1 and 4.2.2 is that, if we know how to compute quickly at time t^{n+1} good approximations of the former positions of the nodes at time t^n, the reminder of the task is just to solve linear problems.

However, if the fluid flow is too complex, it may happen that the computations of the $\mathbf{X}_h^n(\mathbf{a}^i)$ be difficult and expensive and the previous strategy does not work well. Let us indicate some alternatives, where we conserve the notation introduced in Sect. 4.2.

4.3.1 Standard Approximation in Time (Euler or Gear)

It is very natural to introduce the following space-time approximation scheme:

1. Compute $\mathbf{u}_h^0$ and ρ_h^0 with

$$\mathbf{u}_h^0 \approx \mathbf{u}_0 \text{ and } \rho_h^0 \approx \rho_0, \text{ with } (\mathbf{u}_h^0, \rho_h^0) \in S_h \times W_h^1. \tag{4.19}$$

2. Then, given $n \geq 0$, $\mathbf{u}_h^n$ and ρ_h^n, compute $\rho_h^{n+1} \in R_h$ and $(\mathbf{u}^{n+1}, p^{n+1}) \in S_h \times W_h^1$ satisfying

$$\int_\Omega (\rho_h^{n+1} + \tau \mathbf{u}_h^{n+1} \cdot \nabla \rho_h^{n+1})\, \varphi = \int_\Omega \rho_h^n \varphi \quad \forall \varphi \in W_h^1, \tag{4.20}$$

$$\left\{ \begin{array}{l} \displaystyle\int_\Omega (\rho_h^{n+1}(\mathbf{u}_h^{n+1} + \tau(\mathbf{u}_h^{n+1} \cdot \nabla)\mathbf{u}_h^{n+1}) \cdot \mathbf{v} + \tau\mu \nabla \mathbf{u}_h^{n+1} : \nabla \mathbf{v}) - \tau \int_\Omega p_h^{n+1} (\nabla \cdot \mathbf{v}) \\ \qquad\qquad \displaystyle = \int_\Omega \rho_h^{n+1} (\mathbf{u}_h^n + \tau \mathbf{f}^{n+1}) \cdot \mathbf{v} \quad \forall \mathbf{v} \in S_h, \\ \displaystyle\int_\Omega q\, (\nabla \cdot \mathbf{u}_h^{n+1}) = 0 \quad \forall q \in W_h^1. \end{array} \right. \tag{4.21}$$

Obviously, a similar scheme can be deduced with second-order Gear approximations of the time derivatives.

We see now that, at each time step, we must solve a nonlinear finite-dimensional system. This can be achieved (for instance) by introducing Newton or quasi-Newton iterates. At the end, the systems to solve have the form

$$\int_\Omega (\rho_h + \tau(\mathbf{u}_h^n \cdot \nabla \rho_h + \mathbf{u}_h \cdot \nabla \rho_h^n))\, \varphi = \int_\Omega g\, \varphi \quad \forall \varphi \in W_h^1, \tag{4.22}$$

$$\begin{cases} \displaystyle\int_\Omega (\rho_h^n(\mathbf{u}_h+\tau(\mathbf{u}_h\cdot\nabla)\mathbf{u}_h^n+(\mathbf{u}_h^n\cdot\nabla)\mathbf{u}_h-\rho_h\mathbf{k})\cdot\mathbf{v}+\tau\mu\nabla\mathbf{u}_h:\nabla\mathbf{v}) \\ \qquad\qquad \displaystyle -\tau\int_\Omega p_h(\nabla\cdot\mathbf{v}) = \int_\Omega \mathbf{q}\cdot\mathbf{v} \quad \forall\mathbf{v}\in S_h, \\ \displaystyle\int_\Omega q\,(\nabla\cdot\mathbf{u}_h) = 0 \quad \forall q\in W_h^1 \end{cases} \tag{4.23}$$

for some functions g, $\mathbf{k}$ and $\mathbf{q}$, where the unknowns are ρ_h, $\mathbf{u}_h$ and p_h.

Of course, suitable simplified versions of (4.20) and (4.21) make sense and can be enough to furnish satisfactory results. For example, we can change $\mathbf{u}_h^{n+1}$ by $\mathbf{u}_h^n$ in (4.20) and then $(\mathbf{u}_h^{n+1}\cdot\nabla)\mathbf{u}_h^{n+1}$ by $(\mathbf{u}_h^n\cdot\nabla)\mathbf{u}_h^{n+1}$ in (4.21). This leads to the linear system

$$\int_\Omega (\rho_h^{n+1}+\tau\mathbf{u}_h^n\cdot\nabla\rho_h^{n+1})\,\varphi = \int_\Omega \rho_h^n\,\varphi \quad \forall\varphi\in W_h^1, \tag{4.24}$$

$$\begin{cases} \displaystyle\int_\Omega (\rho_h^{n+1}\tau(\mathbf{u}_h^{n+1}+(\mathbf{u}_h^n\cdot\nabla)\mathbf{u}_h^{n+1})\cdot\mathbf{v}+\tau\mu\nabla\mathbf{u}_h^{n+1}:\nabla\mathbf{v})-\tau\int_\Omega p_h^{n+1}(\nabla\cdot\mathbf{v}) \\ \qquad\qquad \displaystyle = \int_\Omega \rho_h^{n+1}(\mathbf{u}_h^n+\tau\mathbf{f}^{n+1})\cdot\mathbf{v} \quad \forall\mathbf{v}\in S_h, \\ \displaystyle\int_\Omega q\,(\nabla\cdot\mathbf{u}_h^{n+1}) = 0 \quad \forall q\in W_h^1, \end{cases} \tag{4.25}$$

that can be solved in cascade in a relatively easy way.

A variant of (4.24) and (4.25) introduced by Guermond and Salgado in [12] is as follows:

$$\int_\Omega (\rho_h^{n+1}+\tau\mathbf{u}_h^n\cdot\nabla\rho_h^{n+1}+\frac{\tau}{2}(\nabla\cdot\mathbf{u}_h^n)\rho_h^{n+1})\,\varphi = \int_\Omega \rho_h^n\,\varphi \quad \forall\varphi\in W_h^1, \tag{4.26}$$

$$\begin{aligned} &\int_\Omega (\rho_h^n(\mathbf{u}_h^{n+1}-\mathbf{u}_h^n)+\tau\rho_h^{n!+!1}(\mathbf{u}_h^n\cdot\nabla)\mathbf{u}_h^{n+1})+\frac{\tau}{4}\rho_h^{n!+!1}(\nabla!\cdot!\mathbf{u}_h^n)\mathbf{u}_h^{n+1})\cdot\mathbf{v} \\ &\quad +\tau\mu\int_\Omega \nabla\mathbf{u}_h^{n+1}:\nabla\mathbf{v}-\tau\int_\Omega p_h^n(\nabla\cdot\mathbf{v})=\int_\Omega \rho_h^{n+1}(\mathbf{u}_h^n+\tau\mathbf{f}^{n+1})\cdot\mathbf{v} \ \ \forall\mathbf{v}\in S_h, \end{aligned} \tag{4.27}$$

$$\int_\Omega \nabla p_h^{n+1}\cdot\nabla\varphi = -\frac{\chi}{\tau}\int_\Omega (\nabla\cdot\mathbf{u}_h^n)\varphi \quad \forall\varphi\in W_h^1, \tag{4.28}$$

where $\chi > 0$.

It is seen that the computations of$\mathbf{u}_h^{n+1}$ and p_h^{n+1} are decoupled and, accordingly, no Stokes-like problem has to be solved.

Thanks to the presence of the terms concerning $\nabla\cdot\mathbf{u}_h^n$ in the left-hand sides of (4.26) and (4.27), the authors were able to prove conditional stability and related convergence results.

Other similar schemes have also been given there; see also [13].

4.3.2 Splitting Methods

Another variant of (4.20) and (4.21) relies on "splitting". The idea is inspired by the previous work of several people for the classical Navier-Stokes and is motivated by the aim of separating (that is, make independent) the numerical treatments of nonlinearity and incompressibility.

In the simplest case, for any given $n \geq 0$, we do as follows:

$$\int_\Omega (\rho_h^{n+1/2} + \frac{\tau}{2}\mathbf{u}_h^n \cdot \nabla \rho_h^{n+1/2})\,\varphi = \int_\Omega \rho_h^n\,\varphi \quad \forall \varphi \in W_h^1, \tag{4.29}$$

$$\left\{\begin{array}{l} \displaystyle\int_\Omega (\rho_h^{n+1/2}(\mathbf{u}_h^{n+1/2} + (\mathbf{u}_h^{n+1/2}\cdot\nabla)\mathbf{u}_h^{n+1/2})\cdot\mathbf{v} + \frac{\tau\mu}{2}\nabla\mathbf{u}_h^{n+1/2} : \nabla\mathbf{v}) \\ \displaystyle\quad = -\frac{\tau}{2}\int_\Omega p_h^{n+1/2}(\nabla\cdot\mathbf{v}) + \int_\Omega \rho_h^{n+1/2}(\mathbf{u}_h^n + \frac{\tau}{2}\mathbf{f}^{n+1/2})\cdot\mathbf{v} \quad \forall \mathbf{v} \in S_h, \end{array}\right. \tag{4.30}$$

$$\int_\Omega (\rho_h^{n+1} + \frac{\tau}{2}\mathbf{u}_h^{n+1/2} \cdot \nabla \rho_h^{n+1})\,\varphi = \int_\Omega \rho_h^{n+1/2}\,\varphi \quad \forall \varphi \in W_h^1, \tag{4.31}$$

$$\left\{\begin{array}{l} \displaystyle\int_\Omega (\rho_h^{n+1}\mathbf{u}_h^{n+1}\cdot\mathbf{v} + \frac{\tau\mu}{2}\nabla\mathbf{u}_h^{n+1} : \nabla\mathbf{v}) - \frac{\tau}{2}\int_\Omega p_h^{n+1}(\nabla\cdot\mathbf{v}) \\ \displaystyle\quad = \int_\Omega \rho_h^{n+1}(\mathbf{u}_h^n + \frac{\tau}{2}\mathbf{f}^{n+1} - (\mathbf{u}_h^{n+1/2}\cdot\nabla)\mathbf{u}_h^{n+1/2})\cdot\mathbf{v} \quad \forall \mathbf{v} \in S_h, \\ \displaystyle\int_\Omega q\,(\nabla\cdot\mathbf{u}_h^{n+1}) = 0 \quad \forall q \in W_h^1. \end{array}\right. \tag{4.32}$$

Thus, the task amounts to solve two linear scalar transport problems, a nonlinear elliptic system (where the incompressibility condition has been forgotten) and a linear stationary quasi-Stokes problem.

For the resolution of (4.30), it is a good idea to reformulate the system as a least squares extremal problem and then use (for example) standard conjugate gradient techniques. On the other hand, (4.32) can be solved arguing as we did for (4.15), i.e., by penalization and algebraic factorization.

Note that, in practice, we go from $t = t^n$ to $t = t^{n+1}$ by first computing a "compressible" approximation of $\mathbf{u}(\cdot\,, t^{n+1/2})$ and then an incompressible $\mathbf{u}^{n+1}$ that is assumed to be close to $\mathbf{u}(\cdot\,, t^{n+1})$. This is related to some well known Peaceman-Rachford alternate direction methods, see to this respect [9].

A more elaborate but similar scheme of the Strang kind consists of solving first a transport-Stokes system (to get an approximation at time $t^{n+1/4}$), then a nonlinear system in $(t^{n+1/4}, t^{n+3/4})$ and then again a linear transport-Stokes system that furnishes an approximation at $t = t^{n+1}$. A well known generalization is the so called θ-scheme. It consists in performing similar computations first in $(t^n, t^{n+\theta})$, then in $(t^{n+\theta}, t^{n+1-\theta})$ and finally in $(t^{n+1-\theta}, t^{n+1})$, where $0 < \theta < 1/2$; see [10] and the references therein for more details.

We will end the presentation of splitting methods with a variant of (4.20) and (4.21) where the two main tasks, to solve a nonlinear problem and to take into account incompressibility and pressure effects, can be carried out in parallel. This parallelization in time process can be very convenient for computational purposes, in particular if it is followed by appropriate parallelization in space techniques; see [1] to this respect.

In the simplest parallelized in time version of (4.20) and (4.21), at the n-th time step we compute ρ_h^{n+1} by solving (4.24), then we compute simultaneously $\mathbf{u}_h^{n+a}$ and $(\mathbf{u}^{n+b}, p^{n+1})$, with

$$\left\{ \begin{aligned} &\int_\Omega (\rho_h^{n+1}(\mathbf{u}_h^{n+a} + (\mathbf{u}_h^{n+a}\cdot\nabla)\mathbf{u}_h^{n+a})\cdot\mathbf{v} + \tau\mu\nabla\mathbf{u}_h^{n+a} : \nabla\mathbf{v}) \\ &\quad = -\tau\int_\Omega p_h^n(\nabla\cdot\mathbf{v}) + \int_\Omega \rho_h^{n+1}(\mathbf{u}_h^n + \tau\mathbf{f}^{n+1})\cdot\mathbf{v} \quad \forall\mathbf{v}\in S_h \end{aligned} \right. \tag{4.33}$$

and

$$\left\{ \begin{aligned} &\int_\Omega (\rho_h^{n+1}\mathbf{u}_h^{n+b}\cdot\mathbf{v} + \tau\mu\nabla\mathbf{u}_h^{n+b} : \nabla\mathbf{v}) - \tau\int_\Omega p_h^{n+1}(\nabla\cdot\mathbf{v}) \\ &\quad = \int_\Omega \rho_h^{n+1}(\mathbf{u}_h^n + \tau\mathbf{f}^{n+1})\cdot\mathbf{v} \quad \forall\mathbf{v}\in S_h, \\ &\int_\Omega q\,(\nabla\cdot\mathbf{u}_h^{n+b}) = 0 \quad \forall q\in W_h^1. \end{aligned} \right. \tag{4.34}$$

and we finally set

$$\mathbf{u}_h^{n+1} = \frac{1}{2}(\mathbf{u}^{n+a} + \mathbf{u}^{n+b}). \tag{4.35}$$

Unfortunately, not much is known on the stability and convergence of these methods; for more details and other schemes, see for instance [7, 11].

4.4 Exercises of Chapter 4

Exercise 4.1* Prove that, for every $f \in L^2(Q)$ and every $\theta_0 \in L^2(\Omega)$ there exists exactly one solution to (4.1), with

$$\theta \in L^2(0,T;H_0^1(\Omega)) \cap C^0([0,T];L^2(\Omega)) \text{ and } \partial_t\theta \in L^2(0,T;H^{-1}(\Omega)).$$

Prove also that, if $\theta_0 \in H_0^1(\Omega)$, then

$$\theta \in L^2(0,T;H^2(\Omega)) \cap C^0([0,T];H_0^1(\Omega)) \text{ and } \partial_t\theta \in L^2(Q).$$

HINT: Argue as in the proof of Theorem 3.1.

Exercise 4.2* Prove that the stationary problems (4.2) are uniquely solvable. Also, prove that the associated θ_τ converge to the unique solution to (4.1) in the following sense:

$$\begin{aligned}
&\theta_\tau \to \theta \text{ weakly in } L^2(0,T;\,H_0^1(\Omega)) \text{ and weakly-}* \text{ in } L^\infty(0,T;\,L^2(\Omega)),\\
&\theta_\tau \to \theta \text{ strongly in } L^2(Q) \text{ and a.e.},\\
&\partial_t\theta_\tau \to \partial_t\theta \text{ weakly in } L^2(0,T;\,H^{-1}(\Omega)).
\end{aligned}$$

HINT: First, prove that the θ_τ and the $\partial_t\theta_\tau$ are bounded in the spaces indicated above. Then, deduce that the (weak) limit of any subsequence is necessarily equal to θ.

Exercise 4.3* Same statement using (4.4) instead of (4.2).

Exercise 4.4* Prove that $\dim \mathbb{P}_\ell(K) = \binom{N+\ell}{\ell}$ for all N and any ℓ.

Exercise 4.5* Prove that the W_h^ℓ are finite-dimensional subspaces of $H^1(\Omega)$. Which is the dimension of W_h^3? Which are the dimensions of the spaces Y_h^ℓ for $\ell = 1, 2, 3$?

Exercise 4.6** Fix n and τ, let us denote by θ the solution to (4.9) and set $G = \theta^n + \tau f^{n+1}$. Consider the family $\{\theta_h\}_{h\in(0,1]}$, with θ_h solving for each h the finite dimensional problem

$$\begin{cases}
\displaystyle\int_\Omega (\theta_h v + \tau\nabla\theta_h\cdot\nabla v)\,d\mathbf{x} = \int_\Omega G v\,d\mathbf{x}\\
\forall v \in Y_h^1,\quad \theta_h \in Y_h^1,
\end{cases}$$

where the Y_h^1 are given by (4.7) and (4.8) for $\ell = 1$ and the corresponding $\{\mathcal{T}_h\}$ is a regular family. Prove the following:

1. $\theta_h \to \theta$ strongly in $H_0^1(\Omega)$ as $h \to 0$ (use that dist.$(\theta, Y_h^1) \to 0$ as $h \to 0$).
2. If $\theta \in H^2(\Omega) \cap H_0^1(\Omega)$, there exists $C > 0$ (independent of h) such that

$$\|\theta - \theta_h\|_{H^1} \le Ch\|\theta\|_{H^2} \quad \forall h \in (0,1].$$

Exercise 4.7* Let the $\theta_{\tau,h}$ be given as in Sect. 4.1. Assume that they are associated to a regular family of triangulations $\{\mathcal{T}_h\}$. Prove that the $\theta_{\tau,h}$ converge as $\tau \to 0$ and $h \to 0$ to the unique solution to (4.1) in the sense indicated in Exercise 4.2.

Exercise 4.8* Same statement using the second-order Gear scheme to approximate in time.

Exercise 4.9* Describe the analog of the time discretization scheme along characteristics in Sect. 4.2.1 with not necessarily equispaced t^n.

Exercise 4.10* Prove the existence and uniqueness of a solution to (4.11). Prove the same for the discretized in time system in Remark 4.2.

Exercise 4.11* Prove that the scheme (4.13)–(4.17) is well defined, i.e., prove the existence and uniqueness of the triplets $\{\rho_h^{n+1}, \mathbf{u}_h^{n+1}, \nabla p_h^{n+1}\}$ for all n and h.

Exercise 4.12** Fix n and τ. Prove results similar to those in Exercise 4.6 for the family $\{\mathbf{u}_h, p_h\}_{h\in(0,1]}$ of solutions to problems of the kind (4.15).

Exercise 4.13* Prove the existence and uniqueness of a solution to (4.18).

Exercise 4.14* Rewrite (4.18) as an algebraic system using the canonical basis in S_h and W_h^1 and applying the `Freefem` "trick".

Exercise 4.15* Prove an existence result for the system (4.20) and (4.21).

Exercise 4.16** Indicate how can be solved (4.20) and (4.21) by applying Newton's method. Prove a local convergence result. Is it also possible to prove convergence for appropriate quasi-Newton variants?

Exercise 4.17* Prove an existence result for the system (4.23). Under which conditions the solution is unique?

Exercise 4.18* Same statement for (4.24).

Exercise 4.19* Prove existence results for the systems in (4.29)–(4.32).

Exercise 4.20* Reformulate (4.30) as a least squares extremal problem and indicate how this problem can be solved by applying preconditioned conjugate gradient techniques.

References

1. Albarreal, I., et al.: Time and space parallelization of the Navier-Stokes equations. Comput. Appl. Math. **24**(3), 417–438 (2005)
2. An, R.: Error analysis of a new fractional-step method for the incompressible Navier-Stokes equations with variable density. J. Sci. Comput. **84**(1), Paper No. 3, 21 (2020)
3. Boukir, K., Maday, Y., Métivet, B.: A high order characteristics method for the incompressible Navier-Stokes equations. ICOSAHOM'92 (Montpellier, 1992). Comput. Methods Appl. Mech. Eng. **116**(1–4), 211–218 (1994)
4. Boyer, F., Fabrie, P.: Mathematical Tools for the Study of the Incompressible Navier-Stokes Equations and Related Models, volume 183 of Applied Mathematical Sciences. Springer, New York (2013)
5. Ciarlet, P.G.: The Finite Element Method for Elliptic Problems. Reprint of the 1978 original [North-Holland, Amsterdam], Classics in Applied Mathematics, vol. 40. Society for Industrial and Applied Mathematics (SIAM), Philadelphia, PA (2002)
6. Davis, T.: UMFPACK version 4.3 user guide. https://www.researchgate.net/, Texas A & M University (2005, Nov.)

7. DiPierro, B., Abid, M.: A projection method for the spectral solution of non-homogeneous and incompressible Navier-Stokes equations. Int. J. Numer. Methods Fluids **71**(8), 1029–1054 (2013)
8. Dong, H., Zhang, Y., Wang, K.: Mass, momentum and energy identical-relation-preserving scheme for the Navier-Stokes equations with variable density. Comput. Math. Appl. **137**, 73–92 (2023)
9. Glowinski, R.: Finite Element Methods for Incompressible Viscous Flow. Handbook of Numerical Analysis, vol. IX, pp. 3–1176. North-Holland, Amsterdam (2003)
10. Glowinski, R., Pan, T.-W.: Numerical Simulation of Incompressible Viscous Flow? Methods and Applications. De Gruyter Series in Applied and Numerical Mathematics, vol. 7. De Gruyter, Berlin (2022)
11. Guermond, J.-L., Minev, P.D.: Efficient parallel algorithms for unsteady incompressible flows. In: Numerical Solution of Partial Differential Equations: Theory, Algorithms, and their Applications, Springer Proc. Math. Stat., vol. 45, pp. 185–201. Springer, New York (2013)
12. Guermond, J.-L., Salgado, A.: A splitting method for incompressible flows with variable density based on a pressure Poisson equation. J. Comput. Phys. **228**, 2834–2846 (2009)
13. Guermond, J.-L., Salgado, A.: Error analysis of a fractional time-stepping technique for incompressible flows with variable density. SIAM J. Numer. Anal. **49**(3), 917–944 (2011)
14. He, X., Axelsson, O., Neytcheva, M.: Numerical solution of the time-dependent Navier-Stokes equation for variable density–variable viscosity. Part I. Math. Model. Anal. **20**(2), 232–260 (2015)
15. Hecht, F.: New developments in freefem++. J. Numer. Math. **20**(3-4), 251–265 (2012)
16. Herbin, R., Latché, J.C., Batteux, L., Gallouët, T., Poullet, P.: Convergence of the MAC scheme for the incompressible Navier-Stokes equations with variable density and viscosity. Math. Comput. **92**(342), 1595–1631 (2023)
17. Liu, C., Walkington, N.J.: Convergence of numerical approximations of the incompressible Navier-Stokes equations with variable density and viscosity. SIAM J. Numer. Anal. **45**(3), 1287–1304 (2007)
18. Quarteroni, A.: Numerical Models for Differential Problems. MS & A. Modeling, Simulation and Applications, 3rd edn., vol. 16. Springer, Cham (2017)
19. Si, Z., Wang, J., Sun, W.: Unconditional stability and error estimates of modified characteristics FEMs for the Navier-Stokes equations. Numer. Math. **134**(1), 139–161 (2016)

Chapter 5
Strong Solutions, Regularity and Uniqueness

In this chapter, we will consider local and global in time strong solutions to the variable density Navier-Stokes equations. Roughly speaking, we mean by this triplets $\{\mathbf{u}, \rho, p\}$ that satisfy the PDEs *almost everywhere* and not only in the weak (distributional) sense. Accordingly, the results that follow may be viewed as regularity properties of the solution(s) furnished by Theorem 3.1.

As in the previous chapter, we will use the notations $Q := \Omega \times (0, T)$ and $\Sigma := \partial\Omega \times (0, T)$, where $T > 0$. Unless otherwise specified, we will assume that $\Omega \subset \mathbb{R}^N$ is a non-empty bounded connected open set whose boundary is (to fix ideas) of class C^2, with $N = 2$ or $N = 3$.

Recall that this means that there exist non-empty open sets $\Omega_0, \Omega_1, \ldots, \Omega_I$, points $\mathbf{x}^1, \ldots, \mathbf{x}^I$ in $\partial\Omega$ and functions $\varphi_0, \varphi_1, \ldots, \varphi_I$ such that

$$\begin{cases} \Omega \subset \bigcup_{i=0}^{I} \Omega_i, \quad \overline{\Omega_0} \subset \Omega, \quad \partial\Omega \subset \bigcup_{i=1}^{I} \Omega_i, \\ \mathbf{x}^i \in \Omega_i, \quad \varphi_i \in \mathcal{D}(\mathbb{R}^N) \text{ and } \operatorname{Supp} \varphi_i \subset \Omega_i \ \ \forall i = 0, 1, \ldots, I, \end{cases}$$

the $\mathbf{x} \in \Omega_i$ can be parametrized by C^2 local coordinates $y_j = Y_j(x_1, \ldots, x_N)$, with

$$\mathbf{x} \in \Omega_i \cap \Omega \text{ (resp . } \mathbf{x} \in \Omega_i \setminus \overline{\Omega}) \ \Leftrightarrow \ y_N > 0 \text{ (resp . } y_N < 0)$$

and $\sum_{i=0}^{I} \varphi_i = 1$ in Ω; more details on the properties of C^2 domains can be found, for instance, in [1, 63].

P. Braz e Silva et al., *Analysis and Control of the Variable Density Incompressible Navier-Stokes Equations*, MS&A 22, https://doi.org/10.1007/978-3-032-14510-9_5

5.1 The Existence of a Strong Solution

Recall that the system under study is

$$\begin{cases} \dfrac{\partial \rho \mathbf{u}}{\partial t} + \nabla \cdot (\rho \mathbf{u} \otimes \mathbf{u}) - \mu \Delta \mathbf{u} + \nabla p = \rho \mathbf{f}, & (\mathbf{x}, t) \in Q, \\ \nabla \cdot \mathbf{u} = 0, \quad (\mathbf{x}, t) \in Q, \\ \dfrac{\partial \rho}{\partial t} + \nabla \cdot (\rho \mathbf{u}) = 0, \quad (\mathbf{x}, t) \in Q, \\ \mathbf{u} = 0, \quad (\mathbf{x}, t) \in \Sigma, \\ \rho|_{t=0} = \rho_0, \quad (\rho \mathbf{u})|_{t=0} = \rho_0 \mathbf{u}_0, \quad \mathbf{x} \in \Omega. \end{cases} \tag{5.1}$$

Also, recall the definitions of the spaces H, V, $D(A)$ and the Stokes operator $A : D(A) \mapsto H$ in Chap. 2: see (2.47), (2.48) and Definition 2.10; in particular, one has

$$H = \{\mathbf{u} \in L^2(\Omega)^N : \nabla \cdot \mathbf{u} = 0, \quad \mathbf{u} \cdot \mathbf{n} = 0 \text{ on } \partial\Omega\}$$

and

$$V = \{\mathbf{u} \in H_0^1(\Omega)^N : \nabla \cdot \mathbf{u} = 0\}.$$

In the sequel, it will be assumed (at least) that

$$\mathbf{f} \in L^2(Q)^N \tag{5.2}$$

and the initial data ρ_0 and $\mathbf{u}_0$ satisfy

$$\mathbf{u}_0 \in V, \ \ \rho_0 \in C^0(\overline{\Omega}) \ \text{ and } \ 0 < \alpha := \min_{\Omega} \leq \max_{\Omega} := \beta \tag{5.3}$$

for some $\alpha, \beta > 0$.

Let us define with precision the kind of solution we are interested in:

Definition 5.1 It will be said that $\{\mathbf{u}, \rho, p\}$ is a *strong solution* to (5.1) in the cylinder $Q = \Omega \times (0, T)$ if

$$\begin{cases} \rho \in C^0(\overline{Q}), \\ \mathbf{u} \in L^2(0, T; D(A)) \cap C^0([0, T]; V), \ \ \frac{\partial \mathbf{u}}{\partial t} \in L^2(0, T; H), \\ p \in L^2(0, T; H^1(\Omega)), \end{cases}$$

and one has

$$\frac{\partial \rho}{\partial t} + \mathbf{u} \cdot \nabla \rho = 0 \text{ in } \mathcal{D}'(Q), \tag{5.4}$$

$$\rho(\mathbf{x}, 0) = \rho_0(\mathbf{x}), \quad \mathbf{x} \in \Omega, \tag{5.5}$$

$$\rho \left(\frac{\partial \mathbf{u}}{\partial t} + (\mathbf{u} \cdot \nabla)\mathbf{u} \right) - \mu \Delta \mathbf{u} + \nabla p = \rho \mathbf{f} \quad \text{a.e. in } Q, \tag{5.6}$$

$$\nabla \cdot \mathbf{u} = 0 \quad \text{a.e. in } Q, \tag{5.7}$$

$$\mathbf{u}\big|_{t=0} = \mathbf{u}_0 \text{ in } V. \tag{5.8}$$

In the following results, we will frequently deal with strong solutions in sub-cylinders of the form $\Omega \times (0, \tilde{T})$ for various $\tilde{T} \in (0, T]$.

The first result concerning strong solutions is the following:

Theorem 5.1 *Let us assume that* (5.2) *and (5.3) hold. Then, there exist $\tilde{T}$ with* $0 < \tilde{T} \leq T$ *and a strong solution to* (5.1) *in* $\Omega \times (0, \tilde{T})$. *Furthermore, if $N = 2$, one has this with $\tilde{T} = T$.*

Proof In this proof and also in the sequel, we will frequently denote the time derivatives of $\mathbf{u}$, ρ, etc. in the form $\mathbf{u}_t$, ρ_t, etc.

Let λ_k and $\mathbf{w}^k$ be the eigenvalues and eigenfunctions of the Stokes operator.

Recall the notation $V_k := [\mathbf{w}^1, \ldots, \mathbf{w}^k]$ and consider the semi-Galerkin approximated problems

$$\begin{cases} \mathbf{u}^k : [0, T] \mapsto V_k, \quad \rho^k : [0, T] \mapsto L^\infty(\Omega), \\ (\rho^k \mathbf{u}_t^k, \mathbf{v}) + (\rho^k (\mathbf{u}^k \cdot \nabla)\mathbf{u}^k, \mathbf{v}) + \mu(A\mathbf{u}^k, \mathbf{v}) = (\rho^k \mathbf{f}, \mathbf{v}) \text{ in } (0, T) \ \forall \mathbf{v} \in V_k, \\ \rho_t^k + \mathbf{u}^k \cdot \nabla \rho^k = 0, \quad (\mathbf{x}, t) \in Q, \\ \mathbf{u}\big|_{t=0} = \mathbf{u}_0^k, \quad \rho^k|_{t=0} = \rho_0, \end{cases} \tag{5.9}$$

where $k = 1, 2, \ldots$ and the $\mathbf{u}_0^k = P_k \mathbf{u}_0$ are the orthogonal projections of $\mathbf{u}_0$ on the finite-dimensional spaces V_k.

In the sequel, we will collect several known and several new estimates for the approximations $\mathbf{u}^k$.

First of all, recall that whenever ρ^k exists one has

$$\alpha = \inf_{\Omega} \rho_0 \leq \rho^k(\mathbf{x}, t) \leq \sup_{\Omega} \rho_0 = \beta, \tag{5.10}$$

as a consequence of the method of characteristics applied to the transport equation in (5.9).

As in Chap. 3, we can take $\mathbf{v} = \mathbf{u}^k$ in (5.9) to get

$$\frac{1}{2}\frac{d}{dt}\|(\rho^k)^{1/2}\mathbf{u}^k\|^2 + \mu\|\nabla\mathbf{u}^k\|^2 = (\rho^k\mathbf{f}, \mathbf{u}^k).$$

Integrating this identity with respect to time in $[0, t]$ and using (5.10), we easily see that

$$\|\mathbf{u}^k(t)\|^2 + \int_0^t \|\nabla\mathbf{u}^k(s)\|^2\, ds \le C + C\int_0^t \|\mathbf{f}(\cdot\,, s)\|^2\, ds,$$

where (as usual) C is a positive constant that depends on the data of the problem but does not depend on k.

Then, Gronwall's Lemma gives

$$\|\mathbf{u}^k(t)\|^2 + \int_0^t \|\nabla\mathbf{u}^k(s)\|^2\, ds \le C \tag{5.11}$$

and this shows that $\mathbf{u}^k$ is uniformly bounded in $L^\infty(0, T; H) \cap L^2(0, T; V)$.

Now, let us present some uniform estimates of $\mathbf{u}^k_t$ and $A\mathbf{u}^k$.

Setting $\mathbf{v} = \mathbf{u}^k_t(\cdot\,, t)$ in (5.9) and using (5.11), we obtain the following:

$$\begin{aligned}\|(\rho^k)^{1/2}\mathbf{u}^k_t\|^2 + \frac{\mu}{2}\frac{d}{dt}\|\nabla\mathbf{u}^k\|^2 &= (\rho^k\mathbf{f}, \mathbf{u}^k_t) - (\rho^k(\mathbf{u}^k\cdot\nabla)\mathbf{u}^k, \mathbf{u}^k_t)\\ &\le \frac{1}{2}\|(\rho^k)^{1/2}\mathbf{u}^k_t\|^2 + \frac{1}{2}\|(\rho^k)^{1/2}(\mathbf{f} - (\mathbf{u}^k\cdot\nabla)\mathbf{u}^k)\|^2.\end{aligned} \tag{5.12}$$

Since $0 < \alpha \le \rho_0 \le \beta$, one has

$$\alpha\|\mathbf{u}^k_t\|^2 + \mu\frac{d}{dt}\|\nabla\mathbf{u}^k\|^2 \le \beta\|(\mathbf{f} - (\mathbf{u}^k\cdot\nabla)\mathbf{u}^k)\|^2. \tag{5.13}$$

On the other hand, setting $\mathbf{v} = \eta A\mathbf{u}^k(\cdot\,, t)$ in (5.9) (with $\eta > 0$ to be chosen later on), one has

$$\begin{aligned}\eta\mu\|A\mathbf{u}^k\|^2 &= \eta(\rho^k(\mathbf{f} - \mathbf{u}^k_t - (\mathbf{u}^k\cdot\nabla)\mathbf{u}^k), A\mathbf{u}^k)\\ &\le \frac{\eta\mu}{2}\|A\mathbf{u}^k\|^2 + \frac{\eta}{2\mu}\left(\beta\|(\rho^k)^{1/2}\mathbf{u}^k_t\|^2 + \beta^2\|\mathbf{f} - (\mathbf{u}^k\cdot\nabla)\mathbf{u}^k\|^2\right),\end{aligned} \tag{5.14}$$

whence

$$\eta\mu\|A\mathbf{u}^k\|^2 \le \frac{\eta}{\mu}\left(\beta\|(\rho^k)^{1/2}\mathbf{u}^k_t\|^2 + \beta^2\|\mathbf{f} - (\mathbf{u}^k\cdot\nabla)\mathbf{u}^k\|^2\right). \tag{5.15}$$

From (5.13) and (5.15), we easily get

$$(\alpha - \frac{\eta\beta}{\mu})\|(\rho^k)^{1/2}\mathbf{u}_t^k\|^2 + \mu\frac{d}{dt}\|\nabla\mathbf{u}^k\|^2 + \eta\mu\|A\mathbf{u}^k\|^2 \le (\beta + \frac{\eta\beta^2}{\mu})\|\mathbf{f} - (\mathbf{u}^k\cdot\nabla)\mathbf{u}^k\|^2$$

and, choosing $\eta = \dfrac{\alpha\mu}{2\beta}$, the following is obtained:

$$\begin{aligned}\frac{\alpha}{2}\|(\rho^k)^{1/2}\mathbf{u}_t^k\|^2 + \mu\frac{d}{dt}\|\nabla\mathbf{u}^k\|^2 + \frac{\alpha\mu^2}{2\beta}\|A\mathbf{u}^k\|^2 &\le \beta\left(1+\frac{\alpha}{2}\right)\|\mathbf{f} - (\mathbf{u}^k\cdot\nabla)\mathbf{u}^k\|^2 \\ &\le \beta\left(1+\frac{\alpha}{2}\right)\left(\|\mathbf{f}\|^2 + \|(\mathbf{u}^k\cdot\nabla)\mathbf{u}^k\|^2\right).\end{aligned}$$

Let us estimate the term $\|(\mathbf{u}^k\cdot\nabla)\mathbf{u}^k\|$.

For $N = 2$, using the interpolation results and Sobolev inequalities in Chap. 2 (see Lemma 2.5), the properties of the Stokes operator (see Theorem 2.11) and the energy estimates (5.11), we find:

$$\begin{aligned}\|(\mathbf{u}^k\cdot\nabla)\mathbf{u}^k\|^2 &\le \|\mathbf{u}^k\|_{L^4}^2\|\nabla\mathbf{u}^k\|_{L^4}^2 \\ &\le C\|\mathbf{u}^k\|\,\|\nabla\mathbf{u}^k\|^2\|A\mathbf{u}^k\| \\ &\le C_\delta\|\mathbf{u}^k\|^2\|\nabla\mathbf{u}^k\|^4 + \delta\|A\mathbf{u}^k\|^2 \\ &\le C_\delta\|\nabla\mathbf{u}^k\|^4 + \delta\|A\mathbf{u}^k\|^2\end{aligned}$$

for any $\delta > 0$ and suitable constants C_δ that only depend on δ.

On the other hand, for $N = 3$, in view of Lemma 2.6, we get:

$$\begin{aligned}\|(\mathbf{u}^k\cdot\nabla)\mathbf{u}^k\|^2 &\le \|\mathbf{u}^k\|_{L^6}^2\|\nabla\mathbf{u}^k\|_{L^3}^2 \\ &\le C\|\nabla\mathbf{u}^k\|^3\|A\mathbf{u}^k\| \\ &\le C_\delta\|\nabla\mathbf{u}^k\|^6 + \delta\|A\mathbf{u}^k\|^2.\end{aligned}$$

In both cases, choosing δ small enough (depending on μ, α and β) and setting $\theta_k(t) := \|\nabla\mathbf{u}^k(t)\|^2$, $\psi_k(t) := \|(\rho^k)^{1/2}\mathbf{u}_t^k(t)\|^2 + \|A\mathbf{u}^k(t)\|^2$ and $\phi(t) := C\|\mathbf{f}(\cdot\,,t)\|^2$, we see that

$$\frac{d}{dt}\theta_k(t) + \psi_k(t) \le \phi(t) + C\theta_k(t)^N \tag{5.16}$$

for all t.

Also, note that $\theta_k(0)$ is bounded independently of k, since $\mathbf{u}^k(0) = P_k\mathbf{u}_0$ and $\mathbf{u}_0 \in V$.

In view of (5.16), Lemma 2.15 and (5.10), there exists a time $\tilde{T} \in (0, T]$ such that

$$\|\nabla\mathbf{u}^k(t)\|^2 \le F_1(t), \tag{5.17}$$

$$\int_0^t \|A\mathbf{u}^k(s)\|^2\,ds \le H_1(t), \tag{5.18}$$

$$\int_0^t \|(\rho^k)^{1/2}\mathbf{u}_t^k(s)\|^2\, ds \le H_2(t), \tag{5.19}$$

for all $t \in [0, \tilde{T})$ for some continuous functions F_1, H_1 and H_2 in $[0, \tilde{T}]$.

It is also clear that, if $N = 2$, we can take $\tilde{T} = T$ since, in that case, θ_k^N is the product of θ_k by a function uniformly bounded in $L^1(0, T)$, thanks to (5.11).

It is possible to extract a subsequence, again indexed by k, such that

$$\begin{array}{lll} \mathbf{u}^k \to \mathbf{u} & \text{weakly in} & L^2(0, \tilde{T}; D(A)) \text{ and weakly } -* \text{ in } L^\infty(0, \tilde{T}; V), \\ \mathbf{u}_t^k \to \mathbf{u}_t & \text{weakly in} & L^2(\Omega \times (0, \tilde{T})), \\ \rho^k \to \rho & \text{weakly } -* \text{ in} & L^\infty(\Omega \times (0, \tilde{T})), \\ \rho_t^k \to \rho_t & \text{weakly in} & L^2(0, \tilde{T}; W^{-1,\infty}(\Omega)). \end{array}$$

Now, we can apply Corollary 2.3 in Chap. 2 to the spatial derivatives of the $\mathbf{u}^k$ and the ρ^k and deduce the following strong convergence properties:

$$\begin{array}{lll} \mathbf{u}^k \to \mathbf{u} & \text{strongly in} & L^2(0, \tilde{T}; V), \\ \rho^k \to \rho & \text{strongly in} & C^0([0, \tilde{T}]; W^{-1,\infty}(\Omega)). \end{array}$$

Also, from the strong convergence of the $\mathbf{u}^k$ and the weak-$*$ convergence of the ρ^k, we get:

$$\rho^k\mathbf{u}^k \to \rho\mathbf{u} \quad \text{weakly} \quad \text{in } L^2(\Omega \times (0, \tilde{T}))^N.$$

These properties of ρ^k and $\mathbf{u}^k$ are sufficient to ensure that there exists p such that $\{\mathbf{u}, \rho, p\}$ is a strong solution.

Indeed, we have $\mathbf{u} \in L^2(0, \tilde{T}; D(A)) \cap C^0([0, \tilde{T}]; V)$, $\mathbf{u}_t \in L^2(0, \tilde{T}; H)$ and of course $\rho \in L^\infty(\Omega \times (0, \tilde{T}))$. Therefore,

$$S := \rho(\mathbf{u}_t + (\mathbf{u}\dot{\nabla})\mathbf{u}) - \mu\Delta\mathbf{u} - \rho\mathbf{f} \in L^2(0, \tilde{T}; L^2(\Omega)^N) \cong L^2(\Omega; L^2(0, \tilde{T}))^N$$

and, since $\langle S, \boldsymbol{\phi}\rangle = 0$ for all $\boldsymbol{\phi} \in \mathcal{V}$, one has $S = -\nabla p$ for some p that belongs to the space $H^1(\Omega; L^2(0, \tilde{T})) \cong L^2(0, \tilde{T}; H^1(\Omega))$, that is,

$$\rho\mathbf{u}_t - \mu\Delta\mathbf{u} + \rho(\mathbf{u}\cdot\nabla)\mathbf{u} - \rho\mathbf{f} = -\nabla p \quad \text{a.e. in } \Omega \times (0, \tilde{T}).$$

The continuity equation in system (5.1) is also satisfied by ρ and $\mathbf{u}$. Indeed, note that

$$\int_0^{\tilde{T}} \left[(\rho^k, \sigma_t) + (\rho^k\mathbf{u}^k, \nabla\sigma)\right] ds = 0$$

for all $\sigma \in C_0^1(0, \tilde{T}; H^1(\Omega))$. Hence, passing to the limit as $k \to \infty$, one obtains

$$\int_0^{\tilde{T}} [(\rho, \sigma_t) + (\rho \mathbf{u}, \nabla \sigma)] \, ds = 0.$$

At this point, we can use the regularity of $\mathbf{u}$, the definition of the inverse Lagrangian coordinates and the identity relating ρ and ρ_0 (or the results by DiPerna-Lions in [24]) to conclude that $\rho \in C^0(\overline{\Omega} \times [0, \tilde{T}])$ and the second PDE in (5.1) is satisfied a.e. in $\Omega \times (0, \tilde{T})$.

Finally, it can be shown, arguing as in the proof of Theorem 3.1, that the initial conditions in (5.1) are also satisfied. Therefore, $\{\mathbf{u}, \rho, p\}$ is a strong solution to (5.1) in $\Omega \times (0, \tilde{T})$ and the result holds. □

In view of this result and its proof, it makes sense to introduce the maximal time of existence of strong solution. By definition, it is given by

$$T_* = \sup \{\tilde{T} \in (0, T] : \exists \text{ strong solution to } (5.1) \text{ in } \Omega \times (0, \tilde{T})\}.$$

If $N = 2$, one has $T_* = T$. On the other hand, if $N = 3$ and $T_* < T$, then $\mathbf{u}(\cdot, t)$ must blow up to infinity as $t \to T_*$ (in a sense that can be made precise; see Exercise 5.2).

Unfortunately, the regularity of the strong solution furnished by Theorem 5.1 is not enough to ensure uniqueness in its interval of definition; see to this respect some partial results in Sect. 5.2.

Remark 5.1 Thanks to Theorem 3.1, the weak convergence of the Galerkin approximations $\mathbf{u}^k$ is ensured in $L^2(0, T; V)$. Consequently, if $Z \subset L^2(0, \tilde{T}; V)$ is a reflexive space and we are able to get uniform estimates of the $\|\mathbf{u}^k\|_Z$, we will also have $\mathbf{u} \in Z$ (the same can be said if Z is the dual of a Banach space). This has been seen in the previous proof for $Z = L^2(0, \tilde{T}; D(A))$. □

Remark 5.2 The main difference in the regularity properties of the solution that can be proved in two and three spatial dimensions can be seen in (5.16). For $N = 2$, these inequalities, together with Gronwall's Lemma and the energy estimates, lead in a straighforward way to strong estimates in the whole $(0, T)$. Contrarily, for $N = 3$, we only get strong estimates locally in time. Whether or not the maximal time T_* is equal to T is a completely open problem. This already happens for the constant density Navier-Stokes equations; for more details, see [44] and the references therein. □

We will now recall a slightly different result, due to Salvi [56], which leads to the existence and uniqueness of a more regular solution under more restrictive assumptions on the external force $\mathbf{f}$ and the initial data $\mathbf{u}_0$ and ρ_0.

Strictly speaking, this result is not needed to understand the role and properties of strong solutions. However, we have preferred to include it, since its proof relies on a series of estimates that deserve to be recalled and explained.

Theorem 5.2 *Let us assume that (5.3) holds and, furthermore,*

$$\begin{cases} \mathbf{u}_0 \in D(A), \quad \rho_0 \in C^1(\overline{\Omega}), \\ \mathbf{f} \in L^2(0,T;H^1(\Omega)^N) \text{ and } \mathbf{f}_t \in L^2(0,T;L^2(\Omega)^N). \end{cases}$$

There exist T_0 with $0 < T_0 \le T$ and a unique strong solution of system (5.1) *in $\Omega \times (0, T_0)$ satisfying*

$$\mathbf{u} \in L^2(0,T_0;H^3(\Omega)^N) \cap C^0([0,T_0];D(A)) \text{ and } \rho \in C^1(\overline{\Omega} \times [0,T_0]). \tag{5.20}$$

Remark 5.3 Actually, the solution furnished by Theorem 5.2 satisfy some additional regularity properties. For instance, when $N = 3$, for any small positive δ one has

$$\mathbf{u}_t \in C^0([0,T_0];H) \cap L^2(0,T_0;V) \cap L^2(\delta,T_0;D(A))$$

and

$$\mathbf{u}_{tt} \in L^2(\delta,T_0;H).$$

To this respect, see Exercise 5.3. □

Theorem 5.2 can also be proved by the semi-Galerkin method. To simplify the presentation, we will indicate the required *a priori* estimates of the approximations in a sequence of Lemmas.

For simplicity, we will only consider the case $N = 3$; when $N = 2$, the argument is similar and in fact easier.

In what follows, it will be assumed that the functions F_i, G_i, H_i are real-valued, continuous and nondecreasing in a time interval of the form $[0, T_0]$. They will be chosen depending only on Ω, T and the data of the problem.

The first estimates are easy:

Lemma 5.1 *There exists T_0 with $0 < T_0 \le T$ such that the semi-Galerkin approximations $\mathbf{u}^k$ and ρ^k satisfy the following for all $t \in [0, T_0]$:*

$$\alpha \le \rho^k \le \beta, \tag{5.21}$$

$$\|\mathbf{u}^k(t)\|^2 + \int_0^t \|\nabla \mathbf{u}^k(s)\|^2 \, ds \le C, \tag{5.22}$$

$$\|\nabla \mathbf{u}^k(t)\|^2 \le F_1(t), \tag{5.23}$$

$$\int_0^t \|A\mathbf{u}^k(s)\|^2 \, ds \le H_1(t), \tag{5.24}$$

$$\int_0^t \|\mathbf{u}^k_t(s)\|^2 \, ds \le H_2(t). \tag{5.25}$$

Proof The first four inequalities have already been proved; see (5.10), (5.11) and (5.17)–(5.19), obtained in the proof of Theorem 5.1.

Moreover, inequality (5.25) follows directly from (5.19). □

Lemma 5.2 *For all $t \in [0, T_0]$, the $\mathbf{u}^k$ satisfy*

$$\|\mathbf{u}_t^k(t)\|^2 + \int_0^t \|\nabla \mathbf{u}_t^k(s)\|^2\,ds \le F_2(t), \tag{5.26}$$

$$\|A\mathbf{u}^k(t)\|^2 \le F_3(t). \tag{5.27}$$

Proof Differentiating with respect to t the first equation in (5.9) and taking $\mathbf{v} = \mathbf{u}_t^k$, one obtains:

$$\begin{aligned}(\rho^k \mathbf{u}_{tt}^k, \mathbf{u}_t^k) + \mu\|\nabla \mathbf{u}_t^k\|^2 &= (\rho_t^k(\mathbf{f} - \mathbf{u}_t^k - ((\mathbf{u}^k \cdot \nabla)\mathbf{u}^k)), \mathbf{u}_t^k)\\ &\quad + (\rho^k(\mathbf{f}_t - (\mathbf{u}_t^k \cdot \nabla)\mathbf{u}^k - (\mathbf{u}^k \cdot \nabla)\mathbf{u}_t^k), \mathbf{u}_t^k).\end{aligned}$$

Taking into account that $\rho_t^k = -\nabla \cdot (\rho^k \mathbf{u}^k)$ and using several times the Sobolev embedding and the interpolation inequalities in Lemma 2.6, we easily find that

$$\begin{aligned}\frac{d}{dt}\|(\rho^k)^{1/2}\mathbf{u}_t^k\|^2 + \mu\|\nabla \mathbf{u}_t^k\|^2 &\le C\|\mathbf{u}_t^k\|^2\left(1 + \|\nabla \mathbf{u}^k\|^4 + \|\nabla \mathbf{u}^k\|\,\|A\mathbf{u}^k\|\right)\\ &\quad + C\|\nabla \mathbf{u}^k\|^4\|A\mathbf{u}^k\|^2\\ &\quad + C\|\mathbf{f}_t(\cdot\,,t)\|^2 + C\|\mathbf{f}(\cdot\,,t)\|_{H^1}^2\|\mathbf{u}^k\|\,\|\nabla \mathbf{u}^k\|.\end{aligned}$$

Integrating with respect to time, we also have that

$$\|\mathbf{u}_t^k(t)\|^2 + \int_0^t \|\nabla \mathbf{u}_t^k(s)\|^2\,ds \le CM_k(t) + C\int_0^t N_k(s)\|\mathbf{u}_t^k(s)\|^2\,ds, \tag{5.28}$$

where

$$\begin{aligned}M_k(t) := \int_0^t \|\mathbf{f}_t(\cdot\,,s)\|^2\,ds &+ \int_0^t \|\mathbf{f}(\cdot\,,s)\|_{H^1}^2\|\mathbf{u}^k(s)\|\,\|\nabla \mathbf{u}^k(s)\|\,ds\\ &+ \int_0^t \|\nabla \mathbf{u}^k(s)\|^4\|A\mathbf{u}^k(s)\|^2\,ds + \|\mathbf{u}_t^k(0)\|^2\end{aligned}$$

and

$$N_k(t) := 1 + \|\nabla \mathbf{u}^k(t)\|^4 + \|A\mathbf{u}^k(t)\|^2.$$

Now, applying Gronwall's Lemma to (5.28), we find that

$$\|\mathbf{u}_t^k(t)\|^2 \leq CM_k(t) + C\int_0^t M_k(s)N_k(s)\exp\left(C\int_s^t N_k(\sigma)\,d\sigma\right)ds \tag{5.29}$$

for all $t \in [0, T_0]$.

Using the hypotheses on $\mathbf{f}$, $\mathbf{f}_t$ and the estimates in Lemma 5.1, we see that

$$M_k(t) \leq G_1(t) + C\|\mathbf{u}_t^k(0)\|^2 \quad \forall t \in [0, T_0), \tag{5.30}$$

for some nondecreasing function G_1 (of course independent of k).

On the other hand, taking $t = 0$ and $\mathbf{v} = \mathbf{u}_t^k(0)$ in the first equation of system (5.9) and recalling that $\mathbf{u}_0 \in D(A)$, we deduce that

$$\begin{aligned}\|\mathbf{u}_t^k(0)\|^2 &\leq C\left(\|(\mathbf{u}^k(0)\cdot\nabla)\mathbf{u}^k(0)\|^2 + \|\mathbf{f}(\cdot\,,0)\|^2 + \|A\mathbf{u}^k(0)\|^2\right)\\ &\leq C\left(\|\nabla\mathbf{u}_0\|^3\|A\mathbf{u}_0\| + \|\mathbf{f}(\cdot\,,0)\|^2 + \|A\mathbf{u}_0\|^2\right).\end{aligned}$$

Thus, $\|\mathbf{u}_t^k(0)\|^2$ is uniformly bounded and

$$M_k(t) \leq G_2(t) \quad \forall t \in [0, T_0)$$

for some nondecreasing function G_2.

Moreover, from the estimates in Lemma 5.1, one has

$$C\int_0^t N_k(s)\,ds \leq G_3(t) \quad \forall t \in [0, T_0]$$

for some G_3 and, in view of (5.29), there exists G_4 such that

$$\|\mathbf{u}_t^k(t)\|^2 \leq G_2(t) + C\int_0^t G_2(s)N_k(s)e^{G_3(t)}\,ds \leq G_4(t).$$

Using (5.28) and again Gronwall's Lemma, we finally obtain the estimate (5.26). In order to prove (5.27), we can take $\mathbf{v} = A\mathbf{u}^k$ in the first equation of (5.9) to get

$$\|A\mathbf{u}^k\| \leq C\left(\|\mathbf{u}_t^k\| + \|\mathbf{f}(\cdot\,,t)\| + \|(\mathbf{u}^k\cdot\nabla)\mathbf{u}^k\|\right). \tag{5.31}$$

Note that

$$\|(\mathbf{u}^k\cdot\nabla)\mathbf{u}^k\| \leq C\|\nabla\mathbf{u}^k\|^{3/2}\|A\mathbf{u}^k\|^{1/2} \leq \varepsilon\|A\mathbf{u}^k\| + C_\varepsilon\|\nabla\mathbf{u}^k\|^3 \tag{5.32}$$

for any $\varepsilon > 0$. Choosing ε small enough and combining the inequalities (5.31) and (5.32), the following is found:

$$\|A\mathbf{u}^k\| \le C\left(\|\mathbf{u}_t^k\| + \|\nabla \mathbf{u}^k\|^3 + \|\mathbf{f}(\cdot\,,t)\|\right).$$

This inequality, together with (5.23) and (5.26), leads to the desired estimates (5.27).

This ends the proof. □

Lemma 5.3 *The semi-Galerkin approximations* $\mathbf{u}^k$ *and* ρ^k *fulfill the following estimates for all* $t \in [0, T_0]$*:*

$$\int_0^t \|\mathbf{u}^k(s)\|^2_{W^{2,6}}\, ds \le \tilde{F}_1(t), \tag{5.33}$$

$$\int_0^t \|\nabla \mathbf{u}^k(s)\|^2_{L^\infty}\, ds \le \tilde{F}_2(t), \tag{5.34}$$

$$\|\nabla \rho^k(\cdot\,,t)\|_{L^\infty} \le F_4(t) \;\; and \;\; \|\rho_t^k(\cdot\,,t)\|_{L^\infty} \le F_5(t). \tag{5.35}$$

Proof Note that, for every $\boldsymbol{\phi} \in \mathcal{V}$, one has

$$\mu(\nabla \mathbf{u}^k, \nabla \boldsymbol{\phi}) = (\boldsymbol{\chi}^k, \boldsymbol{\phi}), \tag{5.36}$$

where we have set

$$\boldsymbol{\chi}^k(t) := P_k(\rho^k(\cdot\,,t)(\mathbf{f}(\cdot\,,t) - \mathbf{u}_t^k(t) - (\mathbf{u}^k(t)\cdot\nabla)\mathbf{u}^k(t))) \quad \forall t \in [0, T_0].$$

It is clear from the previous estimates that $\boldsymbol{\chi}^k$ is uniformly bounded in the space $L^2(0, T'; L^6(\Omega)^3)$ for all $T' < T_0$. Hence, inequality (5.33) follows directly from the results by Amrouche and Girault for the Stokes operator (see [2]; see also Exercise 5.4).

In particular, from the usual Sobolev embedding results, we also get the inequalities (5.34) (recall that $W^{2,6}(\Omega) \hookrightarrow W^{1,\infty}(\Omega)$ with a continuous and even compact embedding, see Theorem 2.6; in fact, in accordance with (2.34)–(2.35), one has $W^{2,6}(\Omega) \hookrightarrow C^{1,1/2}(\overline{\Omega})$ with a continuous embedding).

Finally, the estimates (5.35) follow directly from known results for the solutions; see [43] and Exercise 5.5. □

Lemma 5.4 *The approximations* $\mathbf{u}^k$ *satisfy*

$$\int_0^t \|\mathbf{u}^k(s)\|^2_{H^3}\, ds \le F_6(t) \tag{5.37}$$

for all $t \in [0, T_0]$.

Proof Let us consider again the identity (5.36). From (5.33) and the estimates in Lemmas 5.1 and 5.2, we conclude that the functions $\boldsymbol{\chi}^k$ are uniformly bounded in $L^2(0, T_0; H^1(\Omega)^3)$.

Therefore, using again the regularity results for the solutions to the Stokes problem in [2], we find that $\mathbf{u}^k$ is also uniformly bounded in $L^2(0, T'; H^3(\Omega)^3)$.

This proves (5.37). □

Remark 5.4 Note that, if $h : (a, b) \mapsto \mathbb{R}$ is a positive continuous function with

$$\int_a^b h(s)\, ds < +\infty,$$

then there exists sequences $\{\varepsilon_n\}$ satisfying $\varepsilon_n \to a^+$ and $\varepsilon_n h(a + \varepsilon_n) \to 0$ as $n \to +\infty$. This elementary result will be very useful in the sequel. □

Lemma 5.5 *Let us set $\sigma(t) := min(1, t)$ for all positive t. The approximations $\mathbf{u}^k$ satisfy the following estimates for all $t \in [0, T_0]$:*

$$\int_0^t \sigma(s)\|\mathbf{u}^k_{tt}(\cdot\,, s)\|^2\, ds \le F_7(t), \tag{5.38}$$

$$\sigma(t)\|\nabla \mathbf{u}^k_t(t)\|^2 \le F_8(t), \tag{5.39}$$

$$\sigma(t)\|\mathbf{u}^k(t)\|^2_{H^3} \le F_9(t). \tag{5.40}$$

Proof In view of the estimates in the previous lemmas, differentiating the first equation in system (5.9) with respect to t and setting $\mathbf{v} = \mathbf{u}^k_{tt}$, it is not difficult to deduce after some work that:

$$\|\mathbf{u}^k_{tt}\|^2 + \frac{d}{dt}\|\nabla \mathbf{u}^k_t\|^2 \le G_5(t)\left(1 + \|\nabla \mathbf{u}^k_t(t)\|^2\right). \tag{5.41}$$

Multiplying both sides of (5.41) by σ and integrating with respect to time over (ε, t), we see that

$$\int_\varepsilon^t \sigma(s)\|\mathbf{u}^k_{tt}(s)\|^2\, ds + \int_\varepsilon^t \sigma(s)\frac{d}{dt}\|\nabla \mathbf{u}^k_t(s)\|^2\, ds$$
$$\le \int_\varepsilon^t G_5(s)\sigma(s)\left(1 + \|\nabla \mathbf{u}^k_t(s)\|^2\right) ds.$$

Note that

$$\int_\varepsilon^t \sigma(s)\frac{d}{dt}\|\nabla \mathbf{u}^k_t(s)\|^2\, ds = \sigma(t)\|\nabla \mathbf{u}^k_t(t)\|^2 - \sigma(\varepsilon)\|\nabla \mathbf{u}^k_t(\varepsilon)\|^2$$
$$- \int_\varepsilon^t \sigma'(s)\|\nabla \mathbf{u}^k_t(s)\|^2\, ds.$$

Moreover, Remark 5.4 can be applied with $a = 0$, $b = T_0$ and $h = \|\nabla \mathbf{u}^k_t\|$. Thus, there exist $\varepsilon_1, \varepsilon_2, \ldots$ with $\varepsilon_n \to 0$ and $\sigma(\varepsilon_n)\|\nabla \mathbf{u}^k_t(\varepsilon_n)\|^2 \to 0$.

Therefore,

$$\begin{aligned}
&\sigma(t)\|\nabla \mathbf{u}_t^k(t)\|^2 + \int_0^t \sigma(s)\|\mathbf{u}_{tt}^k(s)\|^2\,ds \\
&\quad\le \int_0^t G_5(s)\sigma(s)\left(1+\|\nabla \mathbf{u}_t^k(s)\|^2\right)ds + \int_0^t \|\nabla \mathbf{u}_t^k(s)\|^2\,ds \\
&\quad\le G_6(t)\int_0^t \left(1+\|\nabla \mathbf{u}_t^k(s)\|^2\right)ds.
\end{aligned}$$

This leads to (5.38) and (5.39).

Finally, using the bound (5.38) and arguing as in the proof of Lemma 5.4, the estimate (5.40) can be easily proved. □

The following lemma can be proved in a similar way:

Lemma 5.6 *The approximations* $\mathbf{u}^k$ *satisfy*

$$\int_0^t \sigma(s)\|A\mathbf{u}_t^k(s)\|^2\,ds \le F_{10}(t),$$

for all $t \in [0, T_0]$.

We are now in position to prove Theorem 5.2.

Proof of Theorem 5.2 The estimates in Lemmas 5.1–5.6 imply the existence of subsequences of $\{\mathbf{u}^k\}$ and $\{\rho^k\}$, again indexed by k, that converge in the following sense for all small positive δ and any $r \in (1, +\infty)$:

(i) $\mathbf{u}^k \to \mathbf{u}$, strongly in $L^r(\delta, T_0; H^{3-\varepsilon}(\Omega))$,
weakly in $L^2(0, T_0; H^3(\Omega))$
weakly-$*$ in $L^\infty(\delta, T_0; H^3(\Omega))$,

(ii) $\mathbf{u}_t^k \to \mathbf{u}_t$, strongly in $L^r(\delta, T_0; H^{1-\varepsilon}(\Omega))$,
weakly in $L^2(0, T_0; H^1(\Omega))$,
weakly-$*$ in $L^\infty(0, T_0; L^2(\Omega))$
weakly in $L^2(\delta, T_0; H^2(\Omega))$,

(iii) $\mathbf{u}_{tt}^k \to \mathbf{u}_{tt}$, weakly in $L^2(\delta, T_0; L^2(\Omega))$,

(iv) $\rho^k \to \rho$ strongly in $L^r(0, T_0; C^{0,\beta}(\bar{(\Omega)}))$
$(0 \le \beta < 1)$
uniformly in $\overline{\Omega} \times [0, T_0]$,

(v) $\nabla\rho^k \to \nabla\rho$ weakly-$*$ in $L^\infty(\Omega \times (0, T_0))^3$,

(vi) $\rho_t^k \to \rho_t$ weakly-$*$ in $L^\infty(\Omega \times (0, T_0))$.

After recalling the arguments used in the proof of Theorem 5.1, it becomes clear that these convergence properties are sufficient to take limits in the equations satisfied by the Galerkin approximations and deduce that there exists p such that $\{\mathbf{u}, \rho, p\}$ is a strong solution to (5.1) in $\Omega \times (0, T_0)$.

The uniqueness of $\{\mathbf{u}, \rho, \nabla p\}$ is a trivial consequence of the results of [18] mentioned at the end of Sect. 3.2.

Let us show that the initial condition for $\mathbf{u}$ is satisfied in the following sense:

$$\lim_{t\to 0^+} \|A\mathbf{u}(\cdot\,, t) - A\mathbf{u}_0\| = 0. \tag{5.42}$$

To this end, we will first prove that

$$\lim_{t\to 0^+} \|\mathbf{u}(\cdot\,, t) - \mathbf{u}_0\|_V = 0, \tag{5.43}$$

i.e., that $\mathbf{u}$ assumes the initial datum continuously in the H_0^1-norm. Indeed, one has

$$\nabla\mathbf{u}^k(t) - \nabla\mathbf{u}_0^k = \int_0^t \nabla\mathbf{u}_t^k(s)\, ds$$

for all t and any $k \geq 1$. Therefore, in view of (5.26), one has

$$\|\nabla\mathbf{u}^k(t) - \nabla\mathbf{u}_0^k\| \leq \int_0^t \|\nabla\mathbf{u}_t^k(s)\|\, ds \leq F_2(T_0)\, t \quad \forall t \in [0, T_0]. \tag{5.44}$$

Since the $\mathbf{u}^k$ are uniformly bounded in $L^\infty(0, T_0; D(A))$ and the $\mathbf{u}_t^k$ are uniformly bounded in $L^\infty(0, T_0; V)$, taking into account Corollary 2.3 in Chap. 2, it follows that $\mathbf{u}^k \to \mathbf{u}$ strongly in $C^0([0, T_0]; V)$.

Accordingly, we can take limits in (5.44) to obtain that

$$\|\nabla\mathbf{u}(t) - \nabla\mathbf{u}_0^k\| \leq F_2(T_0)t \quad \forall t \in [0, T_0].$$

This gives (5.43), as desired.

Now, let us prove (5.42).

We already know that $\|A\mathbf{u}(\cdot\,, t)\|$ is bounded and $\mathbf{u}(\cdot\,, t) \to \mathbf{u}_0$ strongly in V as $t \to 0$. Consequently, $\liminf_{t\to 0^+} \|A\mathbf{u}(\cdot\,, t)\| \geq \|A\mathbf{u}_0\|$ and it suffices to show that

$$\limsup_{t\to 0^+} \|A\mathbf{u}(\cdot\,, t)\| \leq \|A\mathbf{u}_0\|. \tag{5.45}$$

Taking $\mathbf{v} = A\mathbf{u}^k$ in the first equation in (5.9) and integrating in time, one gets

$$\|A\mathbf{u}^k(t)\|^2 - \|A\mathbf{u}_0^k\|^2 = -\int_0^t (\mathbf{m}(s), A\mathbf{u}_t^k(s))\, ds,$$

where

$$\mathbf{m}(t) := \frac{1}{\mu}\,(\rho^k \mathbf{u}_t^k + \rho^k (\mathbf{u}^k \cdot \nabla)\mathbf{u}^k - \rho^k \mathbf{f})(\cdot\,, t).$$

Note that

$$\int_0^t (\mathbf{m}(s), A\mathbf{u}_t^k(s))\,ds = (\mathbf{m}(t), A\mathbf{u}^k(t)) - (\mathbf{m}(0), A\mathbf{u}_0) - \int_0^t (\mathbf{m}'(s), A\mathbf{u}^k(s))\,ds.$$

Therefore, it is clear that

$$\begin{aligned}
&\left|\int_0^t (\mathbf{m}(s), A\mathbf{u}_t^k(s))\,ds\right| \\
&\quad \le |(\rho^k((\mathbf{u}^k\cdot\nabla)\mathbf{u}^k - \mathbf{f}), A\mathbf{u}^k)(t) - (\rho_0((\mathbf{u}_0\cdot\nabla)\mathbf{u}_0 - \mathbf{f}(\cdot\,,0)), A\mathbf{u}_0)| \\
&\quad + G_8(T_0)\left(t + t^{1/4}\right)
\end{aligned}$$

for all $t \in [0, T_0]$ for some continuous nondecreasing G_8.

Hence,

$$\begin{aligned}
\|A\mathbf{u}^k(t)\|^2 &\le \|A\mathbf{u}_0^k\|^2 + G_8(T_0/2)\left(t + t^{1/4}\right) \\
&\quad + |(\rho^k(\mathbf{u}^k\cdot\nabla)\mathbf{u}^k - \rho^k\mathbf{f}, A\mathbf{u}^k)(t) - (\rho_0(\mathbf{u}_0\cdot\nabla)\mathbf{u}_0 - \rho_0\mathbf{f}(\cdot\,,0), A\mathbf{u}_0)|.
\end{aligned}$$

Note that, for each $t \in [0, T_0]$, $A\mathbf{u}^k(t) \to A\mathbf{u}(\cdot\,,t)$ weakly in $L^2(\Omega)^3$ and $\mathbf{u}^k(t) \to \mathbf{u}(t)$ strongly in V.

Also,

$$\lim_{k\to+\infty} (\rho^k(\mathbf{u}^k\cdot\nabla)\mathbf{u}^k - \rho^k\mathbf{f}, A\mathbf{u}^k)(t) = (\rho_0(\mathbf{u}_0\cdot\nabla)\mathbf{u}_0 - \rho_0\mathbf{f}(\cdot\,,0), A\mathbf{u}_0)$$

and

$$\|A\mathbf{u}(\cdot\,,t)\|^2 \le \|A\mathbf{u}_0\|^2 + G_7(T_0)\left(t + t^{1/4}\right).$$

Obviously, this gives the desired inequality (5.45).

This ends the proof. □

Remark 5.5 From (5.42) and the equations satisfied by $\{\mathbf{u}, \rho, p\}$, it is not difficult to prove that

$$\lim_{t\to 0^+} \|\mathbf{u}_t(\cdot\,,t) - \mathbf{u}_t(\cdot\,,0)\| = 0. \tag{5.46}$$

Thus, $\mathbf{u}_t$ may be viewed as a H-valued function that is continuous at $t = 0$. □

Remark 5.6 The argument used in the last part of the proof of Theorem 5.2 also holds for any $t \in (0, T_0)$ and not only at $t = 0$. This implies continuity from the right of $\mathbf{u}$ and $\mathbf{u}_t$ in the appropriate spaces. The proof can also be adapted to show continuity from the left at any $t \in (0, T_0)$. In particular, this leads to the continuity properties mentioned in Remark 5.3. □

5.2 Other Uniqueness Results

Uniqueness by itself is a delicate subject. This has been explained at the end of Chap. 3, where we have recalled the best achievements to date.

For completeness, we will present in this section some additional results concerning strong and regular weak solutions.

For the moment, let us see that it is possible to ensure uniqueness of the strong solution furnished by Theorem 5.2 in a larger class than the space determined by (5.20).

Thus, let the assumptions in Theorem 5.2 be satisfied, let $T' > 0$ be given with $0 < T' \leq T_0$ and let us set

$$\begin{aligned} \mathcal{S}(T') := \{(\mathbf{v}, \xi) : \mathbf{v} \in L^2(0, T'; V),\ \ \mathbf{v}_t \in L^2(0, T'; L^3(\Omega)^3), \\ \xi \in L^\infty(\Omega \times (0, T')),\ \ \nabla\xi \in L^4(0, T'; L^\infty(\Omega)^3)\}. \end{aligned}$$

The following holds:

Theorem 5.3 *Let* $\{\mathbf{u}, \rho, p\}$ *be the strong solution to* (5.1) *furnished by Theorem 5.2. If* $\{\mathbf{v}, \xi, q\}$ *is a solution to (5.1) in* $\Omega\times(0, T')$ *with* $(\mathbf{v}, \xi) \in \mathcal{S}(T')$*, then* $\{\mathbf{v}, \xi, \nabla q\} = \{\mathbf{u}, \rho, \nabla p\}$ *in* $\Omega \times (0, T')$.

Proof Let $\{\mathbf{v}, \xi, q\}$ be a solution to system (5.1) in $\Omega \times (0, T')$ with $(\mathbf{v}, \xi) \in \mathcal{S}$ and let us introduce $\mathbf{w} := \mathbf{u} - \mathbf{v}$ and $\eta := \rho - \xi$. These functions satisfy

$$\begin{cases} P_H(\rho\mathbf{w}_t+\xi(\mathbf{v}\cdot\nabla)\mathbf{w})+\mu A\mathbf{w}=P_H(\eta\mathbf{f}-\eta\mathbf{v}_t-\eta(\mathbf{u}\cdot\nabla)\mathbf{u}-\xi(\mathbf{w}\cdot\nabla)\mathbf{u}), \\ \eta_t+\mathbf{u}\cdot\nabla\eta=-\mathbf{w}\cdot\nabla\xi. \end{cases} \tag{5.47}$$

Multiplying the first equation in system (5.47) by $\mathbf{w}$ and integrating over Ω, one obtains

$$\begin{aligned} \frac{1}{2}\frac{d}{dt}\|\rho^{1/2}\mathbf{w}\|^2 + \mu\|\nabla\mathbf{w}\|^2 &= (\eta\mathbf{f} - \eta\mathbf{v}_t - \eta(\mathbf{u}\cdot\nabla)\mathbf{u} - \xi(\mathbf{w}\cdot\nabla)\mathbf{u}, \mathbf{w}) \\ &\quad + \frac{1}{2}(\rho_t\mathbf{w}, \mathbf{w}) - (\xi(\mathbf{v}\cdot\nabla)\mathbf{w}, \mathbf{w}). \end{aligned}$$

Now, estimating the terms on the right-hand side as before, we find the integral inequality

$$
\begin{aligned}
&\|\mathbf{w}(\cdot\,,t)\|^2 + \int_0^t \|\nabla \mathbf{w}(\cdot\,,s)\|^2\,ds \\
&\quad \le C\int_0^t \Big(\|\mathbf{f}(\cdot\,,s)\|_{H^1}^2 + \|\mathbf{v}_t(\cdot\,,s)\|_{L^3}^2 + \|\nabla \mathbf{u}(\cdot\,,s)\|^2 \|A\mathbf{u}(\cdot\,,s)\|^2\Big)\,\|\eta(\cdot\,,s)\|^2\,ds \\
&\quad\quad + C\int_0^t \Big(\|\nabla \mathbf{v}(\cdot\,,s)\|^2 + \|\nabla \xi(\cdot\,,s)\|_{L^\infty}^4 + \|\nabla \mathbf{u}(\cdot\,,s)\|^4 \\
&\quad\quad + \|\rho_t(\cdot\,,s)\|_{L^\infty}\Big)\|\mathbf{w}(\cdot\,,s)\|^2\,ds.
\end{aligned}
$$

On the other hand, multiplying the second equation in (5.47) by η and integrating with respect to $\mathbf{x}$ and t over $\Omega \times (0, t)$, one has

$$
\begin{aligned}
\|\eta(\cdot\,,t)\|^2 &\le C \int_0^t \|\mathbf{w}(\cdot\,,s)\|\,\|\nabla \xi(\cdot\,,s)\|_{L^\infty}\|\eta(\cdot\,,s)\|\,ds \\
&\le C \int_0^t \|\nabla \xi(\cdot\,,s)\|_{L^\infty}^2 \|\mathbf{w}(\cdot\,,s)\|^2\,ds + \int_0^t \|\eta(\cdot\,,s)\|^2\,ds.
\end{aligned}
$$

Accordingly, we see that

$$
\|\mathbf{w}(\cdot\,,t)\|^2 + \|\eta(\cdot\,,t)\|^2 \le \int_0^t h(s)(\|\mathbf{w}(\cdot\,,s)\|^2 + \|\eta(\cdot\,,s)\|^2)\,ds
$$

for all $t \in [0, T')$, where

$$
\begin{aligned}
h(\cdot\,,t) = C\,\Big(&\|\mathbf{f}(\cdot\,,t)\|_{H^1}^2 + \|\mathbf{v}_t(\cdot\,,t)\|_{L^3}^2 + \|\nabla \mathbf{u}(\cdot\,,t)\|^2\|A\mathbf{u}(\cdot\,,t)\|^2 \\
&+ \|\nabla \mathbf{u}(\cdot\,,t)\|^4 + \|\rho_t(\cdot\,,t)\|_{L^\infty} + \|\nabla \mathbf{v}(\cdot\,,t)\|^2 + \|\nabla \xi(\cdot\,,t)\|_{L^\infty}^4 + 1\Big).
\end{aligned}
$$

Note that h is an integrable function, due to the regularity of $\mathbf{u}$, ρ, $\mathbf{v}$ and ξ. Consequently, we can apply Gronwall's Lemma to get

$$
\|\mathbf{w}(\cdot\,,t)\|^2 + \|\eta(\cdot\,,t)\|^2 \equiv 0.
$$

In other words, $\mathbf{u} = \mathbf{v}$ and $\xi = \rho$, as desired. □

As already said, even in the two-dimensional case, the regularity of the weak solution is not enough to ensure uniqueness, contrarily to what happens in the case of constant density. Additional regularity must be imposed in order to apply a Gronwall-like Lemma with success.

On the other hand, the existence of a solution satisfying the energy identity can lead to uniqueness. The following result and its proof illustrate rather well the situation (see [48]):

Theorem 5.4 *Assume that* $\mathbf{u}_0 \in D(A)$ *and* $\rho_0 \in C^1(\overline{\Omega})$. *Let* $\{\overline{\mathbf{u}}, \overline{\rho}, \overline{p}\}$ *be a solution to problem* (5.1), *with*

$$\begin{cases} \nabla\overline{\rho},\ \overline{\mathbf{u}}_t \in L^2(0,T;L^\infty(\Omega)^N), & \nabla\overline{\mathbf{u}} \in L^2(0,T;L^\infty(\Omega)^{N\times N}), \\ \overline{\rho} \in C^0(\overline{Q}),\ \overline{\mathbf{u}} \in C^0(\overline{Q})^N. \end{cases} \tag{5.48}$$

Then the solution furnished by Theorem 3.1 satisfies $\mathbf{u} = \overline{\mathbf{u}}$ *and* $\rho = \overline{\rho}$ *a.e.*

Proof The energy identity

$$\frac{1}{2}\left(\int_\Omega \overline{\rho}|\overline{\mathbf{u}}|^2\right)(t) + \mu \iint_{\Omega\times(0,t)} |\nabla\overline{\mathbf{u}}|^2 = \frac{1}{2}\int_\Omega \rho_0|\mathbf{u}_0|^2 + \iint_{\Omega\times(0,t)} \overline{\rho}\,\overline{\mathbf{u}}\cdot\mathbf{f} \tag{5.49}$$

follows by (5.48) and the first equation in (5.1) written for $\overline{\mathbf{u}}$, $\overline{\rho}$ and $\overline{p}$.

Also, the inequalities

$$\frac{1}{2}\left(\int_\Omega \rho|\mathbf{u}|^2\right)(t) + \mu \iint_{\Omega\times(0,t)} |\nabla\mathbf{u}|^2 \le \frac{1}{2}\int_\Omega \rho_0|\mathbf{u}_0|^2 + \iint_{\Omega\times(0,t)} \rho\mathbf{u}\cdot\mathbf{f} \tag{5.50}$$

are satisfied, due to the properties of the semi-Galerkin approximations, as shown in Chap. 3.

Additionally, one has

$$\iint_{\Omega\times(0,t)} (\overline{\rho}(\overline{\mathbf{u}}_t + (\overline{\mathbf{u}}\cdot\nabla)\overline{\mathbf{u}} - \mathbf{f})\cdot\mathbf{u} + \mu\nabla\overline{\mathbf{u}}\cdot\nabla\mathbf{u}) = 0 \tag{5.51}$$

and

$$\iint_{\Omega\times(0,t)} (\rho(\mathbf{u}_t + (\mathbf{u}\cdot\nabla)\mathbf{u} - \mathbf{f})\cdot\overline{\mathbf{u}} + \mu\nabla\mathbf{u}\cdot\nabla\overline{\mathbf{u}}) = 0. \tag{5.52}$$

From (5.49)–(5.52), one gets

$$\begin{aligned} &\left(\int_\Omega \rho|\mathbf{u}-\overline{\mathbf{u}}|^2\right)(t) + \mu \iint_{\Omega\times(0,t)} |\nabla(\mathbf{u}-\overline{\mathbf{u}})|^2 \\ &\quad \le \iint_{\Omega\times(0,t)} \left(A(s)|\mathbf{u}-\overline{\mathbf{u}}|^2 + B_\varepsilon(s)|\rho-\overline{\rho}|^2\right) + \varepsilon \iint_{\Omega\times(0,t)} |\mathbf{u}-\overline{\mathbf{u}}|^2 \end{aligned} \tag{5.53}$$

where, due to the regularity hypothesis (5.48), one has $A, B_\varepsilon \in L^\infty(0,T)$.

On the other hand, from the transport equations satisfied by ρ and $\overline{\rho}$, one also finds that

$$\frac{1}{2}\frac{d}{dt}\int_\Omega |\rho-\overline{\rho}|^2 = \int_\Omega \nabla\overline{\rho}\cdot(\mathbf{u}-\overline{\mathbf{u}})\,(\rho-\overline{\rho}).$$

Therefore,

$$\left(\int_\Omega |\rho-\overline{\rho}|^2\right)(t) \le C \iint_{\Omega\times(0,t)} |\mathbf{u}-\overline{\mathbf{u}}|^2$$

and using these inequalities in (5.53) we see that

$$\left(\int_\Omega \rho|\mathbf{u}-\overline{\mathbf{u}}|^2\right)(t) + \mu \iint_{\Omega\times(0,t)} |\nabla(\mathbf{u}-\overline{\mathbf{u}})|^2 \le \iint_{\Omega\times(0,t)} G(s)|\mathbf{u}-\overline{\mathbf{u}}|^2$$

for some $G \in L^\infty(0,T)$.

This suffices to apply Gronwall's Lemma and to conclude that $\mathbf{u} = \overline{\mathbf{u}}$ and $\rho = \overline{\rho}$ a.e. in Q. □

Remark 5.7 The regularity needed to apply this result is guaranteed, for example, under the assumptions in Theorems 5.7 and 5.8, where some smallness is imposed on the data; see Sect. 5.4. □

Remark 5.8 Another strategy for the proof of uniqueness, based on the ideas in [49, 50] and related to "duality", is as follows. Let $\{\overline{\mathbf{u}}, \overline{\rho}, \overline{p}\}$ and $\{\mathbf{u}, \rho, p\}$ be two solutions to (5.1) and let us set $\mathbf{w} := \mathbf{u}-\overline{\mathbf{u}}$, $\beta := \rho-\overline{\rho}$ and $\pi := p-\overline{p}$. The identities satisfied by $\mathbf{w}$, β and π take the form

$$E(\mathbf{u},\rho,\overline{\mathbf{u}},\overline{\rho})(\mathbf{w},\beta,\pi) = 0$$

for some linear differential operator $E(\mathbf{u},\rho,\overline{\mathbf{u}},\overline{\rho})$. Then, the goal is to prove that

$$N(E(\mathbf{u},\rho,\overline{\mathbf{u}},\overline{\rho})) = \{(\mathbf{0},0,0)\}.$$

To this purpose, it is sufficient to check that $R(E(\mathbf{u},\rho,\overline{\mathbf{u}},\overline{\rho})^*)$ is dense. This means that one must be able to solve, in an appropriate space, a linear system of the form

$$E(\mathbf{u},\rho,\overline{\mathbf{u}},\overline{\rho})^*(\mathbf{z},\eta,q) = (\mathbf{g},h),$$

for instance, for any C^∞ compactly supported $(\mathbf{g},h)$. Unfortunately, this also requires adequate regularity properties for the solution. Actually, the result obtained through this argument is not better than Theorem 5.4. □

An interesting open question is the following:

What are the most general "structural" assumptions on ρ_0 implying uniqueness of a weak solution in the two-dimensional case?

For example, we know that if ρ_0 is constant and $N = 2$, the weak solution is unique. What happens if, for instance, ρ_0 is of the form

$$\rho_0 = \eta_1 \mathbb{1}_{\Omega_1} + \eta_2 \mathbb{1}_{\Omega_2},$$

where η_1 and η_2 are positive constants and $\{\Omega_1, \Omega_2\}$ is a partition of Ω?

Also, we can ask about conditions on $\mathbf{u}_0$ to have uniqueness.

More precisely, it can have an interest to know which is the "optimal" functional space X for the initial velocity field that provides uniqueness of a local in time strong solution with regular and strictly positive ρ_0 and (for instance) $\mathbf{f} = \mathbf{0}$?

Here, "optimal" can be understood in the following sense: $X = D(A^\gamma)$, with $\gamma \in [0, 1/2]$ being as small as possible.[1]

This question was answered in the case of the classical Navier-Stokes equations by Farwig and Sohr in [25]. However, there is still no definitive answer in the context of the variable density system.

A partial answer was given by Danchin and Wang in [21] when $\Omega = \mathbb{R}^N$. There, the authors consider an initial density ρ_0 close to a positive constant and an initial velocity $\mathbf{u}_0$ in the critical homogeneous Besov space

$$\dot{B}^{-1+2/p}_{p,1}(\mathbb{R}^2) \ (1 < p < 2) \ \text{ if } \ N = 2$$

and small in the space

$$\dot{B}^{-1+3/p}_{p,1}(\mathbb{R}^3) \ (1 < p < 3) \ \text{ if } \ N = 3;$$

in both cases, a uniqueness statement is established.

Other questions concerning uniqueness are the following:

- Do we have the uniqueness of a solution obtained as in Theorem 3.1, that is, the limit of a semi-Galerkin family of approximations?
- How "large" is the set of uniqueness data?
- Is it reasonable to expect that a suitable "entropy condition" assumption on the solution $\{\mathbf{u}, \rho, p\}$ implies uniqueness?[2] If so, which one?

5.3 An Additional Regularity Result for the Pressure

The pressure corresponding to the strong solution given by Theorem 5.2 satisfies some additional regularity properties. This is shown in the following result:

---

[1] Recall that, if $\mathbf{u}_0 \in H = D(A^0)$, the existence of a global solution is ensured; if $\mathbf{u}_0 \in V = D(A^{1/2})$, a unique local solution exists.

[2] To this respect, see the comments on parabolic regularization below.

Proposition 5.1 *Let the assumptions in Theorem 5.2 be satisfied. Let us denote by* $\{\mathbf{u}, \rho, p\}$ *the strong solution to problem* (5.1) *in* $\Omega \times (0, T_0)$. *Then* $\nabla p \in C^0([0, T_0]; L^2(\Omega)^3)$.

Proof Let $\{\mathbf{u}, \rho, p\}$ be the strong solution furnished by Theorem 5.2. Set

$$\mathbf{k} := \rho(\mathbf{f} - \mathbf{u}_t - (\mathbf{u} \cdot \nabla)\mathbf{u}) + \mu \Delta \mathbf{u}.$$

From the regularity of $\mathbf{u}$ and ρ (see Remark 5.3), one immediately gets that

$$\mathbf{k} \in C^0([0, T_0]; L^2(\Omega)^3) \cap L^2(0, T_0; H^1(\Omega)^3) \tag{5.54}$$

and

$$\mathbf{k}_t \in L^2(\delta, T_0; L^2(\Omega)^3) \quad \forall \delta > 0. \tag{5.55}$$

In view of (5.6), one has:

$$(\mathbf{k}(\cdot\,, t), \mathbf{v}) = 0 \text{ a.e. in } (0, T_0) \quad \forall \mathbf{v} \in \mathcal{V}.$$

Now, we deduce from De Rham's Theorem (Theorem 2.7) and the fact that both (5.54) and (5.55) hold that $\mathbf{k} = \nabla p$ for some p satisfying

$$p \in C^0([0, T_0]; H^1(\Omega)) \cap L^2(0, T_0; H^2(\Omega))$$

and

$$p_t \in L^2(\delta, T_0; H^1(\Omega)) \quad \forall \delta > 0$$

that depends continuously on $\mathbf{u}_0$, ρ_0 and $\mathbf{f}$ in the corresponding norms.

This ends the proof. □

Remark 5.9 In order to obtain additional information on the pressure at $t = 0$, appropriate compatibility conditions for the initial data have to be imposed. This can be carried out as in [31], where the argument is detailed for the classical (constant density) Navier-Stokes equations. □

5.4 Some Results for Small Data When $N = 3$

In this section, we will present some global in time existence results of strong solutions for sufficiently small data in the three-dimensional case. We will also show that, under suitable conditions, the strong solution converges exponentially to zero as $t \to +\infty$.

Recall that, for two-dimensional flows, we have already seen in Theorem 5.1 that one has the global existence of a strong solution without any smallness requirement for the data.

The first result is the following:

Theorem 5.5 *Assume that $N = 3$, the assumptions in Theorem 5.1 are satisfied and, moreover,*

$$\mathbf{u}_0 \in D(A),\ \rho_0 \in C^1(\overline{\Omega}),\ \mathbf{f} \in L^\infty(0, +\infty; H^1(\Omega)^3)\ and\ \mathbf{f}_t \in L^\infty(0, +\infty; L^2(\Omega)^3).$$

Then, if $\|\mathbf{u}_0\|_{H^1}$ and $\|\mathbf{f}\|_{L^\infty(0,+\infty;L^2)}$ are sufficiently small, the strong solution to problem (5.1) *exists globally in time and satisfies*

$$\mathbf{u} \in C^0([0, +\infty); D(A)), \quad \rho \in C^1(\overline{\Omega} \times [0, +\infty)).$$

Furthermore, there exists C depending only on Ω and the norms of $\mathbf{u}_0$, ρ_0 and $\mathbf{f}$ in their respective spaces such that

$$\sup_{t\geq 0} (\|A\mathbf{u}(\cdot\,, t)\| + \|\mathbf{u}_t(\cdot\,, t)\|) \leq C \tag{5.56}$$

and

$$\sup_{t\geq 0} e^{-\gamma t}\int_0^t e^{\gamma s}\left(\|\mathbf{u}(\cdot,s)\|^2_{W^{2,6}}+\|\nabla\mathbf{u}_t(\cdot,s)\|^2\right)\,ds \leq C\left(1+\gamma+\frac{1}{\gamma}\right) \tag{5.57}$$

for all $\gamma > 0$.

Proof As in the results in Sect. 5.1, it will be sufficient to establish estimates like (5.56) and (5.57) for the semi-Galerkin approximations in the time interval where they are defined.

To keep the notation simple, we omit the index k.

We start from the differential inequality (5.16), proved in Sect. 5.1.

This inequality yields

$$\frac{d}{dt}\|\nabla\mathbf{u}(t)\|^2 + C\|A\mathbf{u}(t)\|^2 \leq C_1\|\nabla\mathbf{u}(t)\|^6 + C_2, \tag{5.58}$$

where C, C_1 and $C_2 = C\sup_{t\geq 0}\|\mathbf{f}(\cdot\,, t)\|$ are positive constants.

Now, since the norm in V is controlled by the norm in $D(A)$, we see from (5.58) that

$$\frac{d}{dt}\|\nabla\mathbf{u}\|^2 \leq C_1\|\nabla\mathbf{u}\|^6 - C_3\|\nabla\mathbf{u}\|^2 + C_2 \tag{5.59}$$

for some $C_3 > 0$. With $\psi(t) := \|\nabla\mathbf{u}(t)\|^2$, we get the differential inequality

$$\frac{d\psi}{dt} \le C_1\psi^3 - C_3\psi + C_2,$$

together with

$$\psi(0) = \|\nabla \mathbf{u}_0\|^2.$$

Let us consider the associated Cauchy problem

$$\frac{d\phi}{dt} = F(C_2, \phi) := C_1\phi^3 - C_3\phi + C_2, \quad \phi(0) = \psi(0).$$

It is clear that $\psi(t) \le \phi(t)$ for all t in the interval of existence.

Note that, when $C_2 = 0$, $F(0, \phi) = C_1\phi^3 - C_3\phi$ and, consequently, $F(0\,,\cdot)$ has one simple positive root, given by $r(0) = (C_2/C_1)^{1/2}$.

This root is unstable. Consequently, for small C_2, $F(C_2\,,\cdot)$ also has one unstable simple root $r(C_2)$ that is close to $r(0)$. Thus, whenever $0 < \psi(0) = \phi(0) < r(C_2)$, one necessarily has $0 \le \psi(t) \le \phi(t) \le r(C_2) < +\infty$ for all t in the interval of existence.

Therefore, there exists a constant $M > 0$ such that

$$\sup_{t\ge 0} \|\nabla \mathbf{u}(t)\| = M < +\infty. \tag{5.60}$$

In particular, we find that the maximal time of existence of strong solution introduced just after Theorem 5.1 is $T_* = +\infty$. In other words, the strong solution to (5.1) is defined in the whole interval $[0, +\infty)$.

In order to establish the estimates (5.56) and (5.57), we argue as in [31].

Thus, from the differential inequality (5.16) proved in Sect. 5.1, we get:

$$\frac{d}{dt}\|\nabla \mathbf{u}\|^2 + C\|A\mathbf{u}\|^2 + C\|\rho^{1/2}\mathbf{u}_t\|^2 \le \|\nabla \mathbf{u}\|^6 + C_3\,.$$

Multiplying by $e^{\gamma t}$ with $\gamma > 0$ and integrating in time from 0 to t, we see that

$$\begin{aligned} &e^{\gamma t}\|\nabla \mathbf{u}(t)\|^2 + C\int_0^t e^{\gamma s}\left(\|\mathbf{u}_t(s)\|^2 + \|A\mathbf{u}(s)\|^2\right) ds \\ &\quad \le C\int_0^t e^{\gamma s}\|\nabla \mathbf{u}(s)\|^6\, ds + \gamma\int_0^t e^{\gamma s}\|\nabla \mathbf{u}(s)\|^2\, ds + C_3\int_0^t e^{\gamma s}\, ds. \end{aligned}$$

Now, multiplying by $e^{-\gamma t}$ and recalling that $\|\nabla \mathbf{u}(t)\|$ is uniformly bounded (see (5.60)), we get:

$$e^{-\gamma t}\int_0^t e^{\gamma s}\|A\mathbf{u}(s)\|^2\, ds \text{ and } e^{-\gamma t}\int_0^t e^{\gamma s}\|\mathbf{u}_t(s)\|^2\, ds \text{ are uniformly bounded.}$$

Let us differentiate the first equation in (5.9) with respect to t and let us take $\mathbf{v} = \mathbf{u}_t$. One has:

$$\frac{1}{2}\frac{d}{dt}\|\rho^{1/2}\mathbf{u}_t\|^2 + \mu\|\nabla\mathbf{u}_t\|^2 = (\rho_t\mathbf{f}(\cdot\,,t),\mathbf{u}_t) + (\rho\mathbf{f}_t(\cdot\,,t),\mathbf{u}_t)$$
$$- 2(\rho(\mathbf{u}\cdot\nabla)\mathbf{u}_t,\mathbf{u}_t) - (\rho(\mathbf{u}_t\cdot\nabla)\mathbf{u},\mathbf{u}_t) - (\rho_t(\mathbf{u}\cdot\nabla)\mathbf{u},\mathbf{u}_t).$$

The terms on the right-hand side can be bounded as follows. First,

$$\begin{aligned}|(\rho(\mathbf{u}\cdot\nabla)\mathbf{u}_t,\mathbf{u}_t)| &\le \|\rho\|_{L^\infty}\|\mathbf{u}\|_{L^4}\|\nabla\mathbf{u}_t\|\,\|\mathbf{u}_t\|_{L^4}\\ &\le \|\rho\|_{L^\infty}\|\mathbf{u}\|^{1/4}\|\nabla\mathbf{u}\|^{3/4}\|\mathbf{u}_t\|^{1/4}\|\nabla\mathbf{u}_t\|^{7/4}\\ &\le \varepsilon\|\nabla\mathbf{u}_t\|^2 + C_\varepsilon\|\mathbf{u}_t\|^2\end{aligned}$$

and

$$\begin{aligned}|(\rho(\mathbf{u}_t\cdot\nabla)\mathbf{u},\mathbf{u}_t)| &\le \|\rho\|_{L^\infty}\|\mathbf{u}_t\|_{L^4}\|\nabla\mathbf{u}_t\|\,\|\mathbf{u}_t\|\\ &\le C\|\rho\|_{L^\infty}\|\mathbf{u}_t\|^{1/2}\|\nabla\mathbf{u}_t\|^{3/2}\|\nabla\mathbf{u}\|\\ &\le \varepsilon\|\nabla\mathbf{u}_t\|^2 + C_\varepsilon\|\mathbf{u}_t\|^2.\end{aligned}$$

Then,

$$|(\rho\mathbf{f}_t(\cdot\,,t),\mathbf{u}_t)| \le \|\mathbf{u}_t\|^2 + C\|\mathbf{f}_t(\cdot\,,t)\|^2.$$

Using the identity $\rho_t = -\nabla\cdot(\rho\mathbf{u})$ and integrating by parts, one also obtains:

$$\begin{aligned}|(\rho_t\mathbf{f}(\cdot\,,t),\mathbf{u}_t)| &\le |(\rho(\mathbf{u}\cdot\nabla)\mathbf{f}(\cdot\,,t),\mathbf{u}_t)| + |(\rho(\mathbf{u}\cdot\nabla)\mathbf{u}_t,\mathbf{f}(\cdot\,,t))|\\ &\le \varepsilon\|\nabla\mathbf{u}_t\|^2 + C_\varepsilon\|\mathbf{f}(\cdot\,,t)\|^2_{H^1}.\end{aligned}$$

Finally,

$$\begin{aligned}-2\int_\Omega \rho_t(\mathbf{u}\cdot\nabla)\cdot\mathbf{u}\,\mathbf{u}_t &= 2\int_\Omega \nabla\cdot(\rho\mathbf{u})(\mathbf{u}\cdot\nabla)\mathbf{u}\cdot\mathbf{u}_t\\ &\le C\|\rho\|_{L^\infty}\|\mathbf{u}\|_{L^6}\|\nabla\mathbf{u}\|_{L^6}\|\nabla\mathbf{u}\|\;\|\mathbf{u}_t\|_{L^6}\\ &\quad+C\|\rho\|_{L^\infty}\|\mathbf{u}\|^2_{L^6}\|A\mathbf{u}\|\;\|\mathbf{u}_t\|_{L^6}\\ &\quad+C\|\rho\|_{L^\infty}\|\mathbf{u}\|^2_{L^6}\|\nabla\mathbf{u}\|_{L^6}\;\|\nabla\mathbf{u}_t\|\\ &\le \varepsilon\|\nabla\mathbf{u}_t\|^2 + C_\varepsilon\|A\mathbf{u}\|^2.\end{aligned}$$

Consequently, if we choose $\varepsilon > 0$ small enough, we deduce that

$$\frac{d}{dt}\|\rho^{1/2}\mathbf{u}_t\|^2 + \|\nabla\mathbf{u}_t\|^2 \le C(\|\mathbf{u}_t\|^2 + \|A\mathbf{u}\|^2 + \|\mathbf{f}_t(t)\|^2 + \|\mathbf{f}(\cdot\,,t)\|^2_{H^1}).$$

Multiplying this inequality by $e^{\gamma t}$ and integrating in time from 0 to t, we find that

$$\begin{aligned} e^{\gamma t}\|\rho^{1/2}\mathbf{u}_t\|^2 &+ \int_0^t e^{\gamma s}\|\nabla\mathbf{u}_t(s)\|^2 ds \\ &\le \|\rho^{1/2}(0)\mathbf{u}_t(0)\| + C\int_0^t e^{\gamma s}(\|\mathbf{u}_t(s)\|^2 + \|A\mathbf{u}(s)\|^2)\,ds \\ &\quad +\gamma\int_0^t e^{\gamma s}\|\rho^{1/2}(s)\mathbf{u}_t(s)\|^2\,ds + C\int_0^t e^{\gamma s}(\|\mathbf{f}\|^2_{H^1} + \|\mathbf{f}_t\|^2)\,ds. \end{aligned} \tag{5.61}$$

Therefore,

$$\begin{aligned} \|\rho^{1/2}\mathbf{u}_t(t)\|^2 &+ e^{-\gamma t}\int_0^t e^{\gamma s}\|\nabla\mathbf{u}_t(s)\|^2\,ds \\ &\le C\left(e^{-\gamma t}\|\mathbf{u}_t(0)\|^2 + \gamma e^{-\gamma t}\int_0^t e^{\gamma s}\|\mathbf{u}_t(s)\|^2\,ds \right. \\ &\quad \left. +e^{-\gamma t}\int_0^t e^{\gamma s}(\|\mathbf{u}_t(s)\|^2 + \|A\mathbf{u}(s)\|^2 + 1)\,ds\right) \\ &\le C\left(e^{-\gamma t}\|\mathbf{u}_t(0)\|^2 + 1 + \gamma + \frac{1}{\gamma}\right). \end{aligned}$$

As in the proof of Lemma 5.2, an estimate for $\|\mathbf{u}_t(0)\|$ follows from (5.60) and the hypotheses on $\mathbf{u}_0$, ρ_0 and $\mathbf{f}$. Thus, one also has that

$$\sup_{t\ge 0}\|\mathbf{u}_t(t)\|^2 \le C. \tag{5.62}$$

Now, let us take $\mathbf{v} = A\mathbf{u}(t)$ in the first equation in (5.9). The following is found:

$$\|A\mathbf{u}\|^2 = (\rho\mathbf{f}, A\mathbf{u}) - (\rho\mathbf{u}_t, A\mathbf{u}) - (\rho(\mathbf{u}\cdot\nabla)\mathbf{u}, A\mathbf{u}).$$

Hence,

$$\begin{aligned} \|A\mathbf{u}\|^2 &\le \left(\|\rho\|_{L^\infty}\|\mathbf{f}\| + \|\rho\|_{L^\infty}\|\mathbf{u}_t\| + C\|\rho\|_{L^\infty}\|\mathbf{u}\|_{L^6}\|\nabla\mathbf{u}\|_{L^3}\right)\|A\mathbf{u}\| \\ &\le C(\|\mathbf{f}\|^2 + \|\mathbf{u}_t\|^2 + \|\nabla\mathbf{u}\|^6) + \frac{1}{2}\|A\mathbf{u}\|^2 \end{aligned}$$

and

$$\sup_{t\geq 0} \|A\mathbf{u}(t)\| \leq C. \tag{5.63}$$

The estimates (5.60), (5.62) and (5.63) give (5.56).

On the other hand, recalling that $A\mathbf{u}(t)$ is the projection of

$$\mathbf{F}(\cdot\,,t) := \rho(\cdot\,,t)(\mathbf{u}_t(t) + (\mathbf{u}(t)\cdot\nabla)\mathbf{u}(t) - \mathbf{f}(\cdot\,,t))$$

onto W and taking (5.60) and (5.63) into account, we see that $\|(\mathbf{u}(t)\cdot\nabla)\mathbf{u}(t)\|_{L^6} \leq C\|A\mathbf{u}(t)\| \leq C$. This, together with (5.62), yields

$$e^{-\gamma t}\int_0^t e^{\gamma s}\|\mathbf{F}(s)\|_{L^6}^2\,ds \leq C\left(1+\gamma+\frac{1}{\gamma}\right) \tag{5.64}$$

and, from the regularity of the Stokes operator (see Theorem 2.11; see also [2]) and Sobolev embeddings, we conclude that

$$e^{-\gamma t}\int_0^t e^{\gamma s}\left(\|\mathbf{u}(s)\|_{W^{2,6}}^2 + \|\nabla\mathbf{u}(s)\|_{L^\infty}^2\right)ds \leq C\left(1+\gamma+\frac{1}{\gamma}\right). \tag{5.65}$$

That is, (5.57) also holds. □

The estimate (5.57) asserts in an informal way that the function

$$s \mapsto \|\mathbf{u}(\cdot,s)\|_{W^{2,6}}^2 + \|\nabla\mathbf{u}_t(\cdot,s)\|^2$$

behaves as if it were in L^∞. The "best" right-hand side is obtained for $\gamma = 1$ and the corresponding inequality indicates that

$$\int_0^t e^s\left(\|\mathbf{u}(\cdot,s)\|_{W^{2,6}}^2 + \|\nabla\mathbf{u}_t(\cdot,s)\|^2\right)ds \leq Ce^t \quad \forall t\geq 0.$$

In the next result, we show that, for two-dimensional flows, a result similar to Theorem 5.5 holds without any smallness assumption on the data:

Theorem 5.6 *Let $N=2$. Suppose that the assumptions in Theorem 5.1 are satisfied and, moreover,*

$$\mathbf{u}_0 \in D(A),\quad \rho_0 \in C^1(\overline{\Omega}),\quad \mathbf{f}\in L^\infty(0,+\infty;H^1(\Omega)^2)$$
$$\text{and}\ \ \mathbf{f}_t \in L^\infty(0,+\infty;L^2(\Omega)^2).$$

Then, the strong solution to (5.1) *given by Theorem 5.1 exists globally in time and satisfies*

$$\mathbf{u}\in C^0([0,+\infty);D(A)) \ \ \text{and}\ \ \rho\in C^1(\overline{\Omega}\times[0,+\infty)).$$

Moreover, the estimates (5.57) *in Theorem 5.5 hold for any* $\gamma > 0$.

Proof Again, we start from the inequality (5.16), proved in Sect. 5.1.

In this two-dimensional case, we are led to the differential inequality

$$\frac{d}{dt}\|\nabla \mathbf{u}\|^2 \leq C\|\nabla \mathbf{u}\|^4 - \|\nabla \mathbf{u}\|^2 + C.$$

Consequently, for $\psi(t) := \|\nabla \mathbf{u}(t)\|^2$ and $\eta_0 = \|\nabla \mathbf{u}_0\|^2$, we get $\psi(t) \leq \eta(t)$, where η is the maximal solution to the Cauchy problem

$$\frac{d\eta}{dt} = (C\psi(t) - 1)\eta + C, \quad \eta(0) = \eta_0.$$

Since $\psi \in L^1(0, T)$ for all $T > 0$, the maximal existence interval contains $[0, +\infty)$, independently of the size of $\|\mathbf{u}_0\|$.

Hence, the global in time existence of the strong solution is ensured.

The rest of the proof can be achieved exactly as in the three-dimensional case. The details are left to the reader. □

Remark 5.10 The variable density Navier-Stokes equations have been studied in [51] through a slightly different formulation (basically, as a nonlinear differential equation in H). This approach yields results similar to Theorem 5.2. For the proof, the author uses some techniques from Tanabe [62]; see also [52]. □

We will end this section with some results from [5, 6] that hold when the external forces decay exponentially in time. Under these assumptions, the solutions have a better behavior at infinity.

Specifically, the following holds:

Theorem 5.7 *Suppose that* $N = 3$, *the hypotheses of Theorem 5.1 are satisfied and moreover*

$$\mathbf{u}_0 \in D(A), \quad \rho_0 \in C^1(\overline{\Omega})$$

and

$$e^{\overline{\gamma}t}\mathbf{f} \in L^\infty(0, +\infty; H^1(\Omega)^3), \quad e^{\overline{\gamma}t}\mathbf{f}_t \in L^\infty(0, +\infty; L^2(\Omega)^3)$$

for some $\overline{\gamma} > 0$. *If* $\|\mathbf{u}_0\|_{H^1}$ *and* $\|e^{\overline{\gamma}t}\mathbf{f}\|_{L^\infty(0,+\infty;L^2)}$ *are sufficiently small, the strong solution to system* (5.1) *given by Theorem 5.1 exists globally in time. Furthermore, there exists* $\gamma^* \leq \overline{\gamma}$ *such that, for any* $0 \leq \gamma < \gamma^*$, *the following estimates hold:*

$$\sup_{t\geq 0}\, e^{\gamma^* t}\|\nabla \mathbf{u}(\cdot\,, t)\|^2 < +\infty,$$

$$\sup_{t\geq 0}\, e^{\gamma t}(\|\mathbf{u}_t(\cdot\,, t)\|^2 + \|A\mathbf{u}(\cdot\,, t)\|^2) < +\infty,$$

$$\sup_{t\ge 0}\int_0^t e^{\gamma s}\left(\|\nabla \mathbf{u}_t(\cdot\,,s)\|^2+\|\mathbf{u}(\cdot\,,s)\|^2_{W^{2,6}}+\|\nabla \mathbf{u}(\cdot\,,s)\|^2_{L^\infty}\right)ds<+\infty,$$

$$\sup_{t\ge 0}\left(\|\nabla\rho(\cdot\,,t)\|_{L^\infty}+\|\rho_t(\cdot\,,t)\|_{L^\infty}\right)<+\infty,$$

$$\sup_{t\ge 0}\sigma(t)e^{\gamma t}\|\nabla \mathbf{u}_t(\cdot\,,t)\|^2<+\infty,$$

$$\sup_{t\ge 0}\int_0^t \sigma(s)e^{\gamma s}\left(\|\mathbf{u}_{tt}(\cdot\,,s)\|^2+\|A\mathbf{u}_t(\cdot\,,s)\|^2\right)ds<+\infty,$$

where $\sigma(t)=\min\{1,t\}$.

Proof In this proof, γ is chosen in $(0,\overline{\gamma})$ and the constants C, C_1, etc. below are independent of γ. Again, we will work with the semi-Galerkin approximations and the superscripts will be suppressed.

Multiplying the differential inequality (5.59) by $e^{\gamma t}$ with $0\le\gamma\le\overline{\gamma}$, we get

$$\begin{aligned}\frac{d}{dt}(e^{\gamma t}\|\nabla \mathbf{u}\|^2)&\le C_1e^{\gamma t}\|\nabla \mathbf{u}\|^6-C_2e^{\gamma t}\|\nabla \mathbf{u}\|^2+\gamma e^{\gamma t}\|\nabla \mathbf{u}\|^2+C_3\\&\le C_1e^{3\gamma t}\|\nabla \mathbf{u}\|^6+(\gamma-C_2)e^{\gamma t}\|\nabla \mathbf{u}\|^2+C_3\,,\end{aligned}$$

where

$$C_3=C\sup_{t\ge 0}e^{\gamma t}\|\mathbf{f}(\cdot\,,t)\|\le C\sup_{t\ge 0}e^{\overline{\gamma}t}\|\mathbf{f}(\cdot\,,t)\|.$$

Now, let us take $\gamma^*=\min\{\overline{\gamma},C_2/2\}$ and let us set $\psi(t):=e^{\gamma^* t}\|\nabla \mathbf{u}(t)\|^2$ for all t. Then

$$\frac{d\psi}{dt}\le C_1\psi^3-\overline{C}_2\psi+C_3\ \text{ and }\ \psi(0)=\|\nabla \mathbf{u}_0\|^2,$$

with $\overline{C}_2=C_2/2$.

As in the proof of Theorem 5.5, it is readily seen that one has

$$\sup_{t\ge 0}e^{\gamma^* t}\|\nabla \mathbf{u}(t)\|^2=\sup_{t\ge 0}\psi(t)\le C<+\infty, \tag{5.66}$$

provided the norms $\|\nabla \mathbf{u}_0\|$ and $\|e^{\overline{\gamma}t}\mathbf{f}\|_{L^\infty(0,+\infty;L^2)}$ are small enough.

Now, we multiply the differential inequality (5.16) by $e^{\gamma t}$ with $0\le\gamma<\gamma^*$ and we see that

$$\begin{aligned}&\frac{d}{dt}(e^{\gamma t}\|\nabla \mathbf{u}(t)\|^2)+Ce^{\gamma t}\|A\mathbf{u}(t)\|^2+Ce^{\gamma t}\|\rho(\cdot\,,t)^{1/2}\mathbf{u}(t)\|^2\\&\quad\le Ce^{\gamma t}\|\nabla \mathbf{u}(t)\|^6+\gamma e^{\gamma t}\|\nabla \mathbf{u}(t)\|^2+Ce^{\gamma t}\|\mathbf{f}(\cdot\,,t)\|^2.\end{aligned}$$

Integrating this inequality, we also see that

$$e^{\gamma t}\|\nabla\mathbf{u}(t)\|^2 + C\int_0^t e^{\gamma s}\|A\mathbf{u}(s)\|^2\,ds + C\int_0^t e^{\gamma s}\|\rho(\cdot\,,s)^{1/2}\mathbf{u}_t(s)\|^2\,ds$$

$$\leq \|\nabla\mathbf{u}_0\|^2 + C(\sup_{s\geq 0}\|\nabla\mathbf{u}(s)\|^4)\,(\sup_{s\geq 0} e^{\gamma^* s}\|\nabla\mathbf{u}(t)\|^2)\int_0^t e^{(\gamma-\gamma^*)s}\,ds$$

$$+\gamma\,(\sup_{t\geq 0} e^{\gamma^* t}\|\nabla\mathbf{u}(t)\|^2)\int_0^t e^{(\gamma-\gamma^*)s}\,ds$$

$$+C\,(\sup_{s\geq 0}\|\mathbf{f}(\cdot\,,s)\|)\,(\sup_{s\geq 0} e^{\gamma^* s}\|\mathbf{f}(\cdot\,,s)\|)\int_0^t e^{(\gamma-\gamma^*)s}\,ds$$

for all $t \geq 0$.

Since $\gamma < \gamma^*$, using (5.66) and the hypotheses of the theorem, we obtain:

$$\sup_{t\geq 0} e^{\gamma t}\|\nabla\mathbf{u}(t)\|^2 \leq C, \tag{5.67}$$

$$\sup_{t\geq 0}\int_0^t e^{\gamma s}\|A\mathbf{u}(s)\|^2\,ds \leq C \tag{5.68}$$

and

$$\sup_{t\geq 0}\int_0^t e^{\gamma s}\|\mathbf{u}_t(s)\|^2\,ds \leq C. \tag{5.69}$$

In order to prove the remaining estimates, we use (5.61) and (5.67)–(5.69) and we find that

$$e^{\gamma t}\|\rho(\cdot\,,t)^{1/2}\mathbf{u}_t(t)\|^2 + \int_0^t e^{\gamma s}\|\nabla\mathbf{u}_t(s)\|^2\,ds$$

$$\leq \|\rho_0^{1/2}\mathbf{u}_t(0)\| + C\int_0^t e^{\gamma s}(\|\mathbf{u}_t(s)\|^2 + \|A\mathbf{u}(s)\|^2)\,ds$$

$$+\gamma\int_0^t e^{\gamma s}\|\rho(\cdot\,,s)^{1/2}\mathbf{u}_t(s)\|^2\,ds + C\int_0^t e^{\gamma s}(\|\mathbf{f}(\cdot\,,s)\|^2_{H^1} + \|\mathbf{f}_t(\cdot\,,s)\|^2)\,ds$$

$$\leq \|\rho_0^{1/2}\mathbf{u}_t(0)\| + C + C\int e^{\gamma s}(\|\mathbf{f}(\cdot\,,s)\|^2_{H^1} + \|\mathbf{f}_t(\cdot\,,s)\|^2)\,ds.$$

We can estimate $\|\mathbf{u}_t(0)\|$ as before (see (5.30)). Also,

$$\int_0^t e^{\gamma s}(\|\mathbf{f}(\cdot\,,s)\|^2_{H^1} + \|\mathbf{f}_t(\cdot\,,s)\|^2)\,ds$$

$$
\begin{aligned}
&= \int_0^t e^{\overline{\gamma}s} e^{-(\overline{\gamma}-\gamma)s} (\|\mathbf{f}(\cdot\,,s)\|_{H^1}^2 + \|\mathbf{f}_t(\cdot\,,s)\|^2)\,ds \\
&\le (\sup_{s\ge 0}(e^{\overline{\gamma}s}(\|\mathbf{f}(\cdot\,,s)\|_{H^1}^2 + \|\mathbf{f}_t(\cdot\,,s)\|^2)) \int_0^t e^{-(\overline{\gamma}-\gamma)s}\,ds \\
&\le \frac{C^2}{\overline{\gamma}-\gamma} < +\infty,
\end{aligned}
$$

since $\gamma < \overline{\gamma}$ and $\mathbf{f}$ and $\mathbf{f}_t$ decay exponentially. Thus, we have

$$
\sup_{t\ge 0} e^{\gamma t} \|\mathbf{u}_t(t)\|^2 \le C \tag{5.70}
$$

and

$$
\sup_{t\ge 0} \int_0^t e^{\gamma s} \|\nabla \mathbf{u}_t(s)\|^2\,ds \le C. \tag{5.71}
$$

Also, arguing as in the proof of Theorem 5.5, we get

$$
\sup_{t\ge 0}\, e^{\gamma t} \|A\mathbf{u}(t)\|^2 \le C. \tag{5.72}
$$

With the aim to obtain uniform estimates of the density, we first note that the identity

$$
A\mathbf{u} = P(\rho(-\mathbf{u}_t - (\mathbf{u}\cdot\nabla)\mathbf{u} + \mathbf{f}))
$$

implies

$$
A(e^{\gamma t}\mathbf{u}(t)) = P(e^{\gamma t}(\rho(\cdot\,,t)(-\mathbf{u}_t(t) - (\mathbf{u}(t)\cdot\nabla)\mathbf{u}(t) + \mathbf{f}(\cdot\,,t)),
$$

with $0 < \gamma < \gamma^*$.

Using again the results by Amrouche and Girault in [2], it follows from the estimates (5.66)–(5.72) and the hypotheses on $\mathbf{f}$ that

$$
e^{\gamma t/2}\mathbf{u} \in L^2(0,+\infty; W^{2,6}(\Omega)^3)
$$

and Sobolev's embedding yields

$$
e^{\gamma t/2}\mathbf{u} \in L^2(0,+\infty; W^{1,\infty}(\Omega)^3). \tag{5.73}
$$

We will now use the following formula, that is not difficult to prove arguing as in Lemma 3.1 (see for instance [43], p. 705):

$$
\|\nabla\rho(\cdot\,,t)\|_{L^\infty} \le C\|\nabla\rho_0\|_{L^\infty} \exp\left(\int_0^t \|\nabla\mathbf{u}(\cdot\,,s)\|_{L^\infty}\,ds\right).
$$

This yields

$$\sup_{t\geq 0} \|\nabla\rho(\cdot\,,t)\|_{L^\infty} \leq C < +\infty. \tag{5.74}$$

Indeed, from (5.73) we can get the estimates

$$\begin{aligned}\int_0^t \|\nabla\mathbf{u}(s)\|_{L^\infty}\,ds &= \int_0^\infty e^{\gamma t/2}\|\nabla\mathbf{u}(t)\|_{L^\infty}e^{-\gamma t/2}dt\\ &\leq \left(\int_0^\infty e^{\gamma t}\|\nabla\mathbf{u}(t)\|_{L^\infty}^2 dt\right)^{1/2}\left(\int_0^\infty e^{-\gamma t}dt\right)^{1/2}\\ &\leq C\end{aligned}$$

and, in view of the equation satisfied by ρ, we see that

$$\sup_{t\geq 0} \|\rho_t(\cdot\,,t)\|_{L^\infty} \leq C.$$

In order to obtain higher order estimates, we proceed as follows.

First, we differentiate the first equation in (5.9) with respect to t and take $\mathbf{v} = \mathbf{u}_{tt}$. This gives:

$$\begin{aligned}\frac{1}{2}\frac{d}{dt}\|\nabla\mathbf{u}(t)\|^2 + \|\rho^{1/2}\mathbf{u}_{tt}\|^2 &= (\rho_t\mathbf{f}, \mathbf{u}_{tt}) + (\rho\mathbf{f}_t, \mathbf{u}_{tt}) - (\rho_t(\mathbf{u}\cdot\nabla)\mathbf{u}, \mathbf{u}_{tt})\\ &\quad -(\rho(\mathbf{u}_t\cdot\nabla)\mathbf{u}, \mathbf{u}_{tt}) + (\rho(\mathbf{u}\cdot\nabla)\mathbf{u}_t, \mathbf{u}_{tt})\\ &\quad -(\rho_t\mathbf{u}_t, \mathbf{u}_{tt}),\end{aligned}$$

which leads to the inequality

$$\frac{d}{dt}\|\nabla\mathbf{u}\|^2 + \alpha\|\mathbf{u}_{tt}\|^2 \leq C(\|\mathbf{f}\|^2 + \|\mathbf{f}_t\|^2 + \|\nabla\mathbf{u}\|^2 + \|\nabla\mathbf{u}_t\|^2 + \|\mathbf{u}_t\|^2).$$

Then, we multiply this inequality by $e^{\gamma t}\sigma(t)$ and we integrate in time from ε to t, which gives

$$\begin{aligned}e^{\gamma t}\|\nabla\mathbf{u}(t)\|^2 + \alpha\int_\varepsilon^t e^{\gamma s}\sigma(s)\|\mathbf{u}_{tt}(s)\|^2\,ds &\leq e^{\gamma\varepsilon}\sigma(\varepsilon)\|\nabla\mathbf{u}(\varepsilon)\|^2\\ &\quad + C\int_\varepsilon^t e^{\gamma s}\sigma(s)(\|\mathbf{f}(\cdot\,,s)\|^2 + \|\mathbf{f}_t(\cdot\,,s)\|^2 + \|\mathbf{u}(s)\|_{H^1}^2 + \|\mathbf{u}_t(s)\|_{H^1}^2)\,ds\\ &\leq \left(\int_0^t e^{(\gamma-\gamma')s}\,ds\right)\left(\sup_{s\geq 0} e^{\gamma' s}(\|\mathbf{f}(\cdot\,,s)\|^2 + \|\mathbf{f}_t(\cdot\,,s)\|^2 + \|\nabla\mathbf{u}(s)\|^2)\right)\\ &\quad + \int_0^t e^{\gamma' s}\|\mathbf{u}_t(s)\|_{H^1}^2\,ds\\ &\leq C,\end{aligned}$$

as a consequence of the assumptions and previous estimates.

Hence, taking a sequence $\{\varepsilon_n\}$ with $\varepsilon_n \to 0$ and using Remark 5.4, after some straightforward manipulations, we see that the remaining estimates hold. □

Once more, in the two-dimensional case, a similar result holds with no smallness assumption on $\mathbf{u}_0$ and $\mathbf{f}$:

Theorem 5.8 *Let $N = 2$. Suppose that the assumptions in Theorem 5.1 are satisfied and, moreover,*

$$\mathbf{u}_0 \in D(A), \quad \rho_0 \in C^1(\overline{\Omega})$$

and

$$e^{\overline{\gamma}t}\mathbf{f} \in L^\infty(0, +\infty; H^1(\Omega)^2), \quad e^{\overline{\gamma}t}\mathbf{f}_t \in L^\infty(0, +\infty; L^2(\Omega)^2)$$

for some $\overline{\gamma} > 0$. Then, the solution to problem (5.1) *furnished by Theorem 5.1 exists globally in time and, furthermore, the estimates in Theorem 5.7 hold for any $0 \leq \gamma < \gamma^*$ with $\gamma^* = \overline{\gamma}$.*

Proof We recall that, in this case,

$$\frac{d}{dt}\|\nabla \mathbf{u}\|^2 \leq C\|\nabla \mathbf{u}\|^4 + C\|\mathbf{f}\|^2$$

for a positive constant C. Multiplying this inequality by $e^{\overline{\gamma}t}$, we get

$$\begin{aligned}\frac{d}{dt}(e^{\overline{\gamma}t}\|\nabla \mathbf{u}\|^2) &\leq Ce^{\overline{\gamma}t}\|\nabla \mathbf{u}\|^4 + \overline{\gamma}e^{\overline{\gamma}t}\|\nabla \mathbf{u}\|^2 + Ce^{\overline{\gamma}t}\|\mathbf{f}\|^2 \\ &\leq C_1 e^{2\overline{\gamma}t}\|\nabla \mathbf{u}\|^4 + C_2 + Ce^{\overline{\gamma}t}\|\mathbf{f}\|^2.\end{aligned}$$

Defining $\psi(t) := e^{\overline{\gamma}t}\|\nabla \mathbf{u}(t)\|^2$ and $C_3 := C_2 + C\sup_{t\geq 0} e^{\overline{\gamma}t}\|\mathbf{f}(\cdot\,, s)\|^2$, the above inequality takes the form

$$\frac{d\psi}{dt} \leq C\psi^2 + C$$

and can be analyzed exactly as in Theorem 5.6.

The rest of the proof is completely similar to the proof of Theorem 5.7 and will not be repeated. □

Remark 5.11 Note that the estimates for $\mathbf{u}_{tt}$ given in the last two results hold true in the case of local existence if the supremum is taken in any interval $[0, T']$, with T' smaller than the maximal existence time. This shows that the solution to the local existence theorem actually satisfies $\mathbf{u}_t \in C^0([0, T']; H)$ and therefore $\mathbf{u} \in C^1([0, T']; H)$ for all such T'. □

5.5 Further Results and Open Questions

Many other results and questions can be considered in the context of existence, uniqueness and regularity of a solution to (5.1). Frequently (but not always), they have been motivated by the desire to extend results already known for the constant density Navier-Stokes equations.

Let us mention some of them and let us also indicate some recent advances and related references.[3]

- **Minimal hypotheses for good behavior.**

In view of well known results for the constant density Navier-Stokes equations, it would be interesting to find minimal or sharp assumptions on the weak solution to problem (5.1) of the kind

$$\mathbf{u} \in L^r(0, T; L^s(\Omega)^N), \quad \rho \in C^m(\overline{Q}),$$

leading to "good" properties such as the following:

- The triplet $\{\mathbf{u}, \rho, p\}$ is a strong solution.
- Additional regularity of the data implies additional regularity of $\{\mathbf{u}, \rho, p\}$. Recall that the results in Serrin [60, 61] ensure that a solution $\{\mathbf{u}, p\}$ to the classical Navier-Stokes equations with zero right-hand side such that $\mathbf{u}$ is (for example) locally L^∞ in time and space is in fact of class C^∞. Is this (or something similar) also true for the solutions to (5.1)?
- The energy identity is satisfied, etc.

Note that the answers are essentially unknown even when $N = 2$.

Several sufficient conditions have been given in the constant density case. For example, one has the results in [61] and [36]. These works have had a lot generalizations and improvements in various directions in recent times.

In the variable density framework, several conditions have also been found. Thus, it is proved in [38] that, if $N = 3$ and a strong solution in $(0, T^*)$ blows up at T^*, then

$$\int_0^{T^*} \|\mathbf{u}(\cdot\,, t)\|_{L^r}^s \, dt = +\infty$$

for all exponents r and s with

$$\frac{2}{s} + \frac{3}{r} = 1, \quad r > 3,$$

[3] Let us however indicate that there are several groups working at present on the analysis of variable density Navier-Stokes systems and related PDEs, specially in China. Consequently, it is not simple to summarize the most recent results on the field.

We thank one of the anonymous referees for this very appropriate comment.

that is, it cannot belong to $L^s(0, T^*; L^r(\Omega)^3)$. As immediate consequences, we find a regularity theorem and a global existence criterion for strong solutions.

Let us also mention [39], where the author investigates the three-dimensional variable density Navier-Stokes PDEs in a vacuum and obtains Liouville-type theorems in Lorentz spaces for a class of smooth solutions.

- **Regularity results for $\mu = \mu(\rho)$ and/or $\inf \rho_0 = 0$.**

As mentioned in Chaps. 1 and 2, it is meaningful to consider density dependent viscosity fluids governed by motion equations of the form

$$\frac{\partial \rho \mathbf{u}}{\partial t} + \nabla \cdot (\rho \mathbf{u} \otimes \mathbf{u}) - \nabla \cdot (2\mu(\rho) D(\mathbf{u})) + \nabla p = \rho \mathbf{f}.$$

Here, it also makes sense to assume that ρ_0 is non-negative, but not necessarily bounded from below by a positive constant. The existence and regularity of a strong solution in these cases is more difficult to analyze. Some results can be found in [48]; see also [54], where the authors deal with two-dimensional fluids in $\mathbb{R}^2$ and use some previous ideas by Chemin [11] to establish global existence and uniqueness.

- **Fujita-Kato analysis for variable density fluids.**

Assume that $\Omega = \mathbb{R}^N$.

For the Navier-Stokes equations, a classical way to investigate the existence and uniqueness of a solution is to write the system in the form

$$\begin{cases} \mathbf{u}_t + A\mathbf{u} = B(\mathbf{u}, \mathbf{u}), \\ \mathbf{u}(0) = \mathbf{u}_0, \end{cases} \tag{5.75}$$

where A is the Stokes operator and $B : X \times X \mapsto X'$ is given by

$$\langle B(\mathbf{u}, \mathbf{v}), \mathbf{w} \rangle_{X',X} := \int_\Omega (\mathbf{u} \cdot \nabla)\mathbf{v} \cdot \mathbf{w}$$

for all $\mathbf{u}$, $\mathbf{v}$, $\mathbf{w}$ in an appropriate Hilbert space X.

Then, some general abstract results by Kato and Fujita [37] can be applied and one gets that, for each $T > 0$, there exists $\delta > 0$ such that, whenever $\|\mathbf{u}_0\|_X \le \delta$, problem (5.75) has exactly one "mild" solution in $[0, T]$.

This approach has been revisited by many authors; see [44] for a complete description; see also [10, 12].

More recently, this analysis has been adapted to the variable density Navier-Stokes equations with the aim to produce existence-uniqueness results for (5.1).

This began with the work of Danchin in [13, 14] and it was then followed by [15, 18, 19, 23, 29, 53]; for a complete description of the tools from Fourier Analysis needed to solve problems of this kind (including Littlewood-Paley decomposition and applications to several specific examples), we refer to [3].

In [13] and [14], the equations are solved respectively in the whole space and in a periodic domain. The initial data are taken in appropriate Besov spaces. Local unique solvability is established for large initial velocity and strictly positive and bounded initial density, close to a constant.

It is also proved that the global solution exists whenever the initial velocity is small with respect to the viscosity μ.

These results are generalized and improved in [15] (where fluids in regular bounded N-dimensional domains are considered) and [29] (where the initial density is not necessarily bounded from below by a positive constant and less regularity is needed).

An interesting idea appears in the more recent paper [18], where a Lagrangian viewpoint is introduced and new existence and uniqueness results are obtained. As indicated in Chap. 3, this leads to uniqueness in any dimension for a rather wide class of velocity fields. Also, global existence is established in dimension two if the density is close enough to a positive constant and in dimension three if, moreover, the initial velocity is small.

The Lagrangian formulation is also used in [19] in the context of exterior domains with initial data in Besov spaces.

Let us also mention [53], where the authors consider the Cauchy problem in $\mathbb{R}^3$ with initial data slowly varying in the vertical variable. Their proof of existence is based on a suitable Paley-Littlewood decomposition and the analysis of the propagation of regularity for a transport equation with convection in anisotropic Besov spaces.

Finally, note that the global existence of a weak solution in $\mathbb{R}^3$ with $\rho_0 \in L^\infty(\mathbb{R}^3)$ satisfying $\rho_0 \geq \alpha > 0$ a.e. and $\mathbf{u}_0$ in the critical Besov space is proved in [64]. This result can be viewed as the extension to variable density fluids of Fujita-Kato theory.

- **Do suitable solutions exist?**

In 1981, Caffarelli et al. [8] introduced the concept of *weak suitable solution* for the constant density Navier-Stokes system. According to their definition, a suitable solution is, roughly speaking, a couple $\{\mathbf{u}, p\}$ that satisfies local in space energy estimates.

In [8], the existence of suitable solutions is established. It is also shown that any suitable solution fulfills good regularity properties concerning the "parabolic Haussdorff" measure of the set of singular points.

The results in [8] are a continuation of the work of Scheffer [57–59]. The proofs of these results have been rewritten and simplified in several papers, see [40–42, 47]. The convergence of approximate solutions to suitable weak solutions has been analyzed in [4, 27, 30].

As far as we know, a similar analysis has not been performed so far for the solutions to (5.1) (some specific difficulties related to the lack of regularity of the pressure make this question nontrivial).

- **The behavior of the solutions as $\mu \to 0$.**

For $\mu = 0$, we obtain the Euler equations for nonhomogeneous fluids.

The behavior of the solutions to (5.1) as $\mu \to 0$ has been analyzed when $\Omega = \mathbb{R}^3$ by Itoh [32, 33] in the L^2 context and by Itoh and Tani [35] in the L^p framework. Also, Danchin has proved in [16, 23] that the system is locally well-posed for initial data in appropriate Besov spaces.

Of course, the situation is much more complicated (and interesting) when the spatial domain Ω is a proper subset of $\mathbb{R}^3$. Actually, this is already a nontrivial problem for the standard Navier-Stokes equations; for a review on the subject, see [9].

- **Some recent results of particular interest.**

We will end this chapter by recalling several (more or less) recent achievements on the subject that, in our opinion, have special interest:

- In [6], it is proved among other facts that, under appropriate assumptions, if the initial data are small and smooth and the right-hand side **f** decays exponentially as $t \to +\infty$, the same happens to the strong solution in the H^2 norm; recall Theorems 5.7 and 5.8.

  This question has been reconsidered and solved for the isentropic compressible Navier-Stokes system in [22].
- In [15], a local-in-time solution in a bounded domain in $\mathbb{R}^N$ ($N \geq 2$) with a rough initial data in L^q ($q > N$) is found under the assumption that the initial density is bounded away from zero. The solution is global in time when $N = 2$ for arbitrary initial data and also when $N = 3$ for small initial data.

  This paper extends a result by O.A. Ladyzenskaja and V.A. Solonnikov [43] to a larger class of initial data. See also [7] for some related results.
- The papers [28] and [46] deal with the study of the so called *P.-L. Lions' open problem* on density patches concerning the propagation of regularity in 2D domains. The considered initial density is of the form

$$\rho_0 = et_1 \mathbb{1}_D + \eta_1 \mathbb{1}_{D^c},$$

  where $\eta_1, \eta_2 > 0$ and $D \subset\subset \Omega$ is a bounded, simply connected domain with smooth boundary. It is shown that, if the initial velocity is smooth enough, the problem has a unique local in time strong solution such that

$$\rho(\cdot\,, t) = et_1 \mathbb{1}_{D_t} + \eta_1 \mathbb{1}_{D_t^c},$$

  where the D_t are bounded, simply connected domains that preserve $W^{3,p}$ boundary regularity.

  A similar analysis for 3D domains is performed in [45].
- In [34], the authors establish the existence and uniqueness of a local in time solution to (5.1) in a (bounded or not) domain in $\mathbb{R}^3$ and, then, the convergence as $\mu \to 0$ of the solutions to a solution to a similar problem with $\mu = 0$. The results are proved in the L^p framework, with $p > 3$.

- In [55], the authors were able to improve the uniqueness results for the two-dimensional problem under the conditions that the initial density ρ_0 is bounded from above and from below by positive constants and the initial velocity u_0 possesses H^s regularity with $s > 0$. A similar result was obtained for $N = 3$ under the same condition on ρ_0, H^1 regularity for $\mathbf{u}_0$ and the requirement that $\|\mathbf{u}_0\| \, \|\nabla \mathbf{u}_0\|$ is sufficiently small.
- Finally, recall that [17] deals with the existence and uniqueness of a solution to (5.1) in $\mathbb{R}^N$ for $N = 2$ and 3 via the Lagrangian approach, that is, by rewriting the sums of the time derivative and convective terms as *material derivatives,* i.e., derivatives along trajectories.

The authors prove in [20] prove new existence and uniqueness results. There, the spatial domain is either the torus $\mathcal{T}^N$ or a simply connected bounded domain in $\mathbb{R}^N$ (again $N = 2$ or $N = 3$), the initial density satisfies just $0 \leq \rho_0 < \rho^*$ (ρ^* is given) and the initial velocity field is of class H^1 (again, a Lagrangian approach is used). See also [26], where similar results re established in the case of a bounded domain of C^2 boundary.

5.6 Exercises of Chapter 5

Exercise 5.1* Prove that the function ρ found in the proof of Theorem 5.1 satisfies $\rho \in C^0(\overline{\Omega} \times [0, T_*])$ and the initial condition $\rho(\cdot\,, 0) = \rho_0$.

HINT: Use the expression of ρ in terms of ρ_0 and the Lagrangian coordinates.

Exercise 5.2* With the notation used in Theorem 5.1, prove that either $T_* = T$ or $\limsup_{t \to T_*^-} \|\nabla \mathbf{u}(\cdot\,, t)\| = +\infty$.

HINT: Prove that, if $T_* < T$ and $\|\nabla \mathbf{u}(\cdot\,, t)\|$ is bounded, then there exist sequences $\{t_n\}$ such that $t_n \to T_*^-$ and $\mathbf{u}(\cdot\,, t_n)$ converges weakly in V to some $z \in V$, whence T_* cannot be the maximal time of existence of a strong solution.

Exercise 5.3* Prove that the velocity $\mathbf{u}$ given by Theorem 5.1 satisfies the assertions in Remark 5.3.

HINT: Deduce uniform bounds in the mentioned spaces of the first and second order time derivatives of the approximations $\mathbf{u}^k$.

Exercise 5.4* Prove (5.33) using (5.36) and the regularity properties of the Stokes operator.

Exercise 5.5* Prove (5.35) using (5.33)–(5.34)

HINT: Recall Lemma 3.1 and the argument used in its proof.

Exercise 5.6* Prove the assertion in Remark 5.4.

Exercise 5.7* Prove the estimates (5.40) using (5.38) and arguing as in the proof of Lemma 5.4.

Exercise 5.8* Prove Lemma 5.6.

HINT: Consider again the time derivative of the first equation in (5.9) and take $\mathbf{v} = A\mathbf{u}_t$.

Exercise 5.9* Justify that, in the proof of Theorem 5.2, the inequality (5.45) suffices to prove (5.42).

HINT: Prove that $\liminf_{t\to 0^+} \|A\mathbf{u}(t)\| \geq \|A\mathbf{u}_0\|$.

Exercise 5.10* Using (5.42) and the equations satisfied by $\{\mathbf{u}, \rho, p\}$, prove the identity (5.46).

Exercise 5.11* Prove the assertions in Remark 5.6.

Exercise 5.12* Complete the details of the proof of Proposition 5.1.

Exercise 5.13* Using (5.64) and the regularity properties of the Stokes operator, prove (5.65).

Exercise 5.14* Give the details of the proof of Theorem 5.6.

Exercise 5.15* Complete the details of the proofs of Theorems 5.7 and 5.8.

Exercise 5.16* Prove (5.53) using (5.49)–(5.52).

Exercise 5.17* Give the details of the proof of the uniqueness result asserted in Remark 5.8.

References

1. Adams, R.A.: Sobolev Spaces. Pure and Applied Mathematics, vol. 65. Academic Press [A subsidiary of Harcourt Brace Jovanovich, Publishers], New York-London (1975)
2. Amrouche, C., Girault, V.: On the existence and regularity of the solutions of Stokes problem in arbitrary dimension. Proc. Jpn. Acad. Ser. A Math. Sci. **67**(5), 171–175 (1991)
3. Bahouri, H., Chemin, J.-Y., Danchin, R.: Fourier Analysis and Nonlinear Partial Differential Equations. Grundlehren der Mathematischen Wissenschaften [Fundamental Principles of Mathematical Sciences], vol. 343. Springer, Heidelberg (2011)
4. Beirão da Veiga, H.: On the construction of suitable weak solutions to the Navier-Stokes equations via a general approximation theorem. J. Math. Pures Appl. (9) **64**(3), 321–334 (1985)
5. Boldrini, J.-L., Rojas-Medar, M.A.: Global solutions to the equations for the motion of stratified incompressible fluids. Mat. Contemp. **3**, 1–8 (1992)
6. Boldrini, J.-L., Rojas-Medar, M.A.: Global strong solutions of the equations for the motion of nonhomogeneous incompressible fluids. In: Conca, C., Gatica, G.N. (eds.) Numerical Methods in Mechanics, number 371. Pitman Res. Notes Math. Ser., pp. 35–45. Longman, Harlow (1997)
7. Braz e Silva, P., Rojas-Medar, M., Villamizar-Roa, E.J.: Strong solutions for the nonhomogeneous Navier-Stokes equations in bounded domains. Mat. Methods Appl. Sci. **33**(3), 358–372 (2010)
8. Caffarelli, L., Kohn, R., Nirenberg, L.: Partial regularity of suitable weak solutions of the Navier-Stokes equations. Commun. Pure Appl. Math. **35**(6), 771–831 (1982)
9. Caflisch, R.E., Sammartino, M.: Existence and singularities for the Prandtl boundary layer equations. ZAMM Z. Angew. Math. Mech. **80**(11–12), 733–744 (2000). Special issue on the occasion of the 125th anniversary of the birth of Ludwig Prandtl

10. Cannone, M.: Harmonic analysis tools for solving the incompressible Navier-Stokes equations. In: Handbook of Mathematical Fluid Dynamics, vol. III, pp. 161–244. North-Holland, Amsterdam (2004)
11. Chemin, J.-Y.: Existence globale pour le problème des poches de tourbillon. C. R. Acad. Sci. Paris Sér. I Math. **312**(11), 803–806 (1991)
12. Chemin, J.-Y.: About weak-strong uniqueness for the 3D incompressible Navier-Stokes system. Commun. Pure Appl. Math. **64**(12), 1587–1598 (2011)
13. Danchin, R.: Density-dependent incompressible viscous fluids in critical spaces. Proc. R. Soc. Edinb. Sect. A **133**(6), 1311–1334 (2003)
14. Danchin, R.: Local and global well-posedness results for flows of inhomogeneous viscous fluids. Adv. Differ. Equ. **9**(3–4), 353–386 (2004)
15. Danchin, R.: Density-dependent incompressible fluids in bounded domains. J. Math. Fluid Mech. **8**(3), 333–381 (2006)
16. Danchin, R.: On the well-posedness of the incompressible density-dependent Euler equations in the L^p framework. J. Differ. Equ. **248**(8), 2130–2170 (2010)
17. Danchin, R., Mucha, P.B.: A Lagrangian approach for the incompressible Navier-Stokes equations with variable density. Commun. Pure Appl. Math. **65**(10), 1458–1480 (2012)
18. Danchin, R., Mucha, P.B.: Incompressible flows with piecewise constant density. Arch. Ration. Mech. Anal. **207**(3), 991–1023 (2013)
19. Danchin, R., Mucha, P.B.: Critical functional framework and maximal regularity in action on systems of incompressible flows. Mém. Soc. Math. Fr. (N.S.) (143), vi+151 (2015)
20. Danchin, R., Mucha, P.B.: The incompressible Navier-Stokes equations in vacuum. Commun. Pure Appl. Math. **72**(7), 1351–1385 (2019)
21. Danchin, R., Wang, S.: Global unique solutions for the inhomogeneous Navier-Stokes equations with only bounded density, in critical regularity spaces. Commun. Math. Phys. **399**(3), 1647–1688 (2023)
22. Danchin, R., Wang, S.: Exponential decay for the inhomogeneous viscous flows on the torus. Z. Angew. Math. Phys. **75**(2) (2024)
23. Danchin, R., Zhang, P.: Inhomogeneous Navier–Stokes equations in the half-space, with only bounded density. J. Funct. Anal. **267**(7), 2371–2436 (2014)
24. DiPerna, R.J., Lions, P.-L.: Ordinary differential equations, transport theory and Sobolev spaces. Invent. Math. **98**(3), 511–547 (1989)
25. Farwig, R., Sohr, H.: Optimal initial value conditions for the existence of local strong solutions of the Navier-Stokes equations. Math. Ann. **345**(3), 631–642 (2009)
26. Farwig, R., Qian, Ch., Zhang, P.: Incompressible inhomogeneous fluids in bounded domains of $\mathbb{R}^3$ with bounded density. J. Funct. Anal. **278**(5), 108394, 36 pp. (2020)
27. Fernández-Cara, E., Marín-Gayte, I.: A new proof of the existence of suitable weak solutions and other remarks for the Navier-Stokes equations. Appl. Math. **9**, 383–402 (2018)
28. Gancedo, F., García-Juárez, E.: Global regularity of 2D density patches for inhomogeneous Navier-Stokes. Arch. Ration. Mech. Anal. **229**(1), 339–360 (2018)
29. Germain, P.: Strong solutions and weak-strong uniqueness for the nonhomogeneous Navier-Stokes system. J. Anal. Math. **105**, 169–196 (2008)
30. Guermond, J.-L.: Faedo-Galerkin weak solutions of the Navier-Stokes equations with Dirichlet boundary conditions are suitable. J. Math. Pures Appl. (9) **88**(1), 87–106 (2007)
31. Heywood, J.G., Rannacher, R.: Finite element approximation of the nonstationary Navier-Stokes problem I: regularity of solutions and second order error estimates for spatial discretization. SIAM J. Numer. Anal. **19**(2), 275–311 (1982)
32. Itoh, S.: On the vanishing viscosity in the Cauchy problem for the equations of a nonhomogeneous incompressible fluid. Glasgow Math. J. **36**(1), 123–129 (1994)
33. Itoh, S.: Cauchy problem for the Euler equations of a nonhomogeneous ideal incompressible fluid. II. J. Korean Math. Soc. **32**(1), 41–50 (1995)
34. Itoh, S., Tani, A.: Solvability of nonstationary problems for nonhomogeneous incompressible fluids and the convergence with vanishing viscosity. Tokyo J. Math. **22**(1), 17–42 (1999)

35. Itoh, S., Tani, A.: The initial value problem for the non-homogeneous Navier-Stokes equations with general slip boundary condition. Proc. R. Soc. Edinb. Sect. A **130**(4), 827–835 (2000)
36. Kaniel, S.: A sufficient condition for smoothness of solutions of Navier-Stokes equations. Israel J. Math. **6**(1968), 354–358 (1969)
37. Kato, T., Fujita, H.: On the nonstationary Navier-Stokes system. Rend. Sem. Mat. Univ. Padova **32**, 243–260 (1962)
38. Kim, H.: A blow-up criterion for the nonhomogeneous incompressible Navier-Stokes equations. SIAM J. Math. Anal. **37**, 1417–1434 (2006)
39. Kim, J.-M.: Regularity criteria and Liouville theorems for 3D inhomogeneous Navier-Stokes flows with vacuum. Arch. Math. (Basel) **121**(1), 89–98 (2023)
40. Kukavica, I.: On partial regularity for the Navier-Stokes equations. Discrete Contin. Dyn. Syst. **21**(3), 717–728 (2008)
41. Kukavica, I.: Partial regularity results for solutions of the Navier-Stokes system. In: Partial Differential Equations and Fluid Mechanics, vol. 364. London Math. Soc. Lecture Note Ser., pp. 121–145. Cambridge University Press, Cambridge (2009)
42. Kukavica, I.: Partial regularity for the Navier-Stokes equations with a force in a Morrey space. J. Math. Anal. Appl. **374**(2), 573–584 (2011)
43. Ladyzhenskaya, O.A., Solonnikov, V.A.: The unique solvability of an initial-boundary value problem for viscous incompressible inhomogeneous fluids. In: Boundary Value Problems of Mathematical Physics and Related Questions of the Theory of Functions 8, vol. 52. Zap. Naučn Sem. Leningrad. Otdel. Mat. Inst. Steklov (LOMI), pp. 52–109, 218–219 (1975)
44. Lemarié-Rieusset, P.G.: The Navier-Stokes Problem in the 21st Century. CRC Press, Boca Raton (2016)
45. Liao, X., Liu, Y.: Global regularity of three-dimensional density patches for inhomogeneous incompressible viscous flow. Sci. China Math. **62**(9), 1749–1764 (2019)
46. Liao, X., Zhang, P.: Global regularity of 2D density patches for viscous inhomogeneous incompressible flow with general density: low regularity case. Commun. Pure Appl. Math. **72**(4), 835–884 (2019)
47. Lin, F.: A new proof of the Caffarelli-Kohn-Nirenberg theorem. Commun. Pure Appl. Math. **51**(3), 241–257 (1998)
48. Lions, P.-L.: Mathematical Topics in Fluid Mechanics. Vol I: Incompressible Models. Oxford Lecture Series in Mathematics and Its Applications, vol. 3. The Clarendon Press, Oxford University Press, New York (1996)
49. Lions, P.-L., Masmoudi, N.: Uniqueness of mild solutions of the Navier-Stokes system in L^N. Commun. Partial Differ. Equ. **26**(11–12), 2211–2226 (2001)
50. Masmoudi, N.: Uniqueness results for some PDEs. In: Journées "Équations aux Dérivées Partielles", pp. Exp. No. X, 13. Univ. Nantes, Nantes (2003)
51. Okamoto, H.: On the equation of nonstationary stratified fluid motion: uniqueness and existence of the solutions. J. Fac. Sci. Univ. Tokyo Sect. IA Math. **30**(3), 615–643 (1984)
52. Ortega-Torres, E., Braz e Silva, P., Rojas-Medar, M.A.: Analysis of an iterative method for variable density incompressible fluids. Ann. Univ. Ferrara Sez. VII Sci. Mat. **55**(1), 129–151 (2009)
53. Paicu, M., Zhang, P.: On some large global solutions to 3-D density-dependent Navier-Stokes system with slow variable: well-prepared data. Ann. Inst. H. Poincaré Anal. Non Linéaire **32**(4), 813–832 (2015)
54. Paicu, M., Zhang, P.: Striated regularity of 2-D inhomogeneous incompressible Navier-Stokes systems with variable viscosity. Commun. Math. Phys. **376**(1), 385–439 (2020)
55. Paicu, M., Zhang, P., Zhang, Z.: Global unique solvability of inhomogeneous Navier-Stokes equations with bounded density. Commun. Partial Differ. Equ. **38**(7), 1208–1234 (2013)
56. Salvi, R.: The equations of viscous incompressible nonhomogeneous fluid: on the existence and regularity. J. Austral. Math. Soc. Ser. B **33**(1), 94–110 (1991)
57. Scheffer, V.: Partial regularity of solutions to the Navier-Stokes equations. Pacific J. Math. **66**(2), 535–552 (1976)

58. Scheffer, V.: Turbulence and Hausdorff dimension. In: Turbulence and Navier-Stokes Equations (Proc. Conf., Univ. Paris-Sud, Orsay, 1975). Lecture Notes in Math., vol. 565, pp. 174–183. Springer, Berlin (1976)
59. Scheffer, V.: Hausdorff measure and the Navier-Stokes equations. Commun. Math. Phys. **55**(2), 97–112 (1977)
60. Serrin, J.: On the interior regularity of weak solutions of the Navier-Stokes equations. Arch. Ration. Mech. Anal. **9**, 187–195 (1962)
61. Serrin, J.: The initial value problem for the Navier-Stokes equations. In: Nonlinear Problems (Proc. Sympos., Madison), pp. 69–98. Univ. of Wisconsin Press, Madison (1963)
62. Tanabe, H.: Functional Analytic Methods for Partial Differential Equations. Monographs and Textbooks in Pure and Applied Mathematics, vol. 204. Marcel Dekker, Inc., New York (1997)
63. Vo-Khac, K.: Introduction aux méthodes mathématiques modernes de la physique. Espaces préhilbertiens; série de Fourier; spectre des opérateurs. Distributions, convolutions et transformation de Fourier; espaces de Sobolev. Equations différentielles; transformation de Laplace; fonctions orthogonales. Equations aux dérivées partielles; méthode de Galerkine-Fourier. Centre de Documentation Universitaire, Paris (1968). Faculté des Sciences d'Orléans, Cours de Mathématiques C2 de la Maîtrise de Physique
64. Zhang, P.: Global Fujita-Kato solution of 3-D inhomogeneous incompressible Navier-Stokes system. Adv. Math. **363**, 107007, 43 pp. (2020)

Chapter 6
Control Results

Our aim in this chapter is to present and solve some control problems for systems governed by the variable density Navier-Stokes equations.

The guiding idea in Control Theory is that one or several data (the domain, a right-hand side, a boundary data, etc.) are at our disposal and our interest is to choose these data in order to get solution(s) with "good" or even "the best" properties.

For example, we may be interested in determining the shape of a solid body to get an acceptable behavior of a fluid that flows around; or the inflow velocity we must prescribe in a channel to make the fluid leave at a desired speed; or the way we must move solid walls along a time interval in order to bring the fluid to rest, etc.

First, in Sect. 6.2, we will adopt the viewpoint of optimal control, with several functionals. This means that we try to find controls (the data) such that the associated triplets $\{\mathbf{u}, \rho, p\}$ minimize a cost (or equivalently maximize a payoff).

Then, in Sect. 6.3, we will consider some optimal time control problems, where the goal is to find controls that make $\{\mathbf{u}, \rho, p\}$ satisfy a desired property in a time as short as possible.

Finally, Sect. 6.4 will be devoted to the formulation and solution of some controllability questions. In this case, we will search for controls whose mission is to "drive" the associated solutions (the states) to a prescribed target at final time.

Actually, these strategies or viewpoints can be used in combination, leading to the very interesting subfields of multi-objective and hierarchical control.

In the context of fluid mechanics, we can find many examples of control problems, most of them motivated by real-world questions:

- The control of a fluid in a channel-like domain, with applications to physics, biology, engineering, etc. The action can be performed in several ways: via the inwards velocity, by insulating or suctioning flow through the walls, with an imposed force field of gravitational or magnetic kind, etc.; see for instance [17, 19, 29, 66, 74, 87, 100, 107]. This situation is illustrated in Figs. 6.1 and 6.2,

P. Braz e Silva et al., *Analysis and Control of the Variable Density Incompressible Navier-Stokes Equations*, MS&A 22, https://doi.org/10.1007/978-3-032-14510-9_6

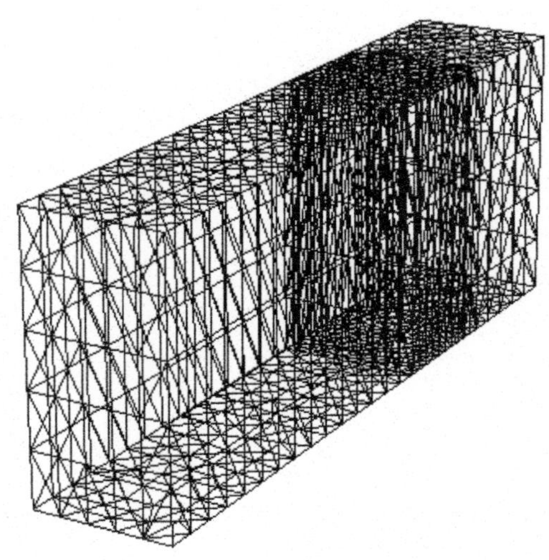

Fig. 6.1 Controlling a fluid in a two-dimensional channel (Poiseuille): a space-time mesh for the computation. The spatial domain is the base and the vertical axis indicates time (number of vertices: 1842; number of elements: 7890; final time $T = 2$). The flow is from back to front. The control is applied on the dark part of the domain

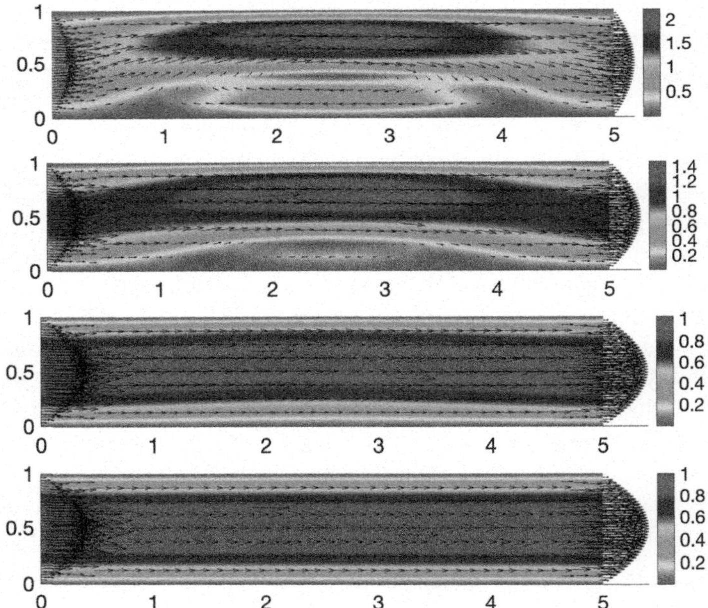

Fig. 6.2 Controlling a fluid in a channel: the velocity. From top to bottom: fields at times $t = 0, 0.6, 1.2, 1.9$

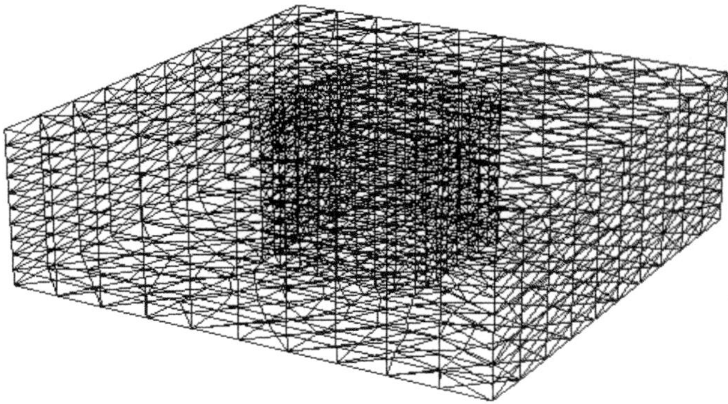

Fig. 6.3 Controlling a fluid in a two-dimensional square (Taylor-Green): a space-time mesh for the computation (number of vertices: 3146; number of elements: 15,900; final time $T = 1$)

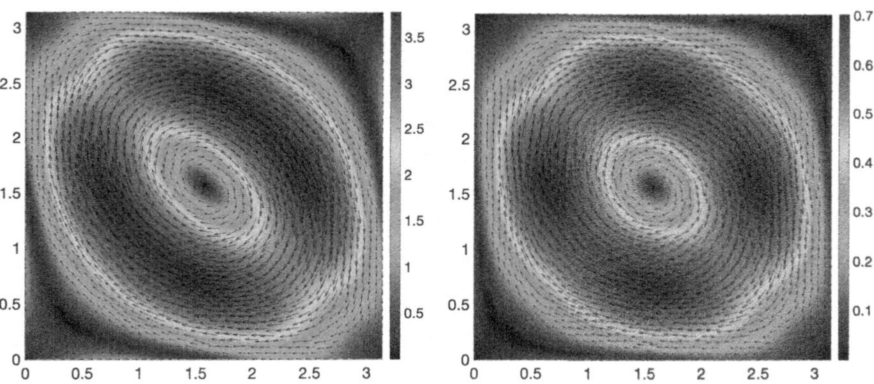

Fig. 6.4 Controlling a fluid in a square (Taylor-Green): the velocity fields at times $t = 0$ (left) and $t = 0.3$ (right)

where a two-dimensional constant density Navier-Stokes flow is driven by the action of a control (an internal force field applied in a small part of the domain) to reach at final time a desired Poiseuille profile; the problem, its approximation and the numerical method are described in [51].

- A similar problem has been exemplified in Figs. 6.3, 6.4, 6.5, and 6.6. There, the fluid particles are assumed to fill a square and, starting from a particular initial state, the system is steered to a so called Taylor-Green flow; the details can also be found in [51].

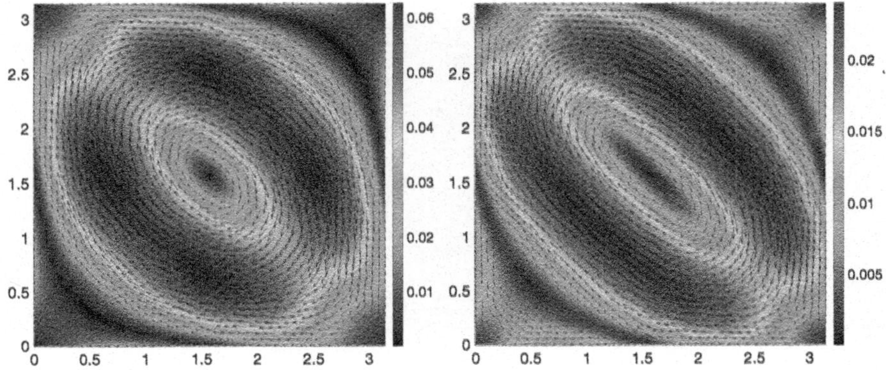

Fig. 6.5 Controlling a fluid in a square (Taylor-Green): the velocity fields at times $t = 0.45$ (left) and $t = 0.55$ (right)

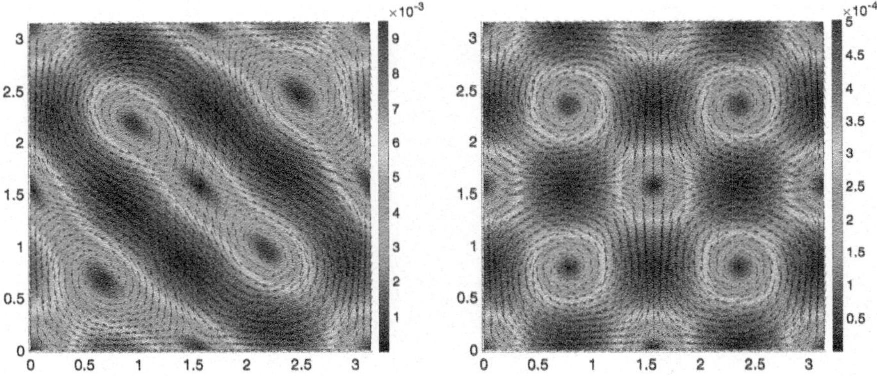

Fig. 6.6 Controlling a fluid in a square (Taylor-Green): the velocity fields at times $t = 0.65$ (left) and $t = 0.95$ (right)

- Controlling a flow in an exterior domain, especially with applications to aeronautics. In this case, there are also several possible control mechanisms to govern the system: the shape of the obstacle(s), a surface heating device, insulating or suctioning flow, etc.; see [54, 57, 62, 67, 70, 88, 92, 103]. An example is exhibited in Fig. 6.7.
- Controlling pollution in fluid domains may be a third source of motivating problems; see [2, 9, 108] for more details.[1] The control of a pollution model is illustrated in Figs. 6.8 and 6.9.

[1] A very interesting and challenging problem is the control of the evolution of the *Great Pacific Garbage Patch*. A complete description of the status and possible solution (the OC001 system) is given in [102]; see also [75].

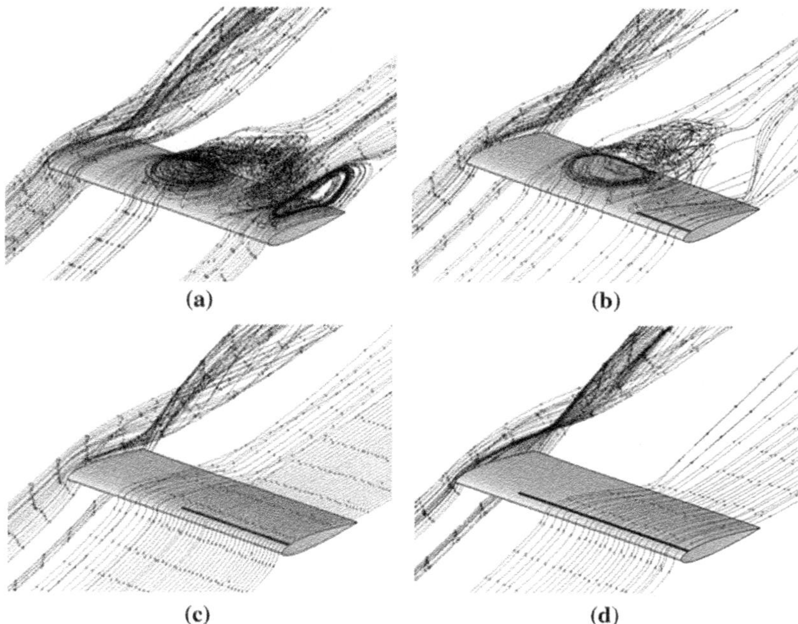

Fig. 6.7 A practical suction control problem: effect of suction jet length on streamlines over finite wing at angle of attack of 18° for center suction; b_s is the length of the active suction segment; C is the half-length of the wing; (**a**) no control; (**b**) $b_s = 0.5\ C$; (**c**) $b_s = 1.0\ C$; (**d**) $b_s = 1.5\ C$. Taken from [111] by courtesy of the authors

For viscous incompressible flows, these tasks lead to a lot of very interesting questions from the theoretical, numerical and practical viewpoints.

Let us end this introductory paragraph by indicating that, from a mathematical point of view, there are at least three natural ways to act that correspond to three different control mechanisms:

1. By fixing the geometrical characteristics of the effective fluid domain.
2. By introducing distributed forces (that is, locally supported right-hand sides in the PDEs).
3. By imposing appropriate data on (a part of) the physical boundary.

Of course, these actions are not incompatible.

We will try to make this clear in the following sections.

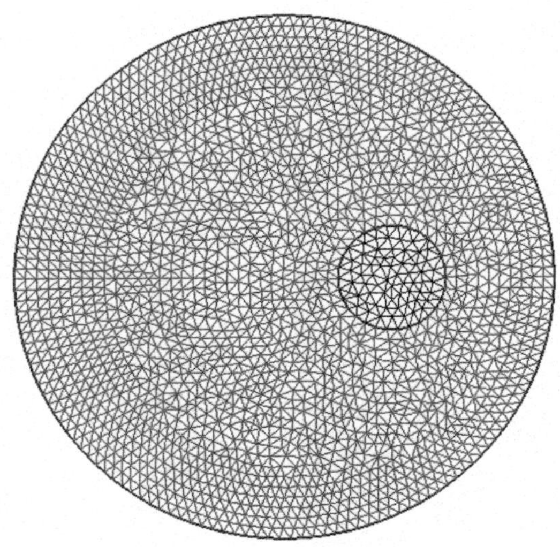

Fig. 6.8 The domain, the mesh and the control region. Number of vertices: 2092. Number of triangles: 4032

6.1 A Brief History of Control Theory

The past results in Control Theory are rich in concepts, arguments and applications. They will be briefly summarized in this section; more details can be found for instance in [12, 47, 82].

There is a well known sentence by Aristotle that can be viewed as a philosophical motivation of Control Theory. It appears in Chapter 3, Book 1, of the monograph "Politics":

> ...if every instrument could accomplish its own work, obeying or anticipating the will of others ...if the shuttle weaved and the pick touched the lyre without a hand to guide them, chief workmen would not need servants, nor masters slaves.

The first elements of Control Theory can be probably found in the ancient Mesopotamia, around 6000 years B.C., in the control of irrigation systems. They can also be recognized in the ancient Egypt around 2500 years B.C., in the building processes of large constructions.

Later, we find in Greece water clocks. They were motivated by the need of accurate time measurements and seem to represent the first realization of feedback control devices.[2]

[2] About 270 B.C., the Greek Ktesibios invented a float regulator, whose mission was to keep constant the water level of a tank. This allowed to fill a second tank at a constant rate and measure time by observing the level in the second tank.

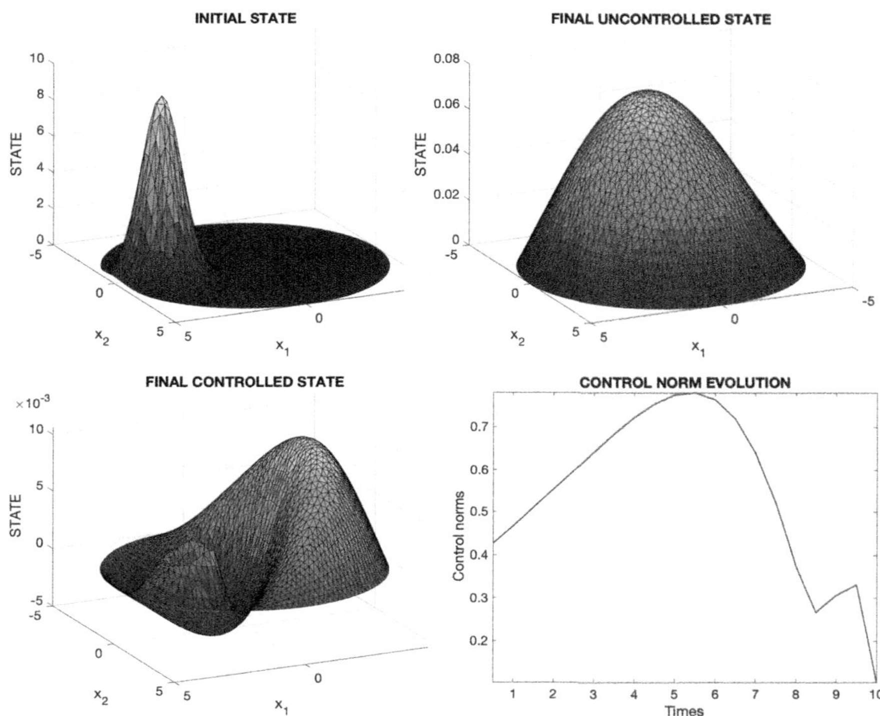

Fig. 6.9 Evolution of the pollution in a medium at rest: from left to right and top to bottom, the initial pollution density, the final time situation without control (maximum value ≈ 0.080 units), the final time controlled pollution density (max. ≈ 0.011 units) and the evolution of the L^2 control norm

We can also mention the Roman aqueducts, starting about 300 years B.C., equipped with ingenious systems of regulating valves and other elements to keep the water level constant between distant points.

Much later, towards the end of the seventeenth century, we find the work by Ch. Huygens and R. Hooke on the *oscillations of the pendulum.* Their aim was to achieve precise measurements of time and location, indispensable for navigation purposes. In particular, Huygens discovered a way to synchronize the motion of two pendulum clocks.

These contributions were followed by advances in the control of the velocity of windmills. The main mechanism invented at that time was based on a system of balls rotating around an axis, with a velocity proportional to the velocity of the windmill. When the rotational velocity increased, the balls got farther from the axis, acting on the wings of the mill through appropriate mechanisms.

The industrial revolution was a period of transition of human economy towards more efficient and stable manufacturing processes. It was mainly characterized by a transformation of hand production into mechanized fabrication systems. Also, it was marked by the invention of advanced devices and very specially steam engines, where speed regulation techniques played a fundamental role.

At this point, it is convenient to mention the contribution of J. Watt. In 1788, he completed the design of a centrifugal machine inspired in the windmill platform. In this case, as the velocity of the balls increased, some valves were opening to let the vapor scape. As the pressure inside the boiler diminished, the velocity began to go down. The goal was of course to keep the velocity as close as possible to a constant.

The regulating system invented by Watts was first analyzed by the astronomer G. Airy and then by J.C. Maxwell, who gave an explanation to the erratic behavior and proposed control mechanisms.

With the developments coming with the industrial revolution, the central ideas of what today is known as Control Theory gained a remarkable impact. Later, in the 1920s and 1930s, engineers began to apply semi-automatic and automatic control techniques everywhere and *Control Engineering* germinated and got the recognition of an individual discipline.

The number of applications increased covering amplifiers in telephone systems, distribution systems in electrical plants, stabilization of airplanes, electrical mechanisms in paper, chemistry, petroleum and steel industry, etc.

During the Second World War and the following years, engineers and scientists improved their experience on the control mechanisms of plane tracking and ballistic missiles and the design of anti-aircraft batteries. This produced an important development of what is called since then frequential methods.

After 1960, all what we have mentioned above began to be considered "Classical Control Theory". By that time, it was observed that the models considered up to the moment were not accurate enough to describe the complexity of real-word. Indeed, it became clear that actually accurate systems are often *nonlinear* and even *non-deterministic*, since they are affected by "noise".

These considerations led to the contributions of R. Bellman in the context of *dynamic programming,* R. Kalman in *filtering techniques* and L. Pontryagin with the *maximum principle* for nonlinear optimal control problems. This allowed to establish the foundations of "Modern Control Theory".

As a consequence of all this, Control Theory can be regarded nowadays as an interdisciplinary area where engineering, mathematics and other sciences melt perfectly and enrich each other. In fact, mathematics has been playing an increasing role in Control Theory since decades. As we have said, the need to consider nonlinear and non-deterministic phenomena increases the need of mathematics and its degree of complexity.

An important component of Control Theory is the concept of *feedback.* This term was incorporated to Control Engineering about 1920 by the engineers of the "Bell Telephone Laboratory". However, at that time, it had already been consolidated in other areas, such as political economics.

In general terms, it is said that we apply a feedback law (or a feedback control definition) if the choice of the control is carried out by observing the state of the system. Nowadays, feedback processes are ubiquitous not only in economics, but also in biology, psychology, etc. Accordingly, in many different related areas, the *cause-effect principle* is not understood as a static rule any more, but it is rather being viewed from a dynamical perspective.

Our society provides every day new problems to Control Theory and this fact is stimulating the creation of new mathematics. Indeed, the range of applications of Control Theory goes from the simplest to the most sophisticated mechanisms we manipulate in everyday life, many of them emerging in new technologies. The books [82] and [97] provide rather complete descriptions of this variety of applications; to this respect, see also section.

6.2 Optimal Control. Problems and Results

We will view the variable density Navier-Stokes system as a state equation, that is, the mathematical model whose solution gives information on the status of the physical system.

For convenience, we will mainly consider distributed controls, locally supported in space, that can be interpreted as force fields. Occasionally, we will also deal with controls of other kinds.

Thus, it will be assumed that the velocity field, the mass density and the pressure of the fluid are governed by the system

$$\begin{cases} \dfrac{\partial \rho \mathbf{u}}{\partial t} + \nabla \cdot (\rho \mathbf{u} \otimes \mathbf{u}) - \mu \Delta \mathbf{u} + \nabla p = \mathbf{v} \mathbb{1}_{\omega}, & (\mathbf{x}, t) \in Q, \\ \nabla \cdot \mathbf{u} = 0, \quad (\mathbf{x}, t) \in Q, \\ \dfrac{\partial \rho}{\partial t} + \nabla \cdot (\rho \mathbf{u}) = 0, \quad (\mathbf{x}, t) \in Q, \\ \mathbf{u} = 0, \quad (\mathbf{x}, t) \in \Sigma, \\ \rho|_{t=0} = \rho_0, \quad (\rho \mathbf{u})|_{t=0} = \rho_0 \mathbf{u}_0, \quad \mathbf{x} \in \Omega, \end{cases} \tag{6.1}$$

where (as in previous chapters) $\Omega \subset \mathbb{R}^N$ is a bounded connected open set whose boundary is of class C^2 ($N=2$ or $N=3$), $Q = \Omega \times (0, T)$, $\Sigma = \partial\Omega \times (0, T)$, $\omega \subset \Omega$ is a (small) non-empty open set and ρ_0 and $\mathbf{u}_0$ are given. In accordance with the usual terms in control theory, $\mathbf{v}$ is the *control* and any associated solution $\{\mathbf{u}, \rho, p\}$ is a *state*. We recall that, for any set G, $\mathbb{1}_G$ denotes the characteristic function of G.

For simplicity, we will assume in this chapter that, at least,

$$\rho_0 \in L^\infty(\Omega) \text{ with } 0 < \alpha \le \rho_0(\mathbf{x}) \le \beta \text{ a.e. in } \Omega \text{ and } \mathbf{u}_0 \in H. \tag{6.2}$$

Consequently, any solution satisfies $\alpha \leq \rho(\mathbf{x}, t) \leq \beta$ a.e. in Q and the initial conditions in (6.1) can also be written in the form

$$\rho|_{t=0} = \rho_0, \quad \mathbf{u}|_{t=0} = \mathbf{u}_0, \quad \mathbf{x} \in \Omega.$$

We must interpret $\mathbf{v}1_\omega$ as a force source. In fact, we should have written $\rho\mathbf{v}1_\omega$ in the right-hand side of (6.1) and then view $\mathbf{v}$ as a control; this is however equivalent to using $\mathbf{v}$ only as long as ρ is bounded from below by α.

6.2.1 Some First Optimal Control Problems

We will deal with a family of problems that share the following structure:

$$\begin{cases} \text{Minimize } \; J(\mathbf{v}, \mathbf{u}, \rho, p) \\ \text{subject to } \; \mathbf{v} \in \mathcal{U}_{\text{ad}}, \; (\mathbf{v}, \mathbf{u}, \rho, p) \text{ satisfies (6.1).} \end{cases} \tag{6.3}$$

Here and henceforth, $\mathcal{U}_{\text{ad}}$ will be a non-empty subset of $L^2(\omega \times (0, T))^N$ and we will consider control-state "quartets" $(\mathbf{v}, \mathbf{u}, \rho, p)$ that "solve" (6.1) at least in the weak sense, i.e., in the sense indicated in Theorem 3.1.

We will denote by $\mathcal{S}_{\text{ad}}$ the following family of these control-states:

$$\mathcal{S}_{\text{ad}} := \{(\mathbf{v}, \mathbf{u}, \rho, p) : \mathbf{v} \in \mathcal{U}_{\text{ad}}, \; \{\mathbf{u}, \rho, p\} \text{is a weak solution to } \; (6.1) \text{ inQ}\}.$$

Several cost functions $J : \mathcal{S}_{\text{ad}} \mapsto \mathbb{R}$ will be considered. The first one is given by

$$\begin{cases} J(\mathbf{v}, \mathbf{u}, \rho, p) = \dfrac{a}{2} \displaystyle\iint_Q |\mathbf{u} - \mathbf{u}_d|^2 + \dfrac{a'}{2} \iint_Q |\rho - \rho_d|^2 \\ \qquad\qquad + \dfrac{b}{2} \displaystyle\iint_{\omega \times (0,T)} |\mathbf{v}|^2, \end{cases} \tag{6.4}$$

where a, a', b are nonnegative constants (obviously, not all them equal to zero) and $\mathbf{u}_d \in L^2(Q)^N$ and $\rho_d \in L^2(Q)$ are prescribed.

A second possible choice of J is the following:

$$J(\mathbf{v}, \mathbf{u}, \rho, p) = \frac{a}{2} \int_\Omega |\mathbf{u}(\mathbf{x}, T) - \mathbf{u}_e(\mathbf{x})|^2 + \frac{b}{2} \iint_{\omega \times (0,T)} |\mathbf{v}|^2, \tag{6.5}$$

where a and b are as before and $\mathbf{u}_e \in L^2(\Omega)^N$ is given.

Recall that, under the conditions (6.2), for every $\mathbf{v} \in L^2(\omega \times (0, T))^3$, any weak solution to (6.1) satisfies $\mathbf{u} \in C^0_w([0, T]; H)$. Consequently, $\mathbf{u}(\cdot\,, T)$ has a sense in H and (6.5) is an acceptable definition of cost function.

Later, we will also deal with the somewhat more complicated functional

$$J(\mathbf{v}, \mathbf{u}, \rho, p) = \frac{1}{2}\, T^*(\mathbf{v}, \mathbf{u}, \rho, p; \mathbf{u}_e, \delta)^2 + \frac{b}{2} \iint_{\omega\times(0,T)} |\mathbf{v}|^2, \tag{6.6}$$

where $\mathbf{u}_e \in H$, $\delta > 0$ and

$$T^*(\mathbf{v}, \mathbf{u}, \rho, p; \mathbf{u}_e, \delta) := \inf\{T > 0 : \|\mathbf{u}(\cdot\,, T) - \mathbf{u}_e\| \le \delta\}. \tag{6.7}$$

Clearly, for a given control-state quartet $(\mathbf{v}, \mathbf{u}, \rho, p)$, this quantity can be viewed as the time needed by the fluid to reach a velocity at distance $\le \delta$ from $\mathbf{u}_e$.

The following comments and interpretations are in order:

- Equation (6.4) provides a balance for two criteria: "getting a velocity field and/or a mass density respectively close to $\mathbf{u}_d$ and ρ_d in Q" and "using a non-expensive (small or at least moderate) control $\mathbf{v}$".
- On the other hand, (6.5) provides a balance for "reaching a final velocity that is close to $\mathbf{u}_e$ in Ω" and, again, "using a small $\mathbf{v}$".
- Finally, (6.6) provides a balance for "being at distance $\le \delta$ from $\mathbf{u}_e$ as soon as possible" and, once more, "using a small control".

Many other cost functionals are possible (and interesting). Let us mention some of them:

- Several observation times can be used. Thus, for instance, instead of (6.5), we can take

$$J(\mathbf{v}, \mathbf{u}, \rho, p) = \frac{1}{2} \sum_{i=1}^{I} a_i \int_\Omega |\mathbf{u}(\mathbf{x}, T_i) - \mathbf{u}^i_e(\mathbf{x})|^2 + \ldots \tag{6.8}$$

  for some T_i with $0 < T_1 < \cdots < T_I \le T$, where the dots may contain one or several terms like those above.
- We can also consider other norms. For instance,

$$J(\mathbf{v}, \mathbf{u}, \rho, p) = \frac{a}{\gamma} \int_\Omega |\mathbf{u}(\mathbf{x}, T) - \mathbf{u}_e(\mathbf{x})|^\gamma + \ldots \tag{6.9}$$

  for some appropriate $\gamma \ge 1$.
- An alternative to minimize (6.6) and (6.7) can be *to maximize the time to be far from a fixed target.*
- Another possible strategy is to observe $L\mathbf{u}$ instead of $\mathbf{u}$, for some $L \in \mathcal{L}(H; Z)$ (Z is another suitable Hilbert space). For instance, Z can be a finite-dimensional subspace of H and $L : H \mapsto Z$ can be the corresponding orthogonal projector.

In that case, we can set

$$J(\mathbf{v}, \mathbf{u}, \rho, p) = \frac{a}{2} \int_{\Omega} |(L\mathbf{u}(\cdot, T))(\mathbf{x}) - \mathbf{z}_e(\mathbf{x})|^2 + \dots \tag{6.10}$$

with $\mathbf{z}_e \in Z$.

- Let $O \subset \Omega$ be a non-empty open set. Then, it is meaningful to observe the behavior of the flow only in O at $t = T$. Thus, we can take the following cost functional:

$$J(\mathbf{v}, \mathbf{u}, \rho, p) = \frac{a}{2} \int_{O} |\mathbf{u}(\mathbf{x}, T) - \mathbf{u}_e(\mathbf{x})|^2 + \dots \tag{6.11}$$

In principle, it is reasonable to expect that "small" sets O are related to "easy-to-solve" optimal control problems. But this has to be clarified and quantified.

- Let us assume that, together with (6.2), one has $N = 2$, $\mathbf{u}_0 \in V$ and $\rho_0 \in C^0(\overline{\Omega})$. Then, in view of Theorem 5.1, it makes sense to take

$$J(\mathbf{v}, \mathbf{u}, \rho, p) = \frac{c}{2} \iint_{Q} |(p - \frac{1}{|\Omega|} \int_{\Omega} p) - p_d|^2 + \dots \tag{6.12}$$

or even

$$J(\mathbf{v}, \mathbf{u}, \rho, p) = \frac{c}{2} \iint_{Q} |\nabla p - \nabla p_d|^2 + \dots \tag{6.13}$$

where $c > 0$ and $p_d \in L^2(0, T; H^1(\Omega))$ with $\int_\Omega p_d = 0$ a.e. in $(0, T)$ are given.
With these functionals, we can formulate control problems where the objective is to govern the behavior of the pressure. Such a goal is rather natural when dealing with aeronautical and navigation problems, but also when considering many biomedical applications; see for instance [30, 85, 104, 109].

Note that, due to the lack of uniqueness, the previous cost functionals cannot be regarded as functions of the control $\mathbf{v}$ alone.

The admissible set $\mathcal{U}_{\text{ad}}$ can also be chosen in many different ways. In the simplest case, we just take

$$\mathcal{U}_{\text{ad}} = L^2(\omega \times (0, T))^N. \tag{6.14}$$

It is not difficult to understand that this corresponds to a rather non-realistic situation where we can dispose of controls of any size and structure in $L^2(\omega \times (0, T))^N$.

A more natural and reasonable choice is

$$\mathcal{U}_{\text{ad}} = \{\mathbf{v} \in L^2(\omega \times (0, T))^N : \mathbf{v} = \sum_{i=1}^{I} \mathbf{v}^i(\mathbf{x}) \mathbb{1}_{(t_i, \tau_i)}(t) \text{ a.e., } \mathbf{v}^i \in L^2(\omega)^N\}, \tag{6.15}$$

where the t_i and τ_i satisfy

$$0 \leq t_1 < \tau_1 < t_2 < \tau_2 < \cdots < t_I < \tau_I \leq T.$$

Clearly, this means that the controls can be active only for $t \in \bigcup_{i=1}^{I}(t_i, \tau_i)$. If the intervals (t_i, τ_i) are small, these controls can be viewed as approximations to instantaneous in time functions of the form $\mathbf{v} = \sum_{i=1}^{I} \mathbf{v}^i(\mathbf{x})\, \delta_{(t=t_i)}$.

It also makes sense to take

$$\mathcal{U}_{\rm ad} = \{\mathbf{v} \in L^2(\omega \times (0,T))^N : |\mathbf{v}| \leq M \text{ a.e.}\}, \tag{6.16}$$

with $M > 0$. This choice has an obvious motivation.

More generally, for any given partially ordered Banach space Z and any continuous convex function $\Phi : L^2(\omega \times (0,T))^N \mapsto Z$, we can take

$$\mathcal{U}_{\rm ad} = \{\mathbf{v} \in L^2(\omega \times (0,T))^N : \Phi(\mathbf{v}) \leq 0\}. \tag{6.17}$$

Control problems arise in many areas with many different objectives and applications. In fluid mechanics they are very natural and have been studied since many years.

Many details on the history and the present state of the art of Control Theory and its applications can be found in [4, 60, 66, 80].

As for many other optimal control problems, three main sets of questions appear in connection with (6.3).

They are the following:

- **Existence and/or uniqueness:**

  Under which conditions on J and $\mathcal{U}_{\rm ad}$ can we ensure that (6.3) has at least one solution $(\mathbf{v}^*, \mathbf{u}^*, \rho^*, p^*)$?

  If it exists, $\mathbf{v}^*$ is called by definition an *optimal control* and any associated solution $\{\mathbf{u}^*, \rho^*, p^*\}$ is called an *optimal state.*

  When can we prove that the solution $(\mathbf{v}^*, \mathbf{u}^*, \rho^*, p^*)$ is unique?

- **Characterization:**

  How can we characterize the solution(s) to (6.3)? In particular, is it possible to find a system necessarily satisfied by any optimal control, an associated state and (maybe) additional variables?

  This is called an optimality system.

  An additional interesting question is: can we find sufficient conditions for optimality (eventually under convenient additional assumptions)?

  Are these conditions necessary?

- **Computation:**

  Can we find iterative algorithms that provide sequences of control-states that converge (in a suitable sense) to a solution to (6.3)?

  If this is the case, can we estimate the convergence rate?

Of course, a comparative analysis of convergence rates versus computational costs is in order.

These issues will be analyzed in the following sections in the particular framework of variable density Navier-Stokes fluids.

Let us mention that there are many other important aspects in Control Theory, more or less connected to those above. One of them is the determination of *feedback laws* and closed loop maps. As already said, in a general control system, a feedback law is a rule that indicates the control needed to achieve a goal in terms of the associated state. It is also relevant to investigate robustness, to deal with singular controls, etc.

For reasons of space, we will not deal with these aspects here; some information can be found, for instance, in [7, 79, 91].

6.2.2 The Existence of Optimal Control-States

Let us emphasize that the mapping *control-to-state* is not well defined in this framework, because of the possible lack of uniqueness of solution. In other words, the values of the cost functional cannot be considered as a function of the control only.

Concerning existence, one has the following general result:

Theorem 6.1 *Assume that the following hypotheses are satisfied:*

1. $\mathcal{U}_{\text{ad}} \subset L^2(\omega \times (0,T))^N$ *is non-empty, closed and convex.*
2. J *is sequentially weakly lower semi-continuous. In other words, if the quartets* $(\mathbf{v}^m, \mathbf{u}^m, \rho^m, p^m)$ *satisfy* (6.1), $\mathbf{v}^m \to \mathbf{v}$ *weakly in* $L^2(\omega \times (0,T))^N$, $\mathbf{u}^m \to \mathbf{u}$ *weakly in* $L^2(0,T;V)$, $\rho^m \to \rho$ *weakly-∗ in* $L^\infty(Q)$ *and* $p^m \to p$ *weakly-∗ in* $W^{-1,\infty}(0,T;L^2(\Omega))$, *then* $(\mathbf{v}, \mathbf{u}, \rho, p)$ *also solves* (6.1) *and*
$$\liminf_{m\to\infty} J(\mathbf{v}^m, \mathbf{u}^m, \rho^m, p^m) \geq J(\mathbf{v}, \mathbf{u}, \rho, p).$$
3. *Either* $\mathcal{U}_{\text{ad}}$ *is bounded or* J *is coercive in* $\mathbf{v}$, *that is, if the* $\mathbf{v}^m \in \mathcal{U}_{\text{ad}}$, *the* $\{\mathbf{u}^m, \rho^m, p^m\}$ *are associated states and* $\|\mathbf{v}^m\|_{L^2(\omega\times(0,T))} \to \infty$, *then*
$$J(\mathbf{v}^m, \mathbf{u}^m, \rho^m, p^m) \to +\infty \ \text{ as } \ m \to +\infty.$$

Then, the optimal control problem (6.3) *possesses at least one solution.*

Proof The argument is classical in control theory; in fact, it stems from Calculus of Variations and, for completeness, will be recalled here.

Let us consider a minimizing sequence $\{(\mathbf{v}^m, \mathbf{u}^m, \rho^m, p^m)\}$. We deduce from hypothesis *(3)* that the $\mathbf{v}^m$ are uniformly bounded in $L^2(\omega \times (0,T))^N$. Consequently, we can assume that they converge weakly to some $\mathbf{v} \in L^2(\omega \times (0,T))^N$.

From hypothesis *(1)*, we have $\mathbf{v} \in \mathcal{U}_{\rm ad}$.

It can be assumed that $\int_\Omega p^m = 0$ in $\mathcal{D}'(0, T)$ for all m. Indeed, it would suffice to change p^m by $p^m - \frac{1}{|\Omega|}\int_\Omega p^m$ if this is needed. Then, from the estimates in Chap. 3, it is clear that the $\{\mathbf{u}^m, \rho^m, p^m\}$ belong to a bounded set in $L^2(0, T; V) \times L^\infty(Q) \times W^{-1,\infty}(0, T; L^2(\Omega))$ and it can also be assumed that

$$\mathbf{u}^m \to \mathbf{u} \ \text{ weakly in } \ L^2(0, T; V), \tag{6.18}$$

$$\rho^m \to \rho \ \text{ weakly-}* \text{ in } L^\infty(Q), \tag{6.19}$$

and

$$p^m \to p \ \text{ weakly-}* \text{ in } \ W^{-1,\infty}(0, T; L^2(\Omega)), \tag{6.20}$$

where $(\mathbf{v}, \mathbf{u}, \rho, p)$ is a control-state quartet, that is, it satisfies (6.1).

Recall that the last convergence property means that

$$\langle p^m, \varphi\rangle_{W^{-1,\infty}(0,T;L^2), W_0^{1,1}(0,T;L^2)} \to \langle p, \varphi\rangle_{W^{-1,\infty}(0,T;L^2), W_0^{1,1}(0,T;L^2)}$$

for every $\varphi \in W_0^{1,1}(0, T; L^2(\Omega))$.

From hypothesis *(2)*, we also have

$$\inf_{\mathcal{U}_{\rm ad}} J = \lim_{m\to\infty} J(\mathbf{v}^m, \mathbf{u}^m, \rho^m, p^m) \geq J(\mathbf{v}, \mathbf{u}, \rho, p).$$

Consequently, $(\mathbf{v}, \mathbf{u}, \rho, p)$ solves (6.3) and the proof is done. □

Note that the particular cost functions (6.4) and (6.5) with $b > 0$ and the admissible sets $\mathcal{U}_{\rm ad}$ given by (6.14), (6.15), (6.16), and (6.17) satisfy the hypotheses in Theorem 6.1.

The unique delicate point to check is hypothesis *(2)*.

In the case of (6.4), this is evident, since the first two terms are convex and continuous respectively in $L^2(Q)^N$ and $L^2(Q)$ and, consequently, lower semicontinuous for the weak convergence in these spaces.

In the case of (6.5), note that, if $\mathbf{v}^m \to \mathbf{v}$ weakly in $L^2(\omega \times (0, T))^N$, we can assume (6.18) and also

$$\mathbf{u}^m \to \mathbf{u} \ \text{ weakly-}* \text{ in } L^\infty(0, T; H) \tag{6.21}$$

and

$$\frac{\partial \mathbf{u}^m}{\partial t} \to \frac{\partial \mathbf{u}}{\partial t} \ \text{ weakly in } \ L^2(0, T; V'). \tag{6.22}$$

Consequently, in view of Lemma 2.19 and Exercise 2.52, one has

$$\mathbf{u}^m(\cdot\,,T) \to \mathbf{u}(\cdot\,,T) \ \text{ weakly in } \ H,$$

whence the weak sequential lower semi-continuity holds too.

Hence, there exist solutions to the optimal control problems corresponding to these choices.

For the cost function (6.6), the argument is more intricate and will be detailed in Sect. 6.3. Other cost functionals considered in Sect. 6.2.1 will be analyzed later, in Sect. 6.2.4.

Observe that Theorem 6.1 does not imply the uniqueness of the minimizer. In practice, uniqueness can only be ensured if the function to minimize is strictly convex and the family of quartets $(\mathbf{v}, \mathbf{u}, \rho, p)$ where we minimize is convex; see for instance [38]. But this is out of question in the context of (6.3), because of the nonlinearities in the PDEs.

6.2.3 Optimality Systems

Now, let us present some optimality results.

As already mentioned, the goal is to deduce a system necessarily satisfied by any optimal control-state together with some additional variables.

This is much in the spirit of *Lagrange multipliers* and is completely natural in the present framework, since (6.3) can (and must) be viewed as a constrained extremal problem: note that we are minimizing $J(\mathbf{v}, \mathbf{u}, \rho, p)$ subject to the requirement $\mathbf{v} \in \mathcal{U}_{\rm ad}$ and the "equality constraints" (6.1).

For convenience, together with the sets $\mathcal{U}_{\rm ad}$ and $\mathcal{S}_{\rm ad}$, we will consider the energy space $E(T)$, formed by the $\{\mathbf{u}, \rho, p\}$ satisfying

$$\mathbf{u} \in L^2(0,T;\,V) \cap L^\infty(0,T;\,H),\quad \rho \in L^\infty(\Omega\times(0,T)),\quad p \in W^{-1,\infty}(0,T;\,L^2(\Omega)).$$

For the moment, we will be concerned with the cost functional (6.4). For simplicity, we will assume that $a' = 0$; see however Exercise 6.3 for a more general case.

The following holds:

Theorem 6.2 *Assume that* $\mathcal{U}_{\rm ad} \subset L^2(\omega\times(0,T))^N$ *is non-empty, closed and convex and* J *is given by* (6.4). *Let* $(\mathbf{v}^*, \mathbf{u}^*, \rho^*, p^*)$ *be an optimal control-state for* (6.3) *and assume that* $\{\mathbf{u}^*, \rho^*, p^*\}$ *satisfies (for instance)*

$$\frac{\partial u_k^*}{\partial t},\ \partial_i u_k^*,\ \partial_i \partial_j u_k^*,\ \frac{\partial \rho^*}{\partial t} \in L^\infty(Q)\ \ \forall i, j, k = 1, \ldots, N. \tag{6.23}$$

Then, there exists $\{\mathbf{w}, \eta, q\}$ *with* $\mathbf{w} \in L^2(0, T; V) \cap L^\infty(0, T; H)$, $\eta \in L^\infty(Q)$ *and* $q \in W^{-1,\infty}(0, T; L^2(\Omega))$ *such that*

$$\begin{cases} \rho^* \dfrac{\partial \mathbf{u}^*}{\partial t} + \rho^*(\mathbf{u}^* \cdot \nabla)\mathbf{u}^* - \mu \Delta \mathbf{u}^* + \nabla p^* = \mathbf{v}^* 1_\omega, \quad (\mathbf{x}, t) \in Q, \\ \nabla \cdot \mathbf{u}^* = 0, \quad (\mathbf{x}, t) \in Q, \\ \dfrac{\partial \rho^*}{\partial t} + \nabla \cdot (\rho^* \mathbf{u}^*) = 0, \quad (\mathbf{x}, t) \in Q, \\ \mathbf{u}^* = 0, \quad (\mathbf{x}, t) \in \Sigma, \\ \rho^*|_{t=0} = \rho_0, \quad \mathbf{u}^*|_{t=0} = \mathbf{u}_0, \quad \mathbf{x} \in \Omega, \end{cases} \tag{6.24}$$

$$\begin{cases} -\rho^* \dfrac{\partial \mathbf{w}}{\partial t} - \rho^*(\mathbf{u}^* \cdot \nabla)\mathbf{w} + \rho^*(\nabla \mathbf{u}^*)\mathbf{w} - \mu \Delta \mathbf{w} + \nabla q + \eta \nabla \rho^* \\ \qquad = a(\mathbf{u}^* - \mathbf{u}_d), \quad (\mathbf{x}, t) \in Q, \\ \nabla \cdot \mathbf{w} = 0, \quad (\mathbf{x}, t) \in Q, \\ -\dfrac{\partial \eta}{\partial t} - \mathbf{u}^* \cdot \nabla \eta + (\dfrac{\partial \mathbf{u}^*}{\partial t} + (\mathbf{u}^* \cdot \nabla)\mathbf{u}^*) \cdot \mathbf{w} = 0, \quad (\mathbf{x}, t) \in Q, \\ \mathbf{w} = 0, \quad (\mathbf{x}, t) \in \Sigma, \\ \eta|_{t=T} = 0, \quad \mathbf{w}|_{t=T} = 0, \quad \mathbf{x} \in \Omega, \end{cases} \tag{6.25}$$

$$\iint_{\omega \times (0,T)} (\mathbf{w} + b\mathbf{v}^*) \cdot (\mathbf{v} - \mathbf{v}^*) \geq 0 \quad \forall \mathbf{v} \in \mathcal{U}_{\text{ad}}. \tag{6.26}$$

Proof In this proof, we will consider the Lagrangian coordinates associated to $\mathbf{u}^*$, that is, the function $\mathbf{X}^* = \mathbf{X}^*(\mathbf{x}, t; s)$ defined by

$$\begin{cases} \dfrac{\partial \mathbf{X}^*}{\partial s} = \mathbf{u}^*(\mathbf{X}^*, s), \quad s \in [0, T], \\ \mathbf{X}^*|_{s=t} = \mathbf{x}. \end{cases}$$

In view of (6.23) and standard regularity results for the solutions to ODEs, the $\mathbf{X}^*_k$ possess first and second-order partial derivatives with respect to the x_i, t and s that belong to $L^\infty(\Omega \times (0, T) \times (0, T))$.

Let us take $\mathbf{v} = \mathbf{v}^* + \varepsilon \mathbf{h}$ with $\varepsilon \in \mathbb{R}_+$ (small), $\mathbf{h} \in L^2(\omega \times (0, T))^N$ and $\mathbf{v} \in \mathcal{U}_{\text{ad}}$.

Let $\{\mathbf{u}, \rho, p\}$ be a state associated to $\mathbf{v}$. After some straightforward computations, we can write that

$$\mathbf{u} = \mathbf{u}^* + \varepsilon \mathbf{y} + \varepsilon \mathbf{y}'_\varepsilon, \quad \rho = \rho^* + \varepsilon \sigma + \varepsilon \sigma'_\varepsilon, \quad p = p^* + \varepsilon \pi + \varepsilon \pi'_\varepsilon,$$

with

$$
\begin{cases}
\rho^*\left(\dfrac{\partial \mathbf{y}}{\partial t}+(\mathbf{u}^*\cdot\nabla)\mathbf{y}+(\mathbf{y}\cdot\nabla)\mathbf{u}^*\right)-\mu\Delta\mathbf{y}+\nabla\pi+\sigma\left(\dfrac{\partial \mathbf{u}^*}{\partial t}+(\mathbf{u}^*\cdot\nabla)\mathbf{u}^*\right)\\
\qquad\qquad =\mathbf{h}1_{\omega},\quad (\mathbf{x},t)\in Q,\\
\nabla\cdot\mathbf{y}=0,\quad (\mathbf{x},t)\in Q,\\
\dfrac{\partial\sigma}{\partial t}+\mathbf{u}^*\cdot\nabla\sigma=-\mathbf{y}\cdot\nabla\rho^*,\quad (\mathbf{x},t)\in Q,\\
\mathbf{y}=0,\quad (\mathbf{x},t)\in\Sigma,\\
\sigma|_{t=0}=0,\quad \mathbf{y}|_{t=0}=0,\quad \mathbf{x}\in\Omega
\end{cases}
\tag{6.27}
$$

and

$$
\begin{cases}
\rho\left(\dfrac{\partial \mathbf{y}'_\varepsilon}{\partial t}+((\mathbf{u}^*+\varepsilon\mathbf{y})\cdot\nabla)\mathbf{y}'_\varepsilon+(\mathbf{y}'_\varepsilon\cdot\nabla)\mathbf{u}\right)-\mu\Delta\mathbf{y}'_\varepsilon+\nabla\pi'_\varepsilon\\
\qquad\qquad =-\varepsilon\sigma\mathbf{y}_t-\varepsilon(\rho^*+\varepsilon\sigma)(\mathbf{y}\cdot\nabla)\mathbf{y},\quad (\mathbf{x},t)\in Q,\\
\nabla\cdot\mathbf{y}'_\varepsilon=0,\quad (\mathbf{x},t)\in Q,\\
\dfrac{\partial\sigma'_\varepsilon}{\partial t}+\mathbf{u}\cdot\nabla\sigma'_\varepsilon=-\mathbf{y}'_\varepsilon\cdot\nabla(\rho^*+\varepsilon\sigma)-\varepsilon\mathbf{y}\cdot\nabla\sigma,\quad (\mathbf{x},t)\in Q,\\
\mathbf{y}'_\varepsilon=0,\quad (\mathbf{x},t)\in\Sigma,\\
\sigma'_\varepsilon|_{t=0}=0,\quad \mathbf{y}'_\varepsilon|_{t=0}=0,\quad \mathbf{x}\in\Omega.
\end{cases}
\tag{6.28}
$$

The functions $\mathbf{y}$ and $\mathbf{y}'_\varepsilon$ must satisfy the same (homogeneous) boundary conditions satisfied by $\mathbf{u}$; on the other hand, $\mathbf{y}$, $\mathbf{y}'_\varepsilon$, σ and σ'_ε must satisfy homogeneous initial conditions at $t=0$.

Note that

$$
\rho^*(\mathbf{x},t)=\rho_0(\mathbf{X}^*(\mathbf{x},t;0))\ \text{ and }\ \sigma(\mathbf{x},t)=-\int_0^t(\mathbf{y}\cdot\nabla\rho^*)(\mathbf{X}^*(\mathbf{x},t;s),s)\,ds\ \text{ in }\ Q.
$$

Let us see that, under the regularity assumptions (6.23), we have:

$$
\mathbf{y}'_\varepsilon\to\mathbf{0}\text{ strongly in }L^2(Q)^N\text{ as }\varepsilon\to0^+.
\tag{6.29}
$$

First, we will get *strong* estimates of $\mathbf{y}$ and σ.

As in previous chapters, it will be sufficient to work with semi-Galerkin approximations. For brevity, we indicate the estimates directly on (6.27).

- Thus, multiplying the first and the third PDE respectively by $\mathbf{y}$ and σ and integrating in space, one has:

$$
\begin{aligned}
\frac{1}{2}\frac{d}{dt}\int_\Omega\rho^*|\mathbf{y}|^2+\mu\|\nabla\mathbf{y}\|^2&=\int_\omega\mathbf{h}\cdot\mathbf{y}\\
&\quad-\int_\Omega\left(\sigma\left(\frac{\partial\mathbf{u}^*}{\partial t}+(\mathbf{u}^*\cdot\nabla)\mathbf{u}^*\right)+\rho^*(\mathbf{y}\cdot\nabla)\mathbf{u}^*\right)\cdot\mathbf{y}
\end{aligned}
$$

and

$$\frac{1}{2}\frac{d}{dt}\int_\Omega |\sigma|^2 = -\int_\Omega (\mathbf{y}\cdot\nabla\rho^*)\sigma.$$

- Arguing as in Theorems 3.1 and 5.1, we see that

$$\frac{1}{2}\frac{d}{dt}\int_\Omega \rho^*|\mathbf{y}|^2 + \mu\|\nabla\mathbf{y}\|^2 \leq \|\mathbf{h}\|^2_{L^2(\omega)} + C\left((1+\|\nabla\mathbf{u}^*\|^4)\|\mathbf{y}\|^2 + \|\frac{\partial\mathbf{u}^*}{\partial t}+(\mathbf{u}^*\cdot\nabla)\mathbf{u}^*\|^2_{L^3}\|\sigma\|^2\right) + \frac{\mu}{4}\|\nabla\mathbf{y}\|^2$$

and

$$\frac{1}{2}\frac{d}{dt}\int_\Omega |\sigma|^2 \leq C\|\nabla\rho^*\|^2_{L^3}\|\sigma\|^2 + \frac{\mu}{4}\|\nabla\mathbf{y}\|^2,$$

and we deduce that $\mathbf{y}\in L^2(0,T;V)\cap L^\infty(0,T;H)$, $\mathbf{y}_t\in L^2(0,T;V')$ and $\sigma\in L^\infty(0,T;L^2(\Omega))$, with the norms in these spaces bounded independently of ε.

- Now, multiplying the first PDE by $\dfrac{\partial\mathbf{y}}{\partial t}$ and integrating in space, the following is found:

$$\int_\Omega \rho^*|\frac{\partial\mathbf{y}}{\partial t}|^2 + \frac{\mu}{2}\frac{d}{dt}\|\nabla\mathbf{y}\|^2 = (\mathbf{F}_\mathbf{y}, \frac{\partial\mathbf{y}}{\partial t}),$$

where

$$\mathbf{F}_\mathbf{y} := \mathbf{h}1_\omega - \sigma(\frac{\partial\mathbf{u}^*}{\partial t}+(\mathbf{u}^*\cdot\nabla)\mathbf{u}^*) - \rho^*((\mathbf{u}^*\cdot\nabla)\mathbf{y} + (\mathbf{y}\cdot\nabla)\mathbf{u}^*)$$

is bounded in $L^2(Q)^N$.

This shows that

$$\mathbf{y}\in L^\infty(0,T;V) \quad\text{and}\quad \frac{\partial\mathbf{y}}{\partial t}\in L^2(0,T;H),$$

with norms bounded independently of ε.

- On the other hand, for t a.e. in $(0, T)$ the couple $\{\mathbf{y}(\cdot\,,t), \pi(\cdot\,,t)\}$ can be regarded as the solution to the Stokes problem

$$\begin{cases} -\mu\Delta\mathbf{z} + \nabla\beta = (\mathbf{h}1\!\!1_\omega)(\cdot\,,t) + \mathbf{F}_\mathbf{y}(\cdot\,,t) & \text{in } \Omega, \\ \nabla\cdot\mathbf{z} = 0 & \text{in } \Omega, \\ \mathbf{z} = 0 & \text{on } \partial\Omega. \end{cases}$$

Therefore, $\mathbf{y} \in L^2(0, T; D(A)) \cap C^0([0, T]; V)$, $\dfrac{\partial\mathbf{y}}{\partial t} \in L^2(0, T; H)$ and $\sigma \in L^\infty(Q)$, again with norms bounded in these spaces.

At this point, let us deduce *energy* estimates for $\mathbf{y}'_\varepsilon$.

Multiplying the first and the third PDE in (6.28) respectively by $\mathbf{y}'_\varepsilon$ and σ'_ε and integrating in space, we see that

$$\begin{aligned} \frac{1}{2}\frac{d}{dt}\int_\Omega \rho|\mathbf{y}'_\varepsilon|^2 + \mu\|\nabla\mathbf{y}'_\varepsilon\|^2 &= -\int_\Omega (\mathbf{y}'_\varepsilon\cdot\nabla)(\mathbf{u}^*+\varepsilon\mathbf{y})\cdot\mathbf{y} \\ &\quad - \varepsilon\int_\Omega \left(\sigma\frac{\partial\mathbf{u}^*}{\partial t} + (\rho^*+\varepsilon\sigma)(\mathbf{y}\cdot\nabla)\mathbf{y}\right)\cdot\mathbf{y}'_\varepsilon. \end{aligned}$$

Using (6.23) and the previous estimates of $\mathbf{y}$ and σ and arguing as before, we easily get that

$$\begin{aligned} \frac{1}{2}\frac{d}{dt}\int_\Omega \rho|\mathbf{y}'_\varepsilon|^2 + \mu\|\nabla\mathbf{y}'_\varepsilon\|^2 &\le C\|\mathbf{y}'_\varepsilon\|^2 \\ &\quad + C\varepsilon^2(\|A\mathbf{y}\|^2 + \|\frac{\partial\mathbf{y}}{\partial t}\|^2) + \frac{\mu}{2}\|\nabla\mathbf{y}'_\varepsilon\|^2. \end{aligned}$$

This, together with Gronwall's Lemma, shows that

$$\|\mathbf{y}'_\varepsilon\|_{L^\infty(0,T;H)} + \|\mathbf{y}'_\varepsilon\|_{L^2(0,T;V)} \le C\varepsilon,$$

whence the previous assertion follows.

By hypothesis, $J(\mathbf{v}, \mathbf{u}, \rho, p) - J(\mathbf{v}^*, \mathbf{u}^*, \rho^*, p^*) \ge 0$. We can write this difference in the form

$$\begin{aligned} &J(\mathbf{v}, \mathbf{u}, \rho, p) - J(\mathbf{v}^*, \mathbf{u}^*, \rho^*, p^*) \\ &\quad = \varepsilon\left(a\iint_Q (\mathbf{u}^*-\mathbf{u}_d)\cdot\mathbf{y} + b\iint_{\omega\times(0,T)} \mathbf{v}^*\cdot\mathbf{h}\right) + \varepsilon Z_\varepsilon, \end{aligned}$$

where

$$Z_\varepsilon = a\iint_Q \left((\mathbf{u}^*-\mathbf{u}_d)\cdot\mathbf{y}'_\varepsilon + \frac{\varepsilon}{2}|\mathbf{y}+\mathbf{y}'_\varepsilon|^2\right) + \frac{\varepsilon b}{2}\iint_{\omega\times(0,T)} |\mathbf{h}|^2.$$

Dividing by ε and taking limits as $\varepsilon \to 0^+$, we see that

$$a \iint_Q (\mathbf{u}^* - \mathbf{u}_d) \cdot \mathbf{y} + b \iint_{\omega\times(0,T)} \mathbf{v}^* \cdot \mathbf{h} \geq 0. \tag{6.30}$$

Now, consider the linear (adjoint) system (6.25). Arguing as in the proof of Theorem 3.1, it can be deduced that (6.25) has at least one weak solution $\{\mathbf{w}, \eta, q\}$ that belongs to the usual energy space, i.e., satisfying

$$\mathbf{w} \in L^2(0,T;V) \cap L^\infty(0,T;H), \quad \eta \in L^\infty(Q), \quad q \in W^{-1,\infty}(0,T;L^2(\Omega)).$$

This solution is unique. Indeed, note that the argument used in the proof of Theorem 5.3 can be applied in this context; see Exercise 6.6. Furthermore, a straightforward integration by parts yields the identity

$$a \iint_Q (\mathbf{u}^* - \mathbf{u}_d) \cdot \mathbf{y} = \iint_{\omega\times(0,T)} \mathbf{w} \cdot \mathbf{h}.$$

This, together with (6.30), gives

$$\iint_{\omega\times(0,T)} (\mathbf{w} + b\mathbf{v}^*) \cdot \mathbf{h} \geq 0.$$

Since this must hold for any $\mathbf{h} = \frac{1}{\varepsilon}(\mathbf{v} - \mathbf{v}^*)$ with $\mathbf{v} \in \mathcal{U}_{\rm ad}$ and $\varepsilon \in (0,1)$, we finally arrive at (6.26).

This ends the proof. □

Remark 6.1 As noticed in the previous proof, in order to establish rigorously (6.30) and then (6.26), we need a regularity assumption on $\{\mathbf{u}^*, \rho^*, p^*\}$. Whether or not Theorem 6.2 remains true without this requirement is an interesting open question. □

Remark 6.2 There are other ways to prove Theorem 6.2. In particular, there exists an "elegant" proof relying on the *Dubovitskii-Milyoutin Formalism,* see [35, 36]. The idea is the following: if $(\mathbf{v}^*, \mathbf{u}^*, \rho^*, p^*)$ is an optimal control-state quartet, at this point there is no descent direction for J at the same time *admissible* for $\mathcal{U}_{\rm ad}$ and (6.1), that is, at most tangential to $\mathcal{U}_{\rm ad}$ and the corresponding solution manifold. Under the regularity assumption (6.23), this implies by duality that a nontrivial linear combination of associated orthogonal directions vanishes. In other words,

$$\begin{cases} \exists \lambda_0, \lambda_1, \lambda_2 \in \mathbb{R} \text{ (not all zero)}, \ \exists \{\mathbf{w}, \eta, q\} \text{ and } \exists \chi \text{ such that:} \\ \quad \{\mathbf{w}, \eta, q\} \text{ solves (6.25)}, \\ \quad \chi \in L^2(\omega \times (0,T))^N \text{ and } \displaystyle\iint_{\omega\times(0,T)} \chi \cdot (\mathbf{v} - \mathbf{v}^*) \leq 0 \ \forall \mathbf{v} \in \mathcal{U}_{\rm ad}, \\ \quad \lambda_0 J'(\mathbf{v}^*, \mathbf{u}^*, \rho^*, p^*) + \lambda_1(\mathbf{0}, \mathbf{w}, \eta, q) + \lambda_2(\chi, \mathbf{0}, 0, 0) = (\mathbf{0}, \mathbf{0}, 0, 0). \end{cases} \tag{6.31}$$

After some work, (6.24)–(6.26) is found again. This argument has been used frequently in the framework of optimal control; some related references are, for instance, [10, 15, 78, 81]. □

Remark 6.3 The inequality (6.26) can be equivalently rewritten in terms of the orthogonal projector $\mathbb{P}_{\mathrm{ad}} : L^2(\omega \times (0, T))^N \mapsto \mathcal{U}_{\mathrm{ad}}$, that assigns to each function in $L^2(\omega \times (0, T))^N$ the corresponding closest control in $\mathcal{U}_{\mathrm{ad}}$. Indeed, (6.26) is equivalent to the identity

$$\mathbf{v}^* = \mathbb{P}_{\mathrm{ad}} \left(-\frac{1}{b} \mathbf{w}|_{\omega \times (0,T)} \right). \tag{6.32}$$

Note that the computation of $\mathbb{P}_{\mathrm{ad}}(\mathbf{z})$ is easy in all the cases considered above. □

In the context of the control of PDE systems, optimality systems like the previous one are very useful. They can be regarded as optimality tests, that is, tools that allow to decide whether or not a given control is optimal. Furthermore, it will be seen later that they serve to introduce iterative algorithms for the computation of optimal control-states.

In fact, it can be still more interesting to deduce second-order optimality conditions, since it is expected that they furnish not only necessary but also sufficient conditions for optimality. However, this is much more complex from the technical viewpoint and will not be contemplated here; for some results in this direction, see [14, 68, 69, 77].

A very similar result can be obtained when the cost functional is given by (6.5):

Theorem 6.3 *Assume that $\mathcal{U}_{\mathrm{ad}} \subset L^2(\omega \times (0, T))^N$ is non-empty, closed and convex and J is given by* (6.5). *Let $(\mathbf{v}^*, \mathbf{u}^*, \rho^*, p^*)$ be an optimal solution to* (6.3) *and assume that $\{\mathbf{u}^*, \rho^*, p^*\}$ satisfies* (6.23). *Then, there exists $\{\mathbf{w}, \eta, q\}$ with $\mathbf{w} \in L^2(0, T; V) \cap L^\infty(0, T; H)$, $\eta \in L^\infty(Q)$ and $q \in W^{-1,\infty}(0, T; L^2(\Omega))$ such that one has* (6.24),

$$\begin{cases} -\rho^* \dfrac{\partial \mathbf{w}}{\partial t} - \rho^*(\mathbf{u}^* \cdot \nabla)\mathbf{w} + \rho^*(\nabla \mathbf{u}^*)\mathbf{w} - \mu \Delta \mathbf{w} + \nabla q + \eta \nabla \rho^* \\ \qquad = \mathbf{0}, \quad (\mathbf{x}, t) \in Q, \\ \nabla \cdot \mathbf{w} = 0, \quad (\mathbf{x}, t) \in Q, \\ -\dfrac{\partial \eta}{\partial t} - \mathbf{u}^* \cdot \nabla \eta + (\dfrac{\partial \mathbf{u}^*}{\partial t} + (\mathbf{u}^* \cdot \nabla)\mathbf{u}^*) \cdot \mathbf{w} = 0, \quad (\mathbf{x}, t) \in Q, \\ \mathbf{w} = 0, \quad (\mathbf{x}, t) \in \Sigma, \\ \eta^*|_{t=T} = 0, \quad \mathbf{w}|_{t=T} = a(\mathbf{u}^*|_{t=T} - \mathbf{u}_e), \quad \mathbf{x} \in \Omega, \end{cases} \tag{6.33}$$

and (6.26).

Remark 6.4 Note that none of these theorems asserts that $\mathbf{v} \mapsto J(\mathbf{v}, \mathbf{u}, \rho, p)$ is differentiable at $\mathbf{v}^*$. In fact, nothing indicates that this function is well defined, since in general it may happen that a control $\mathbf{v}$ close to $\mathbf{v}^*$ produces several associated states. Nevertheless, we have been able to express the variation of J

at $(\mathbf{v}^*, \mathbf{u}^*, \rho^*, p^*)$ in the direction determined by $\mathbf{h}$ in the form

$$\iint_{\omega\times(0,T)} (\mathbf{w} + b\mathbf{v}^*) \cdot \mathbf{h},$$

where $\mathbf{w}$ solves, together with q and η, either (6.25) or (6.33). For this reason, we can interpret $(\mathbf{w} + b\mathbf{v}^*)|_{\omega\times(0,T)}$ as the "gradient" of $\mathbf{v} \mapsto J(\mathbf{v}, \mathbf{u}, \rho, p)$ at $\mathbf{v}^*$. We will take advantage of this interpretation later, in Sect. 6.2.6, where our interest will be to design suitable iterative methods. □

6.2.4 Similar Results with Other Cost Functionals

This section deals with optimal control problems for (6.1) corresponding to other functionals.

It will be seen that existence and characterization results of optimal control-states can be achieved essentially as in Sects. 6.2.2 and 6.2.3.

First, let the T_i be given with $0 = T_0 < T_1 < \cdots < T_I = T$ and set

$$J(\mathbf{v}, \mathbf{u}, \rho, p) := \frac{1}{2}\sum_{i=1}^{I} a_i \int_\Omega |\mathbf{u}(\mathbf{x}, T_i) - \mathbf{u}_e^i(\mathbf{x})|^2 + \frac{b}{2}\iint_{\omega\times(0,T)} |\mathbf{v}|^2, \tag{6.34}$$

where the $a_i > 0$, the $\mathbf{u}_e^i \in L^2(\Omega)^N$ and $b > 0$.

Note that Theorem 6.1 can be applied to the optimal control problem (6.3) with J as in (6.34) and $\mathcal{U}_{\rm ad}$ given by any of the sets (6.14), (6.15), (6.16) or (6.17).

The proof that this cost is well defined and satisfies the weak lower semi-continuity assumption is as in the case of (6.5), see the argument in Sect. 6.2.2. Consequently, there exists at least one related optimal $(\mathbf{v}^*, \mathbf{u}^*, \rho^*, p^*)$.

Also, the following holds:

Theorem 6.4 *Assume that $\mathcal{U}_{\rm ad} \subset L^2(\omega\times(0, T))^N$ is non-empty, closed and convex and J is given by* (6.34). *Let $(\mathbf{v}^*, \mathbf{u}^*, \rho^*, p^*)$ be an associated optimal solution and assume that $\{\mathbf{u}^*, \rho^*, p^*\}$ satisfies* (6.23). *Then, there exists $\{\mathbf{w}, \eta, q\}$ with $\mathbf{w} \in L^2(0, T; V) \cap L^\infty(0, T; H)$ and $\mathbf{w} \in C^0([T_i, T_{i+1}]; H)$ for $1 \le i \le I$, $\eta \in L^\infty(Q)$ and $q \in W^{-1,\infty}(0, T; L^2(\Omega))$ such that one has* (6.24),

$$\begin{cases} -\rho^*\dfrac{\partial \mathbf{w}}{\partial t} - \rho^*(\mathbf{u}^* \cdot \nabla)\mathbf{w} + \rho^*(\nabla\mathbf{u}^*)\mathbf{w} - \mu\Delta\mathbf{w} + \nabla q + \eta\nabla\rho^* \\ \qquad\qquad = \mathbf{0}, \quad (\mathbf{x}, t) \in \Omega \times (T_{i-1}, T_i), \\ \nabla \cdot \mathbf{w} = 0, \quad (\mathbf{x}, t) \in Q, \\ -\dfrac{\partial \eta}{\partial t} - \mathbf{u}^* \cdot \nabla\eta + (\dfrac{\partial \mathbf{u}^*}{\partial t} + (\mathbf{u}^* \cdot \nabla)\mathbf{u}^*) \cdot \mathbf{w} = 0, \quad (\mathbf{x}, t) \in \Omega \times (T_{i-1}, T_i), \\ \mathbf{w} = 0, \quad (\mathbf{x}, t) \in \partial\Omega \times (T_{i-1}, T_i), \\ \eta^*|_{t=T_i^-} = 0, \quad \mathbf{w}|_{t=T_i^-} = a'(\mathbf{u}^*|_{t=T_i} - \mathbf{u}_e^i), \quad \mathbf{x} \in \Omega, \end{cases} \tag{6.35}$$

for $i = 1, \ldots, I$ and (6.26).

Now, assume the cost functional is given by

$$J(\mathbf{v}, \mathbf{u}, \rho, p) := \frac{a}{\gamma} \int_{\Omega} |\mathbf{u}(\mathbf{x}, T) - \mathbf{u}_e(\mathbf{x})|^{\gamma} + \frac{b}{2} \iint_{\omega \times (0,T)} |\mathbf{v}|^2, \tag{6.36}$$

where $\gamma \in (1, 6)$.

Again, Theorem 6.1 holds in this case and an optimality result can be deduced:

Theorem 6.5 *Assume that $\mathcal{U}_{\mathrm{ad}} \subset L^2(\omega \times (0, T))^N$ is as in Theorem 6.4, J is given by* (6.36) *and $(\mathbf{v}^*, \mathbf{u}^*, \rho^*, p^*)$ is an optimal solution such that $\{\mathbf{u}^*, \rho^*, p^*\}$ satisfies* (6.23).. *Then, there exists $\{\mathbf{w}, \eta, q\}$ with $\mathbf{w} \in L^2(0, T; V) \cap L^{\infty}(0, T; H)$, $\eta \in L^{\infty}(Q)$ and $q \in W^{-1,\infty}(0, T; L^2(\Omega))$ such that one has* (6.24),

$$\begin{cases} -\rho^* \dfrac{\partial \mathbf{w}}{\partial t} - \rho^*(\mathbf{u}^* \cdot \nabla)\mathbf{w} + \rho^*(\nabla \mathbf{u}^*)\mathbf{w} - \mu \Delta \mathbf{w} + \nabla q + \eta \nabla \rho^* \\ \qquad = \mathbf{0}, \quad (\mathbf{x}, t) \in Q, \\ \nabla \cdot \mathbf{w} = 0, \quad (\mathbf{x}, t) \in Q, \\ -\dfrac{\partial \eta}{\partial t} - \mathbf{u}^* \cdot \nabla \eta + (\dfrac{\partial \mathbf{u}^*}{\partial t} + (\mathbf{u}^* \cdot \nabla)\mathbf{u}^*) \cdot \mathbf{w} = 0, \quad (\mathbf{x}, t) \in Q, \\ \mathbf{w} = 0, \quad (\mathbf{x}, t) \in \Sigma, \\ \eta^*|_{t=T} = 0, \quad \mathbf{w}|_{t=T} = a\left|\mathbf{u}^*|_{t=T} - \mathbf{u}_e\right|^{\gamma-2}(\mathbf{u}^*|_{t=T} - \mathbf{u}_e), \quad \mathbf{x} \in \Omega, \end{cases} \tag{6.37}$$

and (6.26).

The proof is similar; see Exercise 6.11.

Another interesting optimal control problem appears when we take

$$J(\mathbf{v}, \mathbf{u}, \rho, p) := \frac{a}{2} \|L\mathbf{u}(\mathbf{x}, T) - \mathbf{z}_e\|^2 + \frac{b}{2} \iint_{\omega \times (0,T)} |\mathbf{v}|^2, \tag{6.38}$$

where $L \in \mathcal{L}(H; Z)$ (Z is a Hilbert space with norm $\|\cdot\|_Z$) and $\mathbf{z}_e \in Z$. This functional can be viewed as a generalization of (6.5).

The following result holds:

Theorem 6.6 *Let $\mathcal{U}_{\mathrm{ad}} \subset L^2(\omega \times (0, T))^N$ be as in Theorem 6.4, let J be given by* (6.38) *and let $(\mathbf{v}^*, \mathbf{u}^*, \rho^*, p^*)$ be an optimal solution such that $\{\mathbf{u}^*, \rho^*, p^*\}$ satisfies* (6.23). *Then, there exists $\{\mathbf{w}, \eta, q\}$ with $\mathbf{w} \in L^2(0, T; V) \cap L^{\infty}(0, T; H)$, $\eta \in L^{\infty}(Q)$ and $q \in W^{-1,\infty}(0, T; L^2(\Omega))$ such that one has* (6.24),

$$\begin{cases} -\rho^* \dfrac{\partial \mathbf{w}}{\partial t} - \rho^*(\mathbf{u}^* \cdot \nabla)\mathbf{w} + \rho^*(\nabla \mathbf{u}^*)\mathbf{w} - \mu \Delta \mathbf{w} + \nabla q + \eta \nabla \rho^* \\ \qquad = \mathbf{0}, \quad (\mathbf{x}, t) \in Q, \\ \nabla \cdot \mathbf{w} = 0, \quad (\mathbf{x}, t) \in Q, \\ -\dfrac{\partial \eta}{\partial t} - \mathbf{u}^* \cdot \nabla \eta + (\dfrac{\partial \mathbf{u}^*}{\partial t} + (\mathbf{u}^* \cdot \nabla)\mathbf{u}^*) \cdot \mathbf{w} = 0, \quad (\mathbf{x}, t) \in Q, \\ \mathbf{w} = 0, \quad (\mathbf{x}, t) \in \Sigma, \\ \eta^*|_{t=T} = 0, \quad \mathbf{w}|_{t=T} = aL^*(L\mathbf{u}^*|_{t=T} - \mathbf{z}_e), \quad \mathbf{x} \in \Omega, \end{cases} \tag{6.39}$$

and (6.26).

Other cost functionals can be similarly handled. Consider in particular the control problems corresponding to the functions (6.12) and (6.13), useful when the values taken by the pressure are relevant for our purposes.

6.2.5 A Brief Look at the Turnpike Property

In this short section, we will consider an interesting aspect of optimal control that appears in a very natural way and helps to understand the behavior of a lot of systems in real world.

- **Origin and motivation**

  Roughly speaking, for an optimal control problem where the state is governed by a time-dependent equation or system, it is said that the turnpike property is satisfied if, for most of the time, the optimal solution remains close to the optimal solution to an associated similar stationary (i.e., time-independent) problem.

  This is a very intuitive concept. For example, it is easy to understand that, in many realistic situations, the best way to go by car from one location to other distant point consists of working hard at the beginning and the end of the trip (for instance to get and leave a highway) and, contrarily, keep a constant or nearly constant velocity almost all the time.

  For obvious reasons, it is typical to quantify this property with exponentials. Thus, for a control system in $(0, T)$ where T is large and we denote by $u = u(t)$ (resp. $y = y(t)$) an optimal control (resp. an associated optimal state), we look for estimates of the kind

$$\|y(t) - \overline{y}\|_Y + \|u(t) - \overline{u}\|_U \leq C\left(e^{\lambda t} + e^{\lambda(T-t)}\right) \quad \forall t \in (0, T),$$

  where $(\overline{u}, \overline{y})$ is an optimal stationary control-state pair and Y and U are adequate Banach or Hilbert spaces.

- **What can be done in simple cases (linear and semilinear heat equations)**

  Let us start with a relatively simple situation. Thus, consider the system (6.76) and the optimal control problem

$$\begin{cases} \text{Minimize } \dfrac{a}{2}\displaystyle\iint_Q |\theta - \theta_d|^2 + \dfrac{b}{2}\iint_{\omega\times(0,T)} |h|^2, \\ \text{subject to } h \in L^2(\omega \times (0, T)),\ (h, \theta) \text{ satisfies (6.41)}, \end{cases} \tag{6.40}$$

  where (6.41) reads

$$\begin{cases} \dfrac{\partial\theta}{\partial t} - \Delta\theta = h 1_{\omega}, \quad (\mathbf{x}, t) \in Q, \\ \theta = 0, \quad (\mathbf{x}, t) \in \Sigma, \\ \theta|_{t=0} = \theta_0, \quad \mathbf{x} \in \Omega. \end{cases} \tag{6.41}$$

Here, we assume that $a, b > 0$, $\theta_0, \theta_d \in L^2(\Omega)$ and Ω, ω and T are as in the previous sections.

Also, consider the "stationary" PDE problem

$$\begin{cases} -\Delta\theta = h\mathbb{1}_\omega, & \mathbf{x} \in \Omega, \\ \theta = 0, & \mathbf{x} \in \partial\Omega \end{cases} \tag{6.42}$$

and the time-independent optimal control problem

$$\begin{cases} \text{Minimize } \dfrac{a}{2}\displaystyle\int_\Omega |\theta - \theta_d|^2 + \dfrac{b}{2}\displaystyle\int_\omega |h|^2, \\ \text{subject to } h \in L^2(\omega), \ (h, \theta) \text{ satisfies (6.42).} \end{cases} \tag{6.43}$$

In this particular framework, both (6.40) and (6.43) possess exactly one solution. Let us denote by (h, θ) and $(\overline{h}, \overline{\theta})$ the corresponding optimal control-state pairs. Then the follows hold:

Theorem 6.7 *Let θ_0 and θ_d be fixed in $L^2(\Omega)$. There exist positive constants C and λ independent of θ_0 and θ_d such that, for any sufficiently large $T > 0$, one has*

$$\|\theta(\cdot\,, t) - \overline{\theta}\| + \|h(\cdot\,, t) - \overline{h}\|_{L^2(\omega)} \le C\left(\|\theta_0 - \overline{\theta}\|e^{-\lambda t} + \|\overline{w}\|e^{-\lambda(T-t)}\right) \ \forall t \in (0, T),$$

where $\overline{w} \in H_0^1(\Omega)$ is the adjoint state associated to $\overline{h}$ in the context of the optimal control problem (6.43).

Note that the solution $(\overline{h}, \overline{\theta})$ to (6.43) is, together with $\overline{w}$, the unique solution to the corresponding optimality system

$$\begin{cases} -\Delta\overline{\theta} = \overline{h}\mathbb{1}_\omega, & \mathbf{x} \in \Omega, \\ \overline{\theta} = 0, & \mathbf{x} \in \partial\Omega \end{cases} \tag{6.44}$$

$$\begin{cases} -\Delta\overline{w} = \overline{\theta} - \theta_d, & \mathbf{x} \in \Omega, \\ \overline{w} = 0, & \mathbf{x} \in \partial\Omega \end{cases} \tag{6.45}$$

$$\overline{h} = -\frac{a}{b}\overline{w}\Big|_\omega. \tag{6.46}$$

Several proofs can be given; see [93, 106]. Also, note that the result can be generalized to cover many other situations where the state equation is linear, see [58].

Let us now assume that the time-dependent and the stationary state equations are semilinear:

$$
\begin{cases}
\dfrac{\partial \theta}{\partial t} - \Delta \theta + f(\theta) = h \mathbb{1}_{\omega}, & (\mathbf{x}, t) \in Q, \\
\theta = 0, & (\mathbf{x}, t) \in \Sigma, \\
\theta|_{t=0} = \theta_0, & \mathbf{x} \in \Omega.
\end{cases}
\tag{6.47}
$$

and

$$
\begin{cases}
-\Delta \theta + f(\theta) = h \mathbb{1}_{\omega}, & \mathbf{x} \in \Omega, \\
\theta = 0, & \mathbf{x} \in \partial\Omega,
\end{cases}
\tag{6.48}
$$

where $f \in C^2(\mathbb{R})$, $f(0) = 0$ and $f' \geq 0$.

Again, we consider the optimal control problems

$$
\begin{cases}
\text{Minimize } \dfrac{a}{2} \displaystyle\iint_Q |\theta - \theta_d|^2 + \dfrac{b}{2} \iint_{\omega \times (0,T)} |h|^2, \\
\text{subject to } h \in L^2(\omega \times (0, T)), \ (h, \theta) \text{ satisfies (6.47).}
\end{cases}
\tag{6.49}
$$

and

$$
\begin{cases}
\text{Minimize } \dfrac{a}{2} \displaystyle\int_\Omega |\theta - \theta_d|^2 + \dfrac{b}{2} \int_\omega |h|^2, \\
\text{subject to } h \in L^2(\omega), \ (h, \theta) \text{ satisfies (6.48).}
\end{cases}
\tag{6.50}
$$

Then, as is customary for control problems governed by semilinear or nonlinear PDEs, it is expectable that a result similar to Theorem 6.7 be satisfied in this context under an appropriate smallness condition on θ_0 and θ_d. The corresponding *local* turnpike property is however a little more complicate to state.

Thus, let us consider the optimality systems associated with (6.49) and (6.50). They are respectively given by (6.47) together with

$$
\begin{cases}
-\dfrac{\partial w}{\partial t} - \Delta w + f'(\theta) w = \theta - \theta_d, & (\mathbf{x}, t) \in Q, \\
w = 0, & (\mathbf{x}, t) \in \Sigma, \\
w|_{t=T} = 0, & \mathbf{x} \in \Omega
\end{cases}
\tag{6.51}
$$

and

$$
h = -\frac{a}{b} w \Big|_{\omega \times (0,T)}
\tag{6.52}
$$

and, on the other hand, (6.48) together with

$$
\begin{cases}
-\Delta \overline{w} + f'(\overline{\theta}) \overline{w} = \overline{\theta} - \theta_d, & \mathbf{x} \in \Omega, \\
\overline{w} = 0, & \mathbf{x} \in \partial\Omega
\end{cases}
\tag{6.53}
$$

and

$$\overline{h} = -\frac{a}{b}\overline{w}\Big|_{\omega}. \qquad (6.54)$$

The following holds:

Theorem 6.8 *There exists* $\varepsilon > 0$, $C > 0$ *and* $\lambda > 0$ *such that, if one has*

$$\|\theta_d\| + \|\theta_0 - \overline{\theta}\|_{L^\infty} + \|\overline{w}\|_{L^\infty} \leq \varepsilon,$$

there exist solutions to the optimality systems (6.47), (6.51), (6.52) *and* (6.48), (6.53), (6.54) *satisfying*

$$\|\theta(\cdot\,, t) - \overline{\theta}\|_{L^\infty} + \|w(\cdot\,, t) - \overline{w}\|_{L^\infty} \leq C\left(e^{-\lambda t} + e^{-\lambda(T-t)}\right) \quad \forall t \in [0, T].$$

The proof is given in [94]; other results, additional comments and consequences can be found in [58].

- **Results for 2D classical Navier-Stokes and open questions**
  It is possible to establish results similar to Theorem 6.8 for other semilinear and nonlinear systems.
  In particular, this can be achieved when the state system is the classical Navier-Stokes equations in two dimensions, see [112]. As before, the nonlinearity of the equations makes it necessary a smallness assumption on the data. Moreover, the argument of proof relies strongly on the uniqueness of solution and the continuity of the control-to-state mapping. Accordingly, whether or not the turnpike property holds in the three-dimensional case is an open question.
  For variable density Navier-Stokes systems, the situation is still more delicate. Indeed, recall that, at present, the corresponding stationary problem is not well understood; see the related discussion in Sect. 3.2. Thus, although it is a very interesting question, the turnpike property for control problems like (6.3) and (6.4) remains largely open.

6.2.6 Some Iterative Algorithms

In this section, we will propose some iterative schemes to compute the solutions to the previous optimal control problems.

For simplicity, we will only consider the case where J is given by (6.4) and, consequently, the optimality system is (6.24)–(6.26). The adaptations to (6.5) and the other cost functionals in Sect. 6.2.3 are straightforward, so we do not give an explicit presentation.

The following algorithms rely on the ideas in the proof of Theorem 6.2. Specifically, we note that, if $(\mathbf{v}, \mathbf{u}, \rho, p)$ is given, then for any $\mathbf{h} \in L^2(\omega \times (0, T))^N$, any small $\varepsilon > 0$ and any state $\{\mathbf{u}', \rho', p'\}$ associated to $\mathbf{v} + \varepsilon\mathbf{h}$, it is natural to expect that

$$\frac{1}{\varepsilon}\left(J(\mathbf{v}', \mathbf{u}', \rho', p') - J(\mathbf{v}, \mathbf{u}, \rho, p)\right) = \iint_{\omega\times(0,T)} (\mathbf{w} + b\mathbf{v}) \cdot \mathbf{h} + O(\varepsilon),$$

where $\{\eta, \mathbf{w}, q\}$ solves

$$\begin{cases} -\rho\dfrac{\partial \mathbf{w}}{\partial t} - \rho(\mathbf{u}\cdot\nabla)\mathbf{w} + \rho(\nabla\mathbf{u})\mathbf{w} - \mu\Delta\mathbf{w} + \nabla q + \eta\nabla\rho \\ \qquad\qquad = a(\mathbf{u} - \mathbf{u}_d), \quad (\mathbf{x}, t) \in Q, \\ \nabla\cdot\mathbf{w} = 0, \quad (\mathbf{x}, t) \in Q, \\ -\dfrac{\partial \eta}{\partial t} - \mathbf{u}\cdot\nabla\eta + (\dfrac{\partial \mathbf{u}}{\partial t} + (\mathbf{u}\cdot\nabla)\mathbf{u})\cdot\mathbf{w} = a'(\rho - \rho_d), \quad (\mathbf{x}, t) \in Q, \\ \mathbf{w} = 0, \quad (\mathbf{x}, t) \in \Sigma, \\ \eta|_{t=T} = 0, \quad (\eta\mathbf{w})|_{t=T} = 0, \quad \mathbf{x} \in \Omega, \end{cases} \tag{6.55}$$

and $O(\varepsilon) \to 0$ as $\varepsilon \to 0^+$.

The first proposed algorithm (ALG 1) is given in Table 6.1, at the end of the chapter. If (6.24) has exactly one weak regular solution $\{\mathbf{u}, \rho, p\}$ for each $\mathbf{v} \in \mathcal{U}_{\text{ad}}$, then ALG 1 is just the classical optimal step gradient method with projection. Indeed, in this case, for each $\mathbf{v}^n \in \mathcal{U}_{\text{ad}}$, the corresponding $\mathbf{d}^n := \mathbf{w}^n + b\mathbf{v}^n$ may be viewed as the *gradient* of the function $\mathbf{v} \mapsto J(\mathbf{v}, \mathbf{u}, \rho, p)$ at $\mathbf{v}^n$ and consequently the iterates in ALG 1 read

$$\mathbf{v}^{n+1} = \mathbb{P}_{\text{ad}}(\mathbf{v}^n - r^n\mathbf{d}^n),$$

with r^n as good as possible.

For example, if $\mathcal{U}_{\text{ad}} = L^2(\omega\times(0,T))^N$, in the n-th iterate, we go from $\mathbf{v}^n$ to $\mathbf{v}^{n+1}$ by minimizing J along the straight line through $\mathbf{v}^n$ of direction $\mathbf{d}^n$ (as already said, this is the steepest descent direction at $\mathbf{v}^n$); see more explanations in [18, 60].

Since (6.24) is nonlinear and we have to solve this system by using an iterative scheme, it is reasonable to introduce a variant where we perform mixed loops. This leads to ALG 2 (see Table 6.2), where the number of iterates needed to from $\mathbf{v}^n$ to $\mathbf{v}^{n+1}$ is reduced.

Remark 6.5 A natural choice of the convergence criteria can be

$$\|\mathbf{v}^{n+1} - \mathbf{v}^n\|_{L^2(\omega\times(0,T))} \le \kappa\|\mathbf{v}^{n+1}\|_{L^2(\omega\times(0,T))},$$

for κ small enough. However, note that this can be only partially significant for linear and not superlinear convergence. Consequently, this should be followed by an additional test where we check whether the necessary optimality conditions are (at least) approximately satisfied. On the other hand, since the numerical computation of r^n can be expensive, it may be convenient to simplify ALG 1 and ALG 2 by replacing Step 3 by

3′. Set $\mathbf{d}^n = (\mathbf{w}^n + b\mathbf{v}^n)|_{\omega\times(0,T)}$and $r^n = r$(a prescribed positive constant).

Of course, we can also consider other variants by performing Step 3 only a few times (for instance for $n = 10, 20, 30, \dots$) and keeping in between the same fixed r (equal to the last computed r^n). □

A second and more efficient strategy relies on considering *conjugate gradients*. This leads to algorithms similar to those above, where the main difference is that the descent direction $\mathbf{d}^n$ is not identical to $(\mathbf{w}^n + b\mathbf{v}^n)|_{\omega\times(0,T)}$, but a clever modification that improves their behavior.

The advantages have been checked extensively in a large number of experiments, very especially in the context of fluid control, see [60].

Let us set

$$G_1(\mathbf{f},\mathbf{g}) := \frac{\displaystyle\iint_{\omega\times(0,T)} |\mathbf{f}|^2}{\displaystyle\iint_{\omega\times(0,T)} |\mathbf{g}|^2}, \qquad G_2(\mathbf{f},\mathbf{g}) := \frac{\displaystyle\iint_{\omega\times(0,T)} \mathbf{f}\cdot(\mathbf{f}-\mathbf{g})}{\displaystyle\iint_{\omega\times(0,T)} |\mathbf{g}|^2},$$

for all $\mathbf{f},\mathbf{g} \in L^2(\omega\times(0,T))$ with $\mathbf{g} \neq 0$. The proposed conjugate gradient algorithm with projection (ALG 3) is given in Table 6.3. There, G stands for one of the functions G_1 or G_2; the choice $G = G_1$ (resp. $G = G_2$) corresponds to the Fletcher-Reeves (resp. Polak-Ribière) version; see [19, 60] for more details.

Remark 6.6 Of course, we can modify ALG 3 as we did in Remark 6.5 in order to avoid large computational costs concerning r^n. We can also linearize partially the state systems by simply replacing $\mathbf{u}^n$ by $\mathbf{u}^{n-1}$ in the transport terms. This leads to an analog to ALG 2. For brevity, we omit the details. □

Remark 6.7 Note that, in all these iterative methods, we have to solve at each step a forward (linear or not) problem and then a backward linear similar problem. Thus, it can be useful to parallelize this part of each iterate using information from the previous step. In this way, we can obtain new variants of ALG 1–ALG 3. For clarity, we have described the method in the case of ALG 1 in Table 6.4, at the end of the chapter. □

For reasons of space, we will not discuss here other computational aspects, although they are obviously relevant. For example, we could speak of

- Techniques to *precondition* ALG 3.
- Quasi-Newton, Buckley-Le Nir and other algorithms.
- The role of numerical differentiation.

We refer to [60] for some theoretical and numerical results on these topics.

6.3 Minimizing the Time Needed to Reach a Desired State

In this section, we will consider the optimal control problem (6.3), (6.6), where the time needed to reach a desired state $\mathbf{u}_e$ plays an essential role.

We will prove an existence result. Then, we will deduce the corresponding optimality system.

Time optimal control problems are very interesting from the mathematical point of view. They also appear in connection with many relevant applications; see [3, 6, 110].

In order to discard trivial situations, we will assume that $\|\mathbf{u}_0 - \mathbf{u}_e\| > \delta$; otherwise, $T^*(\mathbf{v}, \mathbf{u}, \rho, p; \mathbf{u}_e, \delta) = 0$ for all $(\mathbf{v}, \mathbf{u}, \rho, p)$.

6.3.1 An Existence Result

Fix $T_0 > 0$ and set $Q_0 := \Omega \times (0, T_0)$. Let us introduce a closed convex set $\mathcal{U}_{\rm ad} \subset L^2(\omega \times (0, T_0))^N$ and set

$$\mathcal{S}_{\rm ad}(T_0) := \{(\mathbf{v}, \mathbf{u}, \rho, p) : \mathbf{v} \in \mathcal{U}_{\rm ad},\ \{\mathbf{u}, \rho, p\} \text{ is a weak solution to (6.1) in } Q_0\}.$$

We have $\mathcal{S}_{\rm ad}(T_0) \subset L^2(\omega \times (0, T_0))^N \times E(T_0)$, where $E(T_0)$ stands for the energy space for the solutions to (6.1) in Q_0. In other words, $E(T_0)$ is the space of triplets $\{\mathbf{u}, \rho, p\}$ satisfying

$$\mathbf{u} \in L^2(0, T_0; V) \cap L^\infty(0, T_0; H),\ \ \rho \in L^\infty(Q_0),\ \ p \in W^{-1,\infty}(0, T_0; L^2(\Omega)).$$

Let us set

$$I(\mathbf{v}, \mathbf{u}, \rho, p) := \frac{1}{2} T^*(\mathbf{v}, \mathbf{u}, \rho, p; \mathbf{u}_e, \delta)^2 + \frac{b}{2} \iint_{\omega \times (0, T_0)} |\mathbf{v}|^2, \tag{6.56}$$

where $\mathbf{u}_e \in H$ and $T^*(\mathbf{v}, \mathbf{u}, \rho, p; \mathbf{u}_e, \delta)$ is given by (6.7), that is, $T^*(\mathbf{v}, \mathbf{u}, \rho, p; \mathbf{u}_e, \delta)$ is the minimal time at which $\mathbf{u}(\cdot\,, t)$ belongs to $\overline{B}(\mathbf{u}_e; \delta)$ (it is not excluded to have $T(\mathbf{v}, \mathbf{u}, \rho, p; \mathbf{u}_e, \delta) = +\infty$).

The time optimal control problem can be written as follows:

$$\begin{cases} \text{Minimize } \ I(\mathbf{v}, \mathbf{u}, \rho, p) \\ \text{subject to } \ (\mathbf{v}, \mathbf{u}, \rho, p) \in \mathcal{S}_{\rm ad}(T_0)\,. \end{cases} \tag{6.57}$$

Our first objective is to establish the existence of an optimal control-state. This is the motivation and goal of the following result:

Theorem 6.9 *Assume that the set*

$$\{(\mathbf{v}, \mathbf{u}, \rho, p) \in \mathcal{S}_{\rm ad}(T_0) : I(\mathbf{v}, \mathbf{u}, \rho, p) < +\infty\} \tag{6.58}$$

is non-empty. Then, there exists at least one solution to (6.57).

Proof The set $\mathcal{U}_{\rm ad}$ is weakly closed in $L^2(\omega \times (0, T_0))^N$ and I is coercive. Accordingly, we only have to check that $\mathcal{S}_{\rm ad}(T_0)$ is weakly-$*$ closed in $L^2(\omega \times (0, T_0))^N \times E(T_0)$ and I is sequentially weakly-$*$ lower semi-continuous for the norm of $L^2(\omega \times (0, T_0))^N \times E(T_0)$.

Let $\{(\mathbf{v}^n, \mathbf{u}^n, \rho^n, p^n)\}$ be a sequence in $\mathcal{S}_{\rm ad}(T_0)$ such that $\mathbf{v}^n \to \mathbf{v}^*$ weakly in $L^2(\omega \times (0, T_0))^N$, $\mathbf{u}^n \to \mathbf{u}^*$ weakly in $L^2(0, T_0; V)$, $\rho^n \to \rho^*$ weakly-$*$ in $L^\infty(Q_0)$ and $p^n \to p^*$ weakly-$*$ in $W^{-1,\infty}(0, T_0; L^2(\Omega))$.

Then, arguing as in the proof of Theorem 6.1, we see that $\{\mathbf{u}^*, \rho^*, p^*\}$ solves the nonlinear system (6.1) in Q_0 for $\mathbf{v} = \mathbf{v}^*$ and $(\mathbf{v}^*, \mathbf{u}^*, \rho^*, p^*) \in \mathcal{S}_{\rm ad}(T_0)$. In particular, one has $\mathbf{u}^* \in C^0_w([0, T_0]; H)$.

We have that

$$\liminf_{n\to+\infty} \iint_{\omega\times(0,T_0)} |\mathbf{v}^n|^2 \geq \iint_{\omega\times(0,T_0)} |\mathbf{v}^*|^2.$$

On the other hand, if we set

$$T^*_n := T^*(\mathbf{v}^n, \mathbf{u}^n, \rho^n, p^n; \mathbf{u}_e, \delta) \quad \text{and} \quad T^* := T^*(\mathbf{v}^*, \mathbf{u}^*, \rho^*, p^*; \mathbf{u}_e, \delta),$$

we also have

$$\liminf_{n\to+\infty} T^*_n \geq T^*. \tag{6.59}$$

Indeed, if this assertion were false, it could be assumed that the T^*_n converge to a time $\tilde{T}$ satisfying

$$\tilde{T} = \lim_{n\to+\infty} T^*_n < T^*. \tag{6.60}$$

We will use the following result. Its proof is given below. □

Lemma 6.1 *Under the assumption* (6.60), *we have*

$$(\mathbf{u}^*(\cdot\,, \tilde{T}) - \mathbf{u}_e, \mathbf{z}) \leq \delta\|\mathbf{z}\| \quad \forall \mathbf{z} \in V. \tag{6.61}$$

Let us suppose that this lemma holds and (6.60) is satisfied. Then, $\|\mathbf{u}^*(\cdot\,, \tilde{T}) - \mathbf{u}_e\| \leq \delta$. But, on the other hand, from the definition of T^* and the fact that $\tilde{T} < T^*$, we must also have $\|\mathbf{u}^*(\cdot\,, \tilde{T}) - \mathbf{u}_e\| > \delta$, which is the opposite inequality.

Thus, we get a contradiction and (6.59) must hold.

This completes the proof of Theorem 6.9.

Proof of Lemma 6.1 Let $\mathbf{u}^n$, T_n^* and $\tilde{T}$ be as in the proof of Theorem 6.9 and let us assume that (6.60) holds.

We can write the following for all n:

$$\begin{aligned} |(\mathbf{u}(\cdot\,,\tilde{T}) - \mathbf{u}_e, \mathbf{z})| &\le |(\mathbf{u}(\cdot\,,\tilde{T}) - \mathbf{u}(\cdot\,,T_n^*), \mathbf{z})| \\ &\quad + |(\mathbf{u}(\cdot\,,T_n^*) - \mathbf{u}^n(\cdot\,,T_n^*), \mathbf{z})| + |(\mathbf{u}^n(\cdot\,,T_n^*) - \mathbf{u}_e, \mathbf{z})|. \end{aligned} \tag{6.62}$$

Let us estimate the three terms in the right-hand side of (6.62).

First, noticing that $\mathbf{u}^n \to \mathbf{u}$ weakly-$*$ in $L^\infty(0,T;H)$ and $\mathbf{u}_t^n \to \mathbf{u}_t$ weakly in $L^\sigma(0,T;V')$, we deduce that $\mathbf{u}^n \to \mathbf{u}$ strongly in $C^0([0,T];V')$. Consequently, for any $\mathbf{z} \in V$, one has

$$\begin{aligned} |(\mathbf{u}(\cdot\,,T_n^*) - \mathbf{u}^n(\cdot\,,T_n^*), \mathbf{z})| &\le C\|\mathbf{u}(\cdot\,,T_n^*) - \mathbf{u}^n(\cdot\,,T_n^*)\|_{V'}\|\mathbf{z}\|_V \\ &\le \|\mathbf{u} - \mathbf{u}^n\|_{C^0([0,T_0];V')}\|\mathbf{z}\|_V, \end{aligned} \tag{6.63}$$

which goes to zero as $n \to +\infty$.

Also, since $T_n^* \to \tilde{T}$ and $\mathbf{u} \in C_w^0([0,T_0];H)$, we have $\mathbf{u}(\cdot\,,T_n^*) \to \mathbf{u}(\cdot\,,\tilde{T})$ weakly in H, whence

$$|(\mathbf{u}(\cdot\,,\tilde{T}) - \mathbf{u}(\cdot\,,T_n^*), \mathbf{z})| \to 0. \tag{6.64}$$

Finally, by the definition of T_n^*,

$$|(\mathbf{u}^n(\cdot\,,T_n^*) - \mathbf{u}_e, \mathbf{z})| \le \|\mathbf{u}^n(\cdot\,,T_n^*) - \mathbf{u}_e\|\,\|\mathbf{z}\| \le \delta\|\mathbf{z}\|. \tag{6.65}$$

From (6.62) and (6.63)–(6.65), we immediately get (6.61).

This ends the proof. □

Remark 6.8 It would be interesting to know whether Theorem 6.9 continues to be true if, in (6.56), $b = 0$. To our knowledge, if $\mathcal{U}_{\rm ad}$ is unbounded, this is unknown. In other words, even under the assumption that there exist a time and a control that drives the system to a state at distance $\le \delta$ from the target $\mathbf{u}_e$, we do not know if there exists a minimal time to do this. □

Remark 6.9 That the set in (6.58) is non-empty or not for any $\mathbf{u}_0$, $\mathbf{u}_e$ and δ is a question that corresponds to *controllability theory*. Issues of this kind will be analyzed below, in Sect. 6.4. □

6.3.2 The Time Optimality Conditions

Now, we will try to characterize the solutions to (6.57) in terms of a suitable optimality system.

Let us introduce the function Φ, with

$$\Phi(T, \mathbf{v}) := \frac{T^2}{2} + \frac{b}{2} \iint_{\omega \times (0, T_0)} |\mathbf{v}|^2, \qquad \forall (T, \mathbf{v}) \in [0, T_0] \times L^2(\omega \times (0, T_0))^N. \tag{6.66}$$

Then, (6.57) can also be written in the form

$$\begin{cases} \text{Minimize} \quad \Phi(T, \mathbf{v}) \\ \text{subject to} \;\; T \in [0, T_0], \\ \qquad\qquad (\mathbf{v}, \mathbf{u}, \rho, p) \in \mathcal{S}_{\rm ad}(T_0), \\ \qquad\qquad \|\mathbf{u}(\cdot\,, T) - \mathbf{u}_e\| \le \delta. \end{cases} \tag{6.67}$$

The following result holds:

Theorem 6.10 *Let the assumptions of Theorem 6.9 be satisfied. Let* $(T^*, \mathbf{v}^*)$ *be a solution to* (6.67)*, with associated state* $\{\mathbf{u}^*, \rho^*, p^*\}$*. Assume that* $\{\mathbf{u}^*, \rho^*, p^*\}$ *satisfies* (6.23),

$$0 < T^* < T_0, \tag{6.68}$$

$$\exists \kappa > 0 \text{ such that the } H\text{-valued function } t \mapsto \mathbf{u}^*(\cdot\,, t) \text{ is } C^1 \text{ in } [T^* - \kappa, T^*] \tag{6.69}$$

and

$$(\mathbf{u}^*(\cdot\,, T^*) - \mathbf{u}_e, \mathbf{u}^*_t(\cdot\,, T^*)) < 0 \tag{6.70}$$

and denote by E^* *the energy space associated to* T^*: $E^* = E(T^*)$*. Then, there exist* $\lambda \in \mathbb{R}$ *and* $\{\mathbf{w}, \eta, q\} \in E^*$ *such that one has:*

$$\begin{cases} \dfrac{\partial \rho^* \mathbf{u}^*}{\partial t} + \nabla \cdot (\rho^* \mathbf{u}^* \otimes \mathbf{u}^*) - \mu \Delta \mathbf{u}^* + \nabla p^* \\ \qquad = \mathbf{v}^* \mathbb{1}_\omega, \;\; (\mathbf{x}, t) \in \Omega \times (0, T^*), \\ \nabla \cdot \mathbf{u}^* = 0, \;\; (\mathbf{x}, t) \in Q, \\ \dfrac{\partial \rho^*}{\partial t} + \nabla \cdot \left(\rho^* \mathbf{u}^*\right) = 0, \;\; (\mathbf{x}, t) \in \Omega \times (0, T^*), \\ \mathbf{u}^* = 0, \;\; (\mathbf{x}, t) \in \partial\Omega \times (0, T^*), \\ \rho^*|_{t=0} = \rho_0, \;\; \mathbf{u}^*|_{t=0} = \mathbf{u}_0, \;\; \mathbf{x} \in \Omega, \end{cases} \tag{6.71}$$

$$
\begin{cases}
-\rho^* \dfrac{\partial \mathbf{w}}{\partial t} - \rho^*(\mathbf{u}^* \cdot \nabla)\mathbf{w} + \rho^*(\nabla \mathbf{u}^*)\mathbf{w} - \mu \Delta \mathbf{w} + \nabla q + \eta \nabla \rho^* \\
\qquad\qquad = \mathbf{0}, \quad (\mathbf{x}, t) \in \Omega \times (0, T^*), \\
\nabla \cdot \mathbf{w} = 0, \quad (\mathbf{x}, t) \in Q, \\
-\dfrac{\partial \eta}{\partial t} - \mathbf{u}^* \cdot \nabla \eta + (\dfrac{\partial \mathbf{u}^*}{\partial t} + (\mathbf{u}^* \cdot \nabla)\mathbf{u}^*) \cdot \mathbf{w} = 0, \quad (\mathbf{x}, t) \in \Omega \times (0, T^*), \\
\mathbf{w} = 0, \quad (\mathbf{x}, t) \in \partial\Omega \times (0, T^*), \\
\eta^*|_{t=T^*} = 0, \quad \mathbf{w}|_{t=T^*} = \lambda(\mathbf{u}|_{t=T^*} - \mathbf{u}_e), \quad \mathbf{x} \in \Omega,
\end{cases}
\tag{6.72}
$$

$$
\iint_{\omega \times (0,T^*)} (\mathbf{w} + b\mathbf{v}^*) \cdot (\mathbf{v} - \mathbf{v}^*) \geq 0 \quad \forall \mathbf{v} \in \mathcal{U}_{\text{ad}}, \quad \mathbf{v}^* \in \mathcal{U}_{\text{ad}}, \tag{6.73}
$$

$$
T^* = P_{[0,T_0]} \left(-\lambda(\mathbf{u}(\cdot\,, T^*) - \mathbf{u}_e, \frac{\partial \mathbf{u}^*}{\partial t}(\cdot\,, T^*)) \right) \tag{6.74}
$$

and

$$
\|\mathbf{u}^*(\cdot\,, T^*) - \mathbf{u}_e\| = \delta. \tag{6.75}
$$

Remark 6.10 The assumption (6.68) on T^* serves to discard trivial cases. On the other hand, (6.69) is a regularity assumption and (6.70) plays the role of a qualification hypothesis; this is explained below. It is a reasonable assumption, at least when $\mathcal{U}_{\text{ad}} = L^2(\omega \times (0, T))^N$; to this respect, see Remark 6.11. □

In order to well understand the situation and interpret Theorem 6.10 appropriately, we will first consider a similar simplified time optimal problem. Thus, let us provisionally replace (6.1) by the simpler system

$$
\begin{cases}
\dfrac{\partial \theta}{\partial t} - \Delta \theta = h 1\!\!1_\omega, \quad (\mathbf{x}, t) \in Q, \\
\theta = 0, \quad (\mathbf{x}, t) \in \Sigma, \\
\theta|_{t=0} = \theta_0, \quad \mathbf{x} \in \Omega.
\end{cases}
\tag{6.76}
$$

Denote by $\mathcal{M}_0$ the family of control-state pairs (h, θ), where $h \in L^2(\omega \times (0, T_0))$ and θ solves (6.76). Let $\theta_e \in L^2(\Omega)$ be given with $\|\theta_0 - \theta_e\| > \delta$ and let (T^*, h^*) be a solution to the problem

$$
\begin{cases}
\text{Minimize} \quad \Psi(T, h) = \dfrac{1}{2} T^2 + \dfrac{b}{2} \displaystyle\iint_{\omega \times (0,T_0)} |h|^2 \\
\text{subject to} \quad T \in [0, T_0] \\
\qquad\qquad (h, \theta) \in \mathcal{M}_0, \\
\qquad\qquad \|\theta(\cdot\,, T) - \theta_e\| \leq \delta
\end{cases}
\tag{6.77}
$$

with $T^* > 0$.

Note that the unique constraint on the control is imposed through the state. Clearly, an existence result similar to Theorem 6.1 can be established for (6.77).

Let θ^* be the state associated to h^* in $\Omega \times (0, T^*)$ and assume that

$$\exists \kappa > 0 \text{ such that } t \mapsto \theta^*(\cdot\,, t) \text{ is } C^1 \text{ in } [T^* - \kappa, T^*]. \tag{6.78}$$

Observe that we must have $\|\theta^*(\cdot\,, T^*) - \theta_e\| = \delta$. Indeed, if we had $\|\theta^*(\cdot\,, T^*) - \theta_e\| < \delta$, there would exist times $T < T^*$ such that the constraint $\|\theta^*(\cdot\,, T) - \theta_e\| \leq \delta$ holds, in contradiction with the definition of T^*. Consequently, we can view (T^*, h^*) as a minimizer of Ψ subject to the equality constraints

$$\begin{aligned} E_1(h, \theta) &:= (\theta_t - \Delta\theta - h 1\!\!1_\omega, \theta(\cdot\,, 0) - \theta_0) = (0, 0), \\ E_2(T, \theta) &:= \frac{1}{2}\|\theta(\cdot\,, T) - \theta_e\|^2 - \frac{\delta^2}{2} = 0. \end{aligned}$$

Note that E_1 and E_2 are well-defined C^1 mappings on appropriate Hilbert spaces and the linear spaces $R(E_1'(h, \theta))$ and $R(E_2'(T, \theta))$ are closed for all h, θ and T. Therefore, due to the classical Lagrange's Theorem, there exist multipliers λ_0, ψ, ζ and λ (not simultaneously equal to zero) with

$$\lambda_0, \lambda \in \mathbb{R}, \quad \psi = \psi(\mathbf{x}, t), \quad \zeta = \zeta(\mathbf{x})$$

and

$$\begin{aligned} 0 &= \lambda_0 \langle \Psi'(T^*, h^*), (S, m) \rangle - \langle (\psi, \zeta), E'(h^*, \theta^*)(m, y) \rangle + \lambda \langle R'(T^*, \theta^*), (S, y) \rangle \\ &= \lambda_0 \left(T^* S + b \iint_{\omega \times (0,T)} h^* m \right) \\ &\quad - \iint_Q \psi(y_t - \Delta y - m 1\!\!1_\omega) - (\zeta, y(\cdot\,, 0)) \\ &\quad + \lambda \left((\theta^*(\cdot\,, T^*) - \theta_e, \theta_t^*(\cdot\,, T^*))S + (\theta^*(\cdot\,, T^*) - \theta_e, y(\cdot\,, T^*)) \right), \end{aligned}$$

for all $S \in \mathbb{R}$ and $(m, y) \in \mathcal{M}_0$.

Since S is arbitrary, the first consequence is that

$$\lambda_0 T^* + \lambda(\theta^*(\cdot\,, T^*) - \theta_e, \theta_t^*(\cdot\,, T^*)) = 0. \tag{6.79}$$

The second consequence is that, for all (m, y), one has

$$\iint_Q \psi(y_t - var Deltay - m 1\!\!1_\omega) - \lambda(\theta^*(\cdot\,, T^*) - \theta_e, y(\cdot\,, T^*)) + (\zeta, y(\cdot\,, 0)) = 0 \tag{6.80}$$

and, after some standard computations involving integrations by parts, this leads to

$$\begin{cases} -\frac{\partial \psi}{\partial t} - \Delta \psi = 0, \quad (\mathbf{x}, t) \in \Omega \times (0, T^*), \\ \psi = 0, \quad (\mathbf{x}, t) \in \partial\Omega \times (0, T^*), \\ \psi(\mathbf{x}, T^*) = \lambda(\theta^*(\mathbf{x}, T^*) - \theta_e(\mathbf{x})), \quad \mathbf{x} \in \Omega \end{cases} \tag{6.81}$$

and

$$\zeta(\mathbf{x}) = \psi(\mathbf{x}, 0) \text{ in } \Omega. \tag{6.82}$$

Finally, we also have

$$\psi + \lambda_0 b h^* = 0 \text{ in } \omega \times (0, T^*). \tag{6.83}$$

We see from (6.79), (6.81), and (6.82) that λ cannot be zero, since otherwise we would also have $\lambda_0 = 0$, $\psi \equiv 0$ and $\zeta \equiv 0$, which is impossible.

The function $t \mapsto \frac{1}{2}\|\theta^*(\cdot\,, t) - \theta_e\|^2$ is non-increasing at $t = T^*$; consequently, $(\theta^*(\cdot\,, T^*) - \theta_e, \theta_t^*(\cdot\,, T^*)) \leq 0$. It is immediate from (6.79) that, if the strict inequality holds, then $\lambda_0 \neq 0$. We can thus assume that $\lambda_0 = 1$ and (6.83) and (6.79) respectively become

$$h^* = -\frac{1}{\lambda_0 b}\psi\big|_{\omega\times(0,T)} \text{ in } \omega \times (0, T^*) \tag{6.84}$$

and

$$T^* = -\lambda(\theta^*(\cdot\,, T^*) - \theta_e, \theta_t^*(\cdot\,, T^*)). \tag{6.85}$$

Finally, note that we must always have

$$\|\theta^*(\cdot\,, T^*) - \theta_e\| = \delta. \tag{6.86}$$

Therefore, we see that, in this simplified case, the optimality system is (6.76), (6.81), (6.84)–(6.86) (with h and θ respectively replaced by h^* and θ^*), which is similar to (6.71)–(6.75).

The minimal time control problem (6.77) for the heat equation was solved both theoretically and numerically in [45].

Remark 6.11 A close look to the simpler problem (6.77) allows to understand the adequacy of the assumptions (6.70) and (6.78). Indeed, assume that (T^*, h^*) solves (6.77), $T^* \in (0, T_0)$ and (6.78) is satisfied. We have already seen that

$$(\theta^*(\cdot\,, T^*) - \theta_e, \theta_t^*(\cdot\,, T^*)) \leq 0.$$

If we had $(\theta^*(\cdot\,, T^*) - \theta_e, \theta_t^*(\cdot\,, T^*)) = 0$, the identities (6.79) and (6.83) would show that $\lambda_0 = 0$ and

$$\zeta = 0 \text{ in } \omega \times (0, T^*).$$

But then $\psi \equiv 0$, because the solutions to systems of the kind (6.81) satisfy the unique continuation property (see, for instance, [101]). From (6.82), we would also have $\eta = 0$. Taking into account the final condition satisfied by ψ and recalling that at least one multiplier must be nonzero, we would deduce that

$$\theta^*(\mathbf{x}, T^*) = \theta_e(\mathbf{x}) \text{ in } \Omega,$$

which is obviously absurd. Consequently, we must necessarily have

$$(\theta^*(\cdot\,, T^*) - \theta_e, \theta_t^*(\cdot\,, T^*)) < 0.$$

This shows that, in Theorem 6.10, inequality (6.70) seems to be a reasonable assumption at least when $\mathcal{U}_{\rm ad} = L^2(\omega \times (0, T_0))^N$. □

Proof of Theorem 6.10 We will give a direct argument relying on the problem satisfied by the adjoint state $\{\mathbf{w}, \eta, q\}$.

Again, we must have $\|\mathbf{u}^*(\cdot\,, T^*) - \mathbf{u}_e\| = \delta$ and $(T^*, \mathbf{v}^*, \mathbf{u}^*, \rho^*, p^*)$ can be regarded as a solution to the problem

$$\begin{cases} \text{Minimize} \quad \Phi(T, \mathbf{v}) \\ \text{subject to } \; T \in [0, T_0], \\ \qquad\qquad (\mathbf{v}, \mathbf{u}, \rho, p) \in \mathcal{S}_{\rm ad}(T_0), \\ \qquad\qquad \|\mathbf{u}(\cdot\,, T) - \mathbf{u}_e\| = \delta. \end{cases} \tag{6.87}$$

Let us introduce $S \in \mathbb{R}$, $\mathbf{m} \in L^2(\omega \times (0, T_0))^N$ and $\varepsilon \in \mathbb{R}_+$ and let us set

$$T := T^* + \varepsilon S \in [0, T_0], \quad \mathbf{v} := \mathbf{v}^* + \varepsilon \mathbf{m} \in \mathcal{U}_{\rm ad}. \tag{6.88}$$

Let $\{\mathbf{u}, \rho, p\}$ be a state associated to $\mathbf{v}$ and let us assume that $\|\mathbf{u}(\cdot\,, T) - \mathbf{u}_e\|^2 = \delta^2$. In view of (6.88), for any small $\varepsilon > 0$ one has

$$\begin{aligned} 0 \le \Phi(T, \mathbf{v}) - \Phi(T^*, \mathbf{v}^*) &= \varepsilon \left(T^* S + b \iint_{\omega\times(0,T_0)} \mathbf{v}^* \cdot \mathbf{m} \right) \\ &+ \frac{\varepsilon^2}{2} \left[S^2 + b \iint_{\omega\times(0,T_0)} |\mathbf{m}|^2 \right]. \end{aligned}$$

Moreover, one must have $\mathbf{v}^* = \mathbf{0}$ for $t \in (T^*, T_0)$, whence

$$T^* S + b \iint_{\omega\times(0,T^*)} \mathbf{v}^* \cdot \mathbf{m} \ge 0. \tag{6.89}$$

As in the proof of Theorem 6.2, we can write that

$$\{\mathbf{u}, \rho, p\} = \{\mathbf{u}^*, \rho^*, p^*\} + \varepsilon\{\mathbf{y}, \sigma, \pi\} + \varepsilon\{\mathbf{y}'_\varepsilon, \sigma'_\varepsilon, p'_\varepsilon\},$$

with $\{\mathbf{y}, \sigma, \pi\}$ and $\{\mathbf{y}'_\varepsilon, \sigma'_\varepsilon, \pi'_\varepsilon\}$ solving systems similar to (6.27) and (6.28), respectively, together with homogeneous boundary and initial conditions.

Arguing as we did in that proof, we see that $\{\mathbf{y}, \sigma, \pi\}, \{\mathbf{y}'_\varepsilon, \sigma'_\varepsilon, p'_\varepsilon\} \in E(T_0)$ and $\|\mathbf{y}'_\varepsilon\|_{L^\infty(0,T;H)} \to 0$ as $\varepsilon \to 0^+$ (among other things). Moreover,

$$\begin{aligned} 0 &= \|\mathbf{u}(\cdot\,, T) - \mathbf{u}_e\|^2 - \delta^2 = \|(\mathbf{u}(\cdot\,, T) - \mathbf{u}^*(\cdot\,, T^*)) + (\mathbf{u}^*(\cdot\,, T^*) - \mathbf{u}_e)\|^2 - \delta^2 \\ &= \|\mathbf{u}(\cdot\,, T) - \mathbf{u}^*(\cdot\,, T^*)\|^2 + 2(\mathbf{u}(\cdot\,, T) - \mathbf{u}^*(\cdot\,, T^*), \mathbf{u}^*(\cdot\,, T^*) - \mathbf{u}_e). \end{aligned}$$

Taking into account that

$$\mathbf{u}(\cdot\,, T) - \mathbf{u}^*(\cdot\,, T^*) = \varepsilon\mathbf{y}(\cdot\,, T^*) + \varepsilon\mathbf{u}^*_t(\cdot\,, T^*)S + \varepsilon\, O(\varepsilon) \;\text{ in }\; L^2(\Omega)$$

where $O(\varepsilon) \to 0$, we easily deduce that

$$-\,(\mathbf{u}^*(\cdot\,, T^*) - \mathbf{u}_e, \mathbf{u}^*_t(\cdot\,, T^*))\, S = (\mathbf{u}^*(\cdot\,, T^*) - \mathbf{u}_e, \mathbf{y}(\cdot\,, T^*)). \tag{6.90}$$

Now, let us introduce $\lambda \in \mathbb{R}$ with

$$-\,(\mathbf{u}^*(\cdot\,, T^*) - \mathbf{u}_e, \mathbf{u}^*_t(\cdot\,, T^*))\, \lambda = T^* \tag{6.91}$$

and let $\{\mathbf{w}, \eta, q\}$ be the solution to system (6.72).

Due to (6.70), λ is well defined.

Furthermore,

$$\begin{aligned} T^*S &= -(\mathbf{u}^*(\cdot\,, T^*) - \mathbf{u}_e, \mathbf{u}^*_t(\cdot\,, T^*))\, \lambda S \\ &= (\lambda(\mathbf{u}^*(\cdot\,, T^*) - \mathbf{u}_e), \mathbf{y}(\cdot\,, T^*)) \\ &= (\mathbf{w}(\cdot\,, T^*), \mathbf{y}(\cdot\,, T^*)) \end{aligned}$$

and, using the equations and boundary and initial conditions satisfied by $\{\mathbf{w}, \eta, q\}$ and $\{\mathbf{y}, \sigma, \pi\}$, we see after some integrations by parts that

$$T^*S = \iint_{\omega\times(0,T_0)} \mathbf{w}\cdot\mathbf{m}.$$

In view of (6.89), this yields

$$\iint_{\omega\times(0,T_0)} (\mathbf{w} + b\mathbf{v}^*)\cdot\mathbf{m} \geq 0.$$

On the other hand, since $T^* \in (0, T_0)$ and λ is given by (6.91), the equality (6.74) is trivially satisfied.

Consequently, the couple $(T^*, \mathbf{v}^*)$, the associated state $\{\mathbf{u}^*, \rho^*, p^*\}$, the multiplier $\lambda \in \mathbb{R}$ and the adjoint state $\{\mathbf{w}, \eta, q\}$ satisfy (6.71)–(6.75).

This ends the proof. □

The optimality system (6.71)–(6.75) can be used to deduce iterative algorithms for the computation of an optimal $(T^*, \mathbf{v}^*, \mathbf{u}^* \rho^*, p^*)$.

Let us indicate more precisely a possible strategy.

First of all, note that, for any given $T > 0$ and $\lambda > 0$, (6.71)–(6.73) is the optimality system corresponding to the optimal control problem

$$\begin{cases} \text{Minimize } \dfrac{\lambda}{2}\displaystyle\int_\Omega |\mathbf{u}(\mathbf{x}, T) - \mathbf{u}_e(\mathbf{x})|^2 + \dfrac{b}{2}\iint_{\omega\times(0,T)} |\mathbf{v}|^2 \\ \text{subject to } \mathbf{v} \in \mathcal{U}_{\text{ad}},\ (\mathbf{v}, \mathbf{u}, \rho, p) \text{ satisfies (6.1).} \end{cases} \tag{6.92}$$

Let us assume that we can solve this problem for each T and λ and we can associate to each couple (T, λ) an optimal control-state $(\mathbf{v}, \mathbf{u}, \rho, p)$ and, therefore, the quantities

$$V_1(T, \lambda) := \frac{1}{2}\|\mathbf{u}(\cdot\,, T) - \mathbf{u}_e\|^2 - \frac{\delta^2}{2}$$

and

$$V_2(T, \lambda) := T - P_{[0,T_0]}\left(-\lambda(\mathbf{u}(\cdot\,, T) - \mathbf{u}_e, \frac{\partial \mathbf{u}}{\partial t}(\cdot\,, T))\right).$$

Then, the following can be done:

1. Choose $\tau, \kappa > 0$, solve (6.92) for $T = \tau, 2\tau, \dots$ and $\lambda = \kappa, 2\kappa, \dots$ and compute the associated $V_1(T, \lambda)$.
2. Using these values, for every $T = \tau, 2\tau, \dots,$ compute a solution to the equation

$$V_1(T, \lambda) = 0, \quad \lambda > 0. \tag{6.93}$$

   Denote this solution by $\lambda(T)$.
3. Using the $\lambda(T)$, solve Eq. $V_2(T, \lambda(T)) = 0$.

Denote by T^* a solution to this last equation.

Set $\lambda^* := \lambda(T^*)$ and let $(\mathbf{v}^*, \mathbf{u}^*, \rho^*, p^*)$ solve the problem (6.92) corresponding to $T = T^*$ and $\lambda = \lambda^*$. Then, $(T^*, \mathbf{v}^*, \mathbf{u}^*, \rho^*, p^*)$ is a solution to (6.71)–(6.75) and, consequently, $(T^*, \mathbf{v}^*)$ is a legitimate candidate to solve the original problem (6.87).

In the case of (6.77), this program was achieved satisfactorily in [45]. There, we proved that the control problems analog to (6.92) are uniquely solvable and, consequently, the $V_1(T, \lambda)$ and the $V_2(T, \lambda)$ are well defined.

It can also be proved that, for each $T > 0$, Eq. (6.93) has at least one solution and satisfactory experiments yield an accurate numerical solution to the time optimal control problem.

6.4 Controllability. Main Concepts, Problems and Results

In this section, we will be concerned with some controllability problems for the variable density Navier-Stokes system (6.1).

The arguments and the results are rather technical and need considerable more work than those in the previous chapters and sections. Accordingly, we will sketch some of them, leaving the details to the reader and indicating appropriate references.

A general *partial controllability* question for (6.1) is the following:

Let Ω, ω and T be given and let us fix a desired requirement for $\mathbf{u}$ at time $t = T$ (for instance, $\mathbf{u}(\cdot\,, T) = 0$ or, more generally, $\mathbf{u}(\cdot\,, T) \in \mathbf{U}_T$ for some $\mathbf{U}_T \subset H$). Then, is it true or not that, for any ρ_0 and $\mathbf{u}_0$ in "good" spaces we can find a control $\mathbf{v} \in L^2(\omega\times(0, T))^N$ such that (6.1) possesses a solution $\{\mathbf{u}, \rho, p\}$ in $Q = \Omega \times (0, T)$ satisfying this desired property?

Accordingly, the goal is to control or govern the system to make the velocity field satisfy a particular property at final time. Note that, in particular, this would permit to stabilize an initially turbulent flow in finite time or even stop (i.e., drive to rest) a flow starting from a disordered (unpleasant) situation.

We say that this is a partial controllability problem, since we are trying to control at final time $\mathbf{u}$ but not necessarily ρ (this will be justified below).

In the previous statement, some particular important sets $\mathbf{U}_T$ are the following:

- A ball in H of small radius. When we are able to find controls that drive associated solutions to (6.1) to arbitrary balls of center $\mathbf{u}_T$, we say that the system is *partially approximately controllable* to $\mathbf{u}_T$. If this is true for any $\mathbf{u}_T \in H$, we simply say that (6.1) is *partially approximately controllable* (at time T).

  This is an interesting property: it means that, after some (possibly hard and expensive) work on $\omega \times (0, T)$, we are always able to get a final velocity as close as desired to a prescribed field.
- $\mathbf{U}_T = \{\mathbf{0}\}$. In this case, we say that we are dealing with a *partial null controllability* problem.

  Solving this problem is still more interesting than controlling approximately. Indeed, this would indicate that we are always able to drive the system to rest at $t = T$.

  In practice, this needs action (i.e., external applied forces) during the bounded interval $(0, T)$ but not for $t > T$. Once the zero velocity field is attained, the solution persists for later times and no additional control is needed to preserve equilibrium.
- $\mathbf{U}_T = \{\overline{\mathbf{u}}(\cdot, T)\}$, where $\overline{\mathbf{u}}$ is, together with some density $\overline{\rho}$ and some pressure $\overline{p}$, a prescribed solution to (6.1) associated with some given control $\overline{\mathbf{v}}$ and some other initial data $\overline{\mathbf{u}}_0$ and $\overline{\rho}_0$. Now, if the corresponding problem can be solved, we say that (6.1) is *partially exactly controllable* to the trajectory determined by $\{\overline{\mathbf{u}}, \overline{\rho}, \overline{p}\}$ at time T.

  Again, this would require control action during $(0, T)$ only; for later times, it would suffice to take $\mathbf{v} = \overline{\mathbf{v}}$.

For the constant density Navier-Stokes equations, the results of some numerical experiments concerning exact controllability to Poiseuille and Taylor-Green flows have been presented above respectively in Figs. 6.1 and 6.2 and Figs. 6.3, 6.4, 6.5, and 6.6.

- A subspace $E \subset H$, possibly of finite dimension. This is the situation where we try to find out whether or not E is *reachable* starting from an arbitrary initial state and applying controls in $L^2(\omega \times (0, T))^N$.

These controllability problems are more difficult to solve than the optimal control problems considered in Sects. 6.2 and 6.3.

In fact, not many things can be said about them and not many satisfactory answers can be given in this context at the present moment.

So far, we have only considered distributed controls $\mathbf{v}$ supported by sets of the form $\omega \times (0, T)$, where $\omega \subset \Omega$ is open and "small". In fact, it is also meaningful (and even maybe more natural from the physical viewpoint) to consider boundary controls. In the related problems, to act on the fluid is to apply an insulating/extracting strategy and/or fix the mass density of the incoming fluid particles.

It can be thought that, in principle, this needs a more delicate analysis, since it concerns the solutions to nonlinear evolution PDEs with nonzero boundary data. However, fortunately, the related controllability problems can be reduced to questions concerning distributed controls via relatively simple domain extension techniques.

Let us be more precise. Consider the system

$$\begin{cases} \dfrac{\partial \rho \mathbf{u}}{\partial t} + \nabla \cdot (\rho \mathbf{u} \otimes \mathbf{u}) - \mu \Delta \mathbf{u} + \nabla p = 0, & (\mathbf{x}, t) \in Q, \\ \nabla \cdot \mathbf{u} = 0, \quad (\mathbf{x}, t) \in Q, \\ \dfrac{\partial \rho}{\partial t} + \nabla \cdot (\rho \mathbf{u}) = 0, \quad (\mathbf{x}, t) \in Q, \\ \mathbf{u} = \mathbf{h} \mathbb{1}_\gamma, \quad \rho = \sigma \mathbb{1}_{\{\mathbf{u} \cdot \mathbf{n} < 0\}}, \quad (\mathbf{x}, t) \in \Sigma, \\ \rho|_{t=0} = \rho_0, \quad \mathbf{u}|_{t=0} = \mathbf{u}_0, \quad \mathbf{x} \in \Omega. \end{cases} \tag{6.94}$$

A general partial boundary controllability problem for (6.94) reads:

Let Ω, γ and T *be given with* $\gamma \subset \partial\Omega$ *being a non-empty open set and let us fix the set* $\mathbf{U}_T \subset H$. *Is it true or not that, for any* ρ_0 *and* $\mathbf{u}_0$, *we can find boundary data* $\mathbf{h}$ *and* σ *respectively for* $\mathbf{u}$ *and* ρ *such that* (6.94) *has a solution* $\{\mathbf{u}, \rho, p\}$ *in* Q *satisfying* $\mathbf{u}(\cdot\,, T) \in \mathbf{U}_T$?

This can be reduced to a partial controllability problem for (6.1) with distributed controls. It suffices to implement the following steps:

1. Extend the spatial domain Ω to a strictly larger open regular set $\tilde{\Omega}$ such that $\gamma = \partial\Omega \cap \tilde{\Omega}$.
2. Extend the initial data $\mathbf{u}_0$ and ρ_0 and the functions in $\mathbf{U}_T$ as suitable functions defined in the whole set $\tilde{\Omega}$.

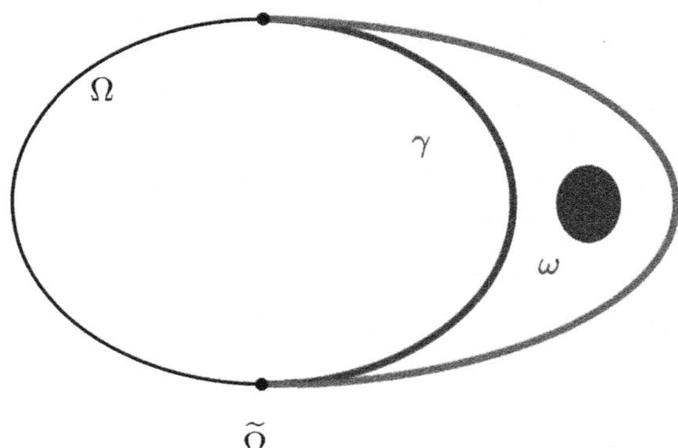

Fig. 6.10 The open sets Ω, ω and $\tilde{\Omega}$

3. Control the resulting system in $\tilde{Q} := \tilde{\Omega} \times (0, T)$ with a distributed control $\tilde{\mathbf{v}} \in L^2(\omega \times (0, T))^N$, where ω is a non-empty open subset of $\tilde{\Omega} \setminus \overline{\Omega}$.
4. Consider the restriction to Q of the "extended state" $\{\tilde{\mathbf{u}}, \tilde{\rho}, \tilde{p}\}$.

See Fig. 6.10 for a graphic representation.

Obviously, the boundary controls we are looking for are the lateral traces of $\mathbf{u}$ and ρ, respectively on $\gamma \times (0, T)$ and Σ_{in}, where

$$\Sigma_{in} := \{(\mathbf{x}, t) \in \gamma \times (0, T) : \mathbf{u}(\mathbf{x}, t) \cdot \mathbf{n}(x) < 0\}$$

(Σ_{in}) is obviously the set of points on $\gamma \times (0, T)$ where the flow is *inward directed.*

Note that this process allows to reduce distributed (or internal) controllability to boundary controllability without any need of supporting the control result on the existence of a state for every admissible control.

Furthermore, if the argument is successful, we are led to the existence of *smooth* controls (much more regular that L^2) that achieve the imposed task $\mathbf{u}(\cdot\,, T) \in \mathbf{U}_T$.

6.4.1 *The Constant Density Case*

In order to well understand the situation, we will first consider the homogeneous Navier-Stokes system

$$\begin{cases} \rho_0 \left(\dfrac{\partial \mathbf{u}}{\partial t} + (\mathbf{u} \cdot \nabla)\mathbf{u} \right) - \mu \Delta \mathbf{u} + \nabla p = \mathbf{v} 1_{\omega}, \quad (\mathbf{x}, t) \in Q, \\ \nabla \cdot \mathbf{u} = 0, \quad (\mathbf{x}, t) \in Q, \\ \mathbf{u} = \mathbf{0} \quad (\mathbf{x}, t) \in \Sigma, \\ \mathbf{u}(\mathbf{x}, 0) = \mathbf{u}_0(\mathbf{x}) \quad \mathbf{x} \in \Omega \end{cases} \tag{6.95}$$

and the similar linear Stokes-like problem

$$\begin{cases} \rho_0\left(\dfrac{\partial \mathbf{u}}{\partial t}+(\tilde{\mathbf{u}}\cdot\nabla)\mathbf{u}+(\mathbf{u}\cdot\nabla)\tilde{\mathbf{u}}\right)-\mu\Delta\mathbf{u}+\nabla p=\mathbf{v}1\!\!1_{\omega}, \quad (\mathbf{x},t)\in Q, \\ \nabla\cdot\mathbf{u}=0, \quad (\mathbf{x},t)\in Q, \\ \mathbf{u}=\mathbf{0} \quad (\mathbf{x},t)\in\Sigma, \\ \mathbf{u}(\mathbf{x},0)=\mathbf{u}_0(\mathbf{x}) \quad \mathbf{x}\in\Omega, \end{cases} \tag{6.96}$$

where $\rho_0 > 0$ is a constant and $\tilde{\mathbf{u}} = \tilde{\mathbf{u}}(\mathbf{x}, t)$ are given and satisfies adequate regularity assumptions.

For the sake of clarity, before considering controllability problems for (6.95) and (6.96), we will recall some well known existence, uniqueness and regularity results. Their proofs can be found arguing as in the proofs of the results in Chaps. 3 and 5. The details can be found, together with many other results, in [20, 105]:

- For any $\mathbf{u}_0 \in H$ and any $\mathbf{v} \in L^2(\omega \times (0,T))^N$, there exists at least one weak solution $\{\mathbf{u}, p\}$ to (6.95), with

$$\mathbf{u} \in L^2(0,T;V)\cap C^0_w([0,T];H), \quad \frac{\partial \mathbf{u}}{\partial t} \in L^\sigma(0,T;V') \quad (\sigma = 4/N),$$
$$p \in L^2(0,T;H^{-1}(\Omega))\cap W^{-1,\infty}(0,T;L^2(\Omega)).$$

  If $N = 2$, the solution is unique (p is unique up to an additive function only depending on t) and, among other properties, one has

$$\mathbf{u} \in C^0([0,T];H).$$

- If $\mathbf{u}_0 \in V$, the solution is strong for small t. That is, there exists $T_* \in (0,T)$ such that

$$\mathbf{u} \in L^2(0,T_*;D(A))\cap C^0([0,T_*];V), \quad \frac{\partial \mathbf{u}}{\partial t} \in L^2(0,T_*;H),$$
$$p \in L^2(0,T_*;H^1(\Omega)).$$

  If $N = 2$, we have this with $T_* = T$.
- If $\tilde{\mathbf{u}}$ is regular enough (for example $\tilde{\mathbf{u}} \in L^2(0,T;V)\cap L^\infty(Q)^N$), there exists a unique weak solution to (6.96). Moreover, if $\mathbf{u}_0 \in V$, the solution is again strong.
- On the other hand, the regularity of the solutions to (6.95) and (6.96) improve with the data. For instance, if $\mathbf{u}_0 \in D(A) = H^2(\Omega)^N \cap V$ and the right-hand side satisfies

$$\mathbf{v} \in L^2(0,T;H^1(\omega)^N) \quad \text{and} \quad \frac{\partial \mathbf{v}}{\partial t} \in L^2(\omega\times(0,T))^N,$$

  there exists $T_* \in (0,T)$ such that the solution to (6.95) satisfies

$$\mathbf{u} \in L^2(0, T_*; H^3(\Omega)^N) \cap C^0([0, T_*]; D(A)), \quad \frac{\partial \mathbf{u}}{\partial t} \in L^2(0, T_*; V),$$
$$p \in L^2(0, T_*; H^2(\Omega)).$$

Again, if $N = 2$, this holds with $T_* = T$.

The results that follow have been taken mainly from [26, 41, 48, 49, 56, 63, 72]. They concern approximate controllability, null controllability and local exact controllability to the trajectories for (6.95) and (6.96). Other important contributions are [21, 23, 27, 28, 52, 61, 65].

It will be said that (6.95) is *approximately controllable* (AC) at time T if, for any $\mathbf{u}_0 \in H$, any $\mathbf{u}_T \in H$ and any $\varepsilon > 0$, there exist $\mathbf{v} \in L^2(\omega \times (0, T))^N$ and an associated solution $\{\mathbf{u}, p\}$ satisfying

$$\|\mathbf{u}(\cdot, T) - \mathbf{u}_T\| \leq \varepsilon. \tag{6.97}$$

It will be said that (6.95) (resp. (6.96)) is *null-controllable* (NC) at time T if, for any $\mathbf{u}_0 \in H$, there exists $\mathbf{v} \in L^2(\omega \times (0, T))^N$ and an associated solution $\{\mathbf{u}, p\}$ satisfying

$$\mathbf{u}(\mathbf{x}, T) = \mathbf{0} \ \text{ in } \ \Omega. \tag{6.98}$$

On the other hand, it will be said that (6.95) is *exactly controllable to the trajectories* (ECT) at time T if, for any $\mathbf{u}_0 \in H$ and any weak solution $\{\overline{\mathbf{u}}, \overline{p}\}$ to (6.95) corresponding to a control $\overline{\mathbf{v}}$ and an initial state $\overline{\mathbf{u}}_0$, there exists $\mathbf{v} \in L^2(\omega \times (0, T))^N$ and an associated solution $\{\mathbf{u}, p\}$ satisfying

$$\mathbf{u}(\mathbf{x}, T) = \overline{\mathbf{u}}(\mathbf{x}, T) \ \text{ in } \ \Omega. \tag{6.99}$$

Similar definitions make sense for the approximate, null and exact to the trajectories controllability of (6.96).

At present, the ECT and even the NC and AC of the solutions to (6.95) are open questions. Nevertheless, it has been possible to establish a collection of partial results, all of them in the positive direction, that make it reasonable to believe that these properties are satisfied.

A description of the state of the art and the known (partial) results are exhibited below, at the end of this section.

Observe that the NC and ECT hold for the linear system (6.96), provided $\tilde{\mathbf{u}}$ satisfies appropriate regularity conditions. In fact, in this linear framework, AC is a consequence of NC, that is equivalent to ECT:

$$\text{NC} \Leftrightarrow \text{ECT} \Rightarrow \text{AC} \tag{6.100}$$

(see Remark 6.17 below).

It is expected that these implications (or at least the fact that ECT implies AC) also hold for (6.95). But this is unknown.

For completeness, let us recall the main ideas used in the proof of the null controllability property of the Stokes-like problem (6.96). This will allow to clarify what can be done for (6.95) and later for (6.94).

The outline of the argument is as follows:

- First, we prove an observability inequality for the solutions to a system similar to (6.96) (more precisely, its adjoint). By this we mean an estimate that allows to get a bound of the *evolved* or *final time* values of the solution in terms of what is *observed*.
  The "evolved" values of the solution to the adjoint, since it is backwards in time, correspond to $t = 0$ and the observation reduces to the values of the velocity field in the set $\omega \times (0, T)$; see the estimate (6.104) below.
  To this purpose, a global estimate of the Carleman kind is previously established. Then, standard energy inequalities lead to the desired observability property. The details are given below.
- Secondly, it is shown that the null controllability of (6.96), together with an estimate of the control in terms of initial data, is equivalent to the *observability* of the solutions to the adjoint systems.

We give now more details.

The adjoint of (6.96) is the backwards in time system

$$\begin{cases} -\rho_0\left(\dfrac{\partial \mathbf{w}}{\partial t}+(\tilde{\mathbf{u}}\cdot\nabla)\mathbf{w}-(D\mathbf{w})\tilde{\mathbf{u}}\right)-\mu\Delta\mathbf{w}+\nabla\pi=\mathbf{g}, \quad (\mathbf{x},t)\in Q, \\ \nabla\cdot\mathbf{w}=0, \quad (\mathbf{x},t)\in Q, \\ \mathbf{w}=\mathbf{0}, \quad (\mathbf{x},t)\in\Sigma, \\ \mathbf{w}(\mathbf{x},T)=\mathbf{w}_T(\mathbf{x}), \quad \mathbf{x}\in\Omega. \end{cases} \tag{6.101}$$

Again, if $\tilde{\mathbf{u}}$ is regular enough (for instance $\tilde{\mathbf{u}} \in L^\infty(Q)^N$), for any $\mathbf{w}_T \in H$ and any $\mathbf{g} \in L^2(Q)^N$, (6.101) possesses exactly one weak solution that is strong if $\mathbf{w}_T \in V$.

In order to get (partial) local exact controllability to a trajectory $\overline{\mathbf{u}} = \overline{\mathbf{u}}(\mathbf{x}, t)$, the following hypotheses are needed:

$$\overline{\mathbf{u}} \in L^\infty(Q)^N, \quad \overline{\mathbf{u}}_t \in L^2(0, T; L^\sigma(\Omega))^N \quad \begin{pmatrix} \sigma > 6/5 \text{ if } N = 3 \\ \sigma > 1 \quad \text{if } N = 2 \end{pmatrix}. \tag{6.102}$$

Observe that the previous assumption on $\overline{\mathbf{u}}_t$ is automatically satisfied if $\overline{\mathbf{u}}$ is, together with some $\overline{p}$, a weak solution to problem (6.95) that belongs to $L^\infty(Q)^N$ and corresponds to an initial velocity $\overline{\mathbf{u}}_0 \in V$ and a control $\overline{\mathbf{v}} \in L^\infty(\omega \times (0, T))^N$.

The fundamental tool for the observability of the solutions to (6.101) is an estimate of the Carleman kind that is given in the following result:

Theorem 6.11 *Let us assume that (6.102) holds. There exist positive functions* $\xi, \xi_*, \xi_i \in C^0(\overline{\Omega} \times [0,T])$ *(*$i = 0, 1, 2$*) that grow to* $+\infty$ *as* $t \to T^-$ *and a constant* C_0 *only depending on* Ω*,* ω*,* T *and the norms of* $\tilde{\mathbf{u}}$ *and* $\tilde{\mathbf{u}}_t$ *in the spaces in* (6.102) *such that, for any* $\mathbf{g} \in L^2(Q)^N$ *and any* $\mathbf{w}_T \in H$*, the associated solution to (6.101) satisfies*

$$\begin{cases} \displaystyle\iint_Q \left[\xi_2^{-2}(|\mathbf{w}_t|^2 + |\Delta \mathbf{w}|^2) + \xi_1^{-2}|\nabla \mathbf{w}|^2 + \xi_0^{-2}|\mathbf{w}|^2\right] \\ \qquad \displaystyle\le C_0 \left(\iint_Q \xi^{-2}|\mathbf{g}|^2 + \iint_{\omega\times(0,T)} \xi_*^{-2}|\mathbf{w}|^2 \right). \end{cases} \tag{6.103}$$

This result is proved in [48]. Many ideas in the proof have been taken from [72]; see also [95, 96] and Exercise 6.18 for detailed explanations and improvements.

The weights ξ, ξ_* and ξ_i furnished by Theorem 6.11 are of the form

$$\eta(\mathbf{x}, t)\, e^{\beta(\mathbf{x})/(T-t)^r}$$

for some positive bounded functions η and β and some $r > 1$; here, we can have $\eta(\mathbf{x}, t) \to 0$ polynomially as $t \to T^-$.

It is important to observe that the Carleman estimate (6.103) is also satisfied by the solutions to the *modified* systems

$$\begin{cases} -\rho\left(\dfrac{\partial \mathbf{w}}{\partial t} + (\tilde{\mathbf{u}}\cdot\nabla)\mathbf{w} - (D\mathbf{w})\tilde{\mathbf{u}}\right) - \mu\Delta\mathbf{w} + \nabla\pi = \mathbf{g}, & (\mathbf{x}, t) \in Q, \\ \nabla \cdot \mathbf{w} = 0, \quad (\mathbf{x}, t) \in Q, \\ \mathbf{w} = \mathbf{0}, \quad (\mathbf{x}, t) \in \Sigma, \\ \mathbf{w}(\mathbf{x}, T) = \mathbf{w}_T(\mathbf{x}), \quad \mathbf{x} \in \Omega. \end{cases}$$

where $\rho \in C^1(\overline{Q})$, $0 < \alpha \le \rho \le \beta$ and the $\mathbf{w}_T$ are chosen in H, with a constant C_0 also depending on α, β and $\|\rho\|_{C^1(\overline{Q})}$. This will be used later, in Sect. 6.4.2, in the context of the variable density Navier-Stokes system.

As already mentioned, used in combination with usual energy estimates for the solutions to (6.101), Theorem 6.11 leads to an observability inequality.

Specifically, the following can be proved:

$$\|\mathbf{w}(\cdot\,, 0)\|^2 \le C \iint_{\omega\times(0,T)} \xi_*^{-2}|\mathbf{w}|^2, \tag{6.104}$$

for a positive constant C only depending on Ω, ω, T and the norms of $\tilde{\mathbf{u}}$ and $\tilde{\mathbf{u}}_t$ in the spaces in (6.102)

In order to get (6.104) from the Carleman estimate (6.103), it suffices to argue as follows:

- First, (6.103) and the properties of the weights yield easily

$$\iint_{\Omega\times(T/4,3T/4)} |\mathbf{w}|^2 \le C \iint_{\omega\times(0,T)} \xi_*^{-2}|\mathbf{w}|^2. \tag{6.105}$$

- Secondly, one has

$$\|\mathbf{w}(\cdot,t)\|^2 \le C \iint_{\Omega\times(T/4,3T/4)} |\mathbf{w}|^2 \quad \forall t \in [0,T/4] \tag{6.106}$$

for any solution to (6.101). Indeed, note that there exists $C > 0$ such that

$$\|\mathbf{w}(\cdot\,,t_1)\|^2 \le C\|\mathbf{w}(\cdot\,,t_2)\|^2 \quad \forall t_1,t_2 \in [0,T] \text{ with } t_1 < t_2.$$

Using (6.105) and (6.106), we easily deduce (6.104) at once.

As announced, a consequence of (6.104) is the null controllability of the linear system (6.96). This is well known and obeys to a principle that remains valid for many linear equations and systems; see [98, 99].

A short explanation is the following.

Let $\xi_* = \xi_*(\mathbf{x},t)$ be a positive continuous function in $\Omega\times[0,T]$, bounded from below and possibly blowing up to $+\infty$ as $t \to T^-$ and let us introduce the space $\mathcal{U}_* := \{\mathbf{v} : \xi_*\mathbf{v} \in L^2(\omega\times(0,T))^N\}$. This is a Hilbert space for the natural norm

$$\|\mathbf{v}\|_{\mathcal{U}_*} := \left(\iint_{\omega\times(0,T)} \xi_*^2|\mathbf{v}|^2\right)^{1/2}.$$

Consider the linear mappings $S_1 \in \mathcal{L}(H;H)$ and $S_2 \in \mathcal{L}(\mathcal{U}_*;H)$, with

- $S_1\mathbf{u}_0 = \mathbf{u}(\cdot\,,T)$, where $\mathbf{u}$ is, together with some p, the solution to (6.96) with $\mathbf{v} = \mathbf{0}$.
- $S_2\mathbf{v} = \mathbf{u}(\cdot\,,T)$, where $\mathbf{u}$ is now, together with another p, the solution to (6.96) with $\mathbf{u}_0 = \mathbf{0}$.

Then, (6.96) is null-controllable if and only if

$$R(S_1) \subset R(S_2). \tag{6.107}$$

Note that the adjoints of S_1 and S_2 are respectively given by

$$S_1^*\mathbf{w}_T = \mathbf{w}(\cdot\,,0) \quad \forall \mathbf{w}_T \in H$$

and

$$S_2^* \mathbf{w}_T = \mathbf{w}\big|_{\omega\times(0,T)} \quad \forall \mathbf{w}_T \in H,$$

where $\mathbf{w}$ is, together with some π, the solution to (6.101) with $\mathbf{g} = \mathbf{0}$. In particular, S_2^* is a continuous linear operator on H with values in

$$\mathcal{U}'_* := \{\mathbf{z} : \xi_*^{-1}\mathbf{z} \in L^2(\omega \times (0,T))^N\}.$$

The following result holds:

Lemma 6.2 *With the previous notation, one has*

$$\forall \mathbf{u}_0 \in H \ \exists \mathbf{v} \in \mathcal{U}_* \text{ such that } S_1\mathbf{u}_0 + S_2\mathbf{v} = 0 \text{ and } \|\mathbf{v}\|_{\mathcal{U}_*} \le C\|\mathbf{u}_0\| \tag{6.108}$$

if and only if there exists $C_0 > 0$ such that

$$\|S_1^*\mathbf{w}_T\| \le C_0\|S_2^*\mathbf{w}_T\|_{\mathcal{U}'_*} \quad \forall \mathbf{w}_T \in H. \tag{6.109}$$

Proof Let us first assume that (6.108) holds. Then consider for any $\mathbf{w} \in H$ the continuous linear form on H

$$\mathbf{u}_0 \mapsto (S_1\mathbf{u}_0, \mathbf{w}).$$

For each $\mathbf{u}_0 \in H$ with $\|\mathbf{u}_0\| \le 1$, there exists $\mathbf{v} \in \mathcal{U}_*$ such that $S_1\mathbf{u}_0 = -S_2\mathbf{v}$ and $\|\mathbf{v}\|_{\mathcal{U}_*} \le C$. Consequently,

$$\|S_1^*\mathbf{w}_T\| = \sup_{\|\mathbf{u}_0\|\le 1} |(S_1\mathbf{u}_0, \mathbf{w}_T)| \le \sup_{\|\mathbf{v}\|_{\mathcal{U}_*}\le C} \left| \iint_{\omega\times(0,T)} \mathbf{v}\cdot S_2^*\mathbf{w}_T \right| \le C\|S_2^*\mathbf{w}_T\|_{\mathcal{U}'_*}.$$

Since $\mathbf{w}_T$ is arbitrary in H, we get (6.109).

Conversely, let us assume that (6.109) holds. We will present a proof of (6.108) that relies on the construction of a family of controls $\{\mathbf{v}^\varepsilon\}$ such that the associated solutions to (6.96) satisfy

$$\|\mathbf{u}^\varepsilon(\cdot\,,T)\| \le \varepsilon.$$

Let $\varepsilon > 0$ be given and let us consider the following extremal problem:

$$\begin{cases} \text{Minimize } B_\varepsilon(\boldsymbol{\varphi}_T) := \dfrac{1}{2}\|S_2^*\boldsymbol{\varphi}_T\|^2_{\mathcal{U}'_*} + \varepsilon\|\boldsymbol{\varphi}_T\| + (\mathbf{u}_0, S_1^*\boldsymbol{\varphi}_T) \\ \text{subject to } \boldsymbol{\varphi}_T \in H. \end{cases} \tag{6.110}$$

Due to (6.109), we are minimizing here a continuous, strictly convex and coercive function in H. Indeed, one has

$$\begin{aligned} B_\varepsilon(\varphi_T) &\geq \frac{1}{2}\|S_2^*\varphi_T\|_{\mathcal{U}'_*}^2 + \varepsilon\|\varphi_T\| - \|\mathbf{u}_0\|\,\|S_1^*\varphi_T\| \\ &\geq \frac{1}{2}\left(\|S_2^*\varphi_T\|_{\mathcal{U}'_*}^2 - \frac{1}{C_0^2}\|S_1^*\varphi_T\|^2\right) + \varepsilon\|\varphi_T\| - \frac{C_0^2}{4}\|\mathbf{u}_0\|^2 \\ &\geq \varepsilon\|\varphi_T\| - \frac{C_0^2}{4}\|\mathbf{u}_0\|^2, \end{aligned}$$

whence we see that $B_\varepsilon(\varphi_T) \to +\infty$ as $\|\varphi_T\| \to +\infty$.

On the other hand, B_ε is differentiable at any $\varphi_T \neq \mathbf{0}$, with

$$(B'_\varepsilon(\tilde{\varphi}_T), \varphi_T) = (S_2^*\tilde{\varphi}_T, S_2^*\varphi_T)_{\mathcal{U}'_*} + \frac{\varepsilon}{\|\tilde{\varphi}_T\|}(\tilde{\varphi}_T, \varphi_T) + (\mathbf{u}_0, S_1^*\varphi_T)$$

for all $\varphi_T \in H$ and any nonzero $\tilde{\varphi}_T \in H$. Hence, there exists a unique minimizer φ_T^ε of B_ε and, moreover, we have

$$\begin{cases} \text{Either } \varphi_T^\varepsilon = \mathbf{0} \text{ or} \\ (S_2^*\varphi_T^\varepsilon, S_2^*\varphi_T)_{\mathcal{U}'_*} + \dfrac{\varepsilon}{\|\varphi_T^\varepsilon\|}(\varphi_T^\varepsilon, \varphi_T) = -(S_1\mathbf{u}_0, \varphi_T) \ \ \forall \varphi_T \in H. \end{cases} \tag{6.111}$$

In both cases, we find that

$$\|S_2^*\varphi_T^\varepsilon\|_{\mathcal{U}'_*} \leq C_0\|\mathbf{u}_0\|$$

and, in particular, φ_T^ε is bounded independently of ε.

Let $\beta : \mathcal{U}'_* \mapsto \mathcal{U}_*$ be the canonical isomorphism from $\mathcal{U}'_*$ into $\mathcal{U}_*$. One has

$$\beta\mathbf{z} = \xi_*^{-2}\mathbf{z} \quad \forall \mathbf{z} \in \mathcal{U}'_*.$$

Then, $\mathbf{v}^\varepsilon := \beta S_2^*\varphi_T^\varepsilon$ is also uniformly bounded in $\mathcal{U}_*$ and it can be assumed that it converges weakly in $\mathcal{U}_*$ to some $\mathbf{v}$ that satisfies

$$\|\mathbf{v}\|_{\mathcal{U}_*} \leq C_0\|\mathbf{u}_0\|.$$

From (6.111), we have that, if $\varphi_T^\varepsilon \neq \mathbf{0}$, then

$$S_2(\beta S_2^*\varphi_T^\varepsilon) + \frac{\varepsilon}{\|\varphi_T^\varepsilon\|}\varphi_T^\varepsilon = -S_1\mathbf{u}_0.$$

Consequently, taking limits as $\varepsilon \to 0$, we see that $S_2\mathbf{v} + S_1\mathbf{u}_0 = 0$ and this proves (6.108). □

Observe that the estimates (6.109) and (6.104) are the same; this is readily seen from the definitions of S_1 and S_2 and their adjoints. Thus, since (6.104) holds and (6.108) is strictly stronger than (6.107), we certainly have the null controllability of (6.96) with controls $\mathbf{v} \in \mathcal{U}_*$ satisfying

$$\iint_{\omega\times(0,T)} \xi_*^2|\mathbf{v}|^2 \leq C\|\mathbf{u}_0\|^2, \tag{6.112}$$

for some C.

For clarity, this is explicitly claimed in the following statement:

Theorem 6.12 *For any* $\mathbf{u}_0 \in H$, *there exist controls* $\mathbf{v} \in \mathcal{U}_*$ *such that the associated solutions to* (6.96) *satisfy* (6.98) *and* (6.112), *for some* C *only depending on* Ω, ω, T *and* $\tilde{\mathbf{u}}$.

Remark 6.12 Obviously, the result in Lemma 6.2 also holds, with the same proof, in a general setting:

Let E, F and G be Hilbert spaces with norms $\|\cdot\|_E$, $\|\cdot\|_F$ and $\|\cdot\|_G$; assume that $S_1 \in \mathcal{L}(E;G)$ and $S_2 \in \mathcal{L}(F;G)$; then, the following assertions are equivalent:

- $\forall e \in E\ \exists f \in F$ such that $S_1e + S_2f = 0$, $\|f\|_F \leq C\|e\|_E$.
- $\exists C > 0$ such that $\|S_1^*g\|_E \leq C\|S_2^*g\|_F\ \ \forall g \in G$.

With the language of control theory, this result shows that "controllability" is equivalent to "observability"; for other "classical" properties concerning range inclusion, see for example [8, 34, 53]. □

Remark 6.13 Observe that Theorem 6.12 holds for all $T > 0$ and for any non-empty control open set $\omega \subset \Omega$, with no restriction on its size or shape. Of course, the "cost" of null controllability, that is, the minimal norm of a control able to drive the system to zero depends on Ω, ω and T and it is expected to grow to infinity as $T \to 0$ and also as $|\omega| \to 0$. □

Remark 6.14 Since (6.96) is linear, it is reasonable to expect that the null controllability property in Theorem 6.12 implies approximate controllability. This is true. Indeed, a consequence of the previous Carleman estimates is the following unique continuation property that must hold for the solutions to (6.101) with $\mathbf{g} = \mathbf{0}$:

If $\mathbf{w} \in L^2(0,T;V)\cap L^\infty(0,T;H)$ solves together with some p the linear problem (6.101) in Q with $\mathbf{g} = \mathbf{0}$ and $\mathbf{w} = 0$ in $\omega \times (0,T)$, then $\mathbf{w} \equiv \mathbf{0}$.

This is equivalent to the approximate controllability of (6.96), since it indicates that $N(S_2^*) = \{\mathbf{0}\}$, i.e., that $R(S_2)$ is dense in H; see [42, 43] for more details on the approximate controllability of linear and semi-linear PDEs. □

Remark 6.15 As already said, we find in the proof of Lemma 6.2 a practical (constructive) method for the computation of a control $\mathbf{v}$ that drives the solution to (6.96) to zero: compute for small $\varepsilon > 0$ the solution $\hat{\boldsymbol{\varphi}}_T^\varepsilon$ to the unconstrained extremal problem (6.110), then take $\mathbf{v}_\varepsilon := \beta S_2^*\hat{\boldsymbol{\varphi}}_T^\varepsilon$ and finally see what happens as $\varepsilon \to 0^+$. This is the basic idea of a family of numerical methods that have been investigated in depth in [11, 71] in the context of parabolic PDEs. □

Remark 6.16 From Theorem 6.11, we can also deduce the null controllability of linear systems of the form

$$\begin{cases} \rho_0\left(\dfrac{\partial \mathbf{u}}{\partial t}+(\tilde{\mathbf{u}}\cdot\nabla)\mathbf{u}+(\mathbf{u}\cdot\nabla)\tilde{\mathbf{u}}\right)-\mu\Delta\mathbf{u}+\nabla p=\mathbf{f}+\mathbf{v}1_\omega, \quad (\mathbf{x},t)\in Q, \\ \nabla\cdot\mathbf{u}=0, \quad (\mathbf{x},t)\in Q, \\ \mathbf{u}=\mathbf{0} \quad (\mathbf{x},t)\in\Sigma, \\ \mathbf{u}(\mathbf{x},0)=\mathbf{u}_0(\mathbf{x}) \quad \mathbf{x}\in\Omega, \end{cases} \tag{6.113}$$

for some appropriate nonzero right-hand sides **f**. There are several ways to prove this. One of them consists of applying the so called Fursikov-Imanuvilov strategy, see [55, 56], that we explain now.

Thus, let us introduce the notation

$$L(\mathbf{u},p) := \rho_0\left(\frac{\partial \mathbf{u}}{\partial t}+(\tilde{\mathbf{u}}\cdot\nabla)\mathbf{u}+(\mathbf{u}\cdot\nabla)\tilde{\mathbf{u}}\right)-\mu\Delta\mathbf{u}+\nabla p$$

and

$$L^*(\mathbf{w},\pi) := -\rho_0\left(\frac{\partial \mathbf{w}}{\partial t}+(\tilde{\mathbf{u}}\cdot\nabla)\mathbf{w}-(D\mathbf{w})\tilde{\mathbf{u}}\right)-\mu\Delta\mathbf{w}+\nabla\pi.$$

The idea of the method is to search for a solution to the problem

$$\begin{cases} \text{Minimize } \dfrac{1}{2}\displaystyle\iint_Q \xi^2|\mathbf{u}|^2+\dfrac{1}{2}\iint_{\omega\times(0,T)}\xi_*^2|\mathbf{v}|^2 \\ \text{subject to } \mathbf{v}\in L^2(\omega\times(0,T))^N, \ (\mathbf{v},\mathbf{u},\rho,p) \text{ satisfies (6.113).} \end{cases} \tag{6.114}$$

It can be proved that (6.114) possesses exactly one solution, given by

$$\mathbf{u}=\xi^{-2}L^*(\mathbf{w},\pi), \quad \mathbf{v}=-\xi_*^{-2}\mathbf{w}\big|_{\omega\times(0,T)},$$

where $\{\mathbf{w},\pi\}$ solves the variational equality

$$\begin{cases} \displaystyle\iint_Q\left(\xi^{-2}L^*(\mathbf{w},\pi)\cdot L^*(\bar{\mathbf{w}},\bar{\pi})+\xi_*^{-2}1_\omega\mathbf{w}\cdot\bar{\mathbf{w}}\right)=\iint_Q\mathbf{f}\cdot\bar{\mathbf{w}}+(\rho_0\mathbf{u}_0,\bar{\mathbf{w}}(\mathbf{x},0)) \\ \forall(\bar{\mathbf{w}},\bar{\pi})\in Z, \quad (\mathbf{w},\pi)\in Z. \end{cases} \tag{6.115}$$

Here, Z is an appropriate space of couples $(\mathbf{w},\pi)$ satisfying

$$\iint_Q\left(\xi^{-2}|L^*(\mathbf{w},\pi)|^2+\xi_*^{-2}1_\omega|\mathbf{w}|^2\right)<+\infty$$

and $\mathbf{w} = \mathbf{0}$ on Σ. Using Theorem 6.11, it is not difficult to check that, under the assumption

$$\iint_Q \xi_*^2 |\mathbf{f}|^2 < +\infty,$$

there exists a unique solution $\{\mathbf{w}, \pi\}$ to (6.115). Furthermore, it is possible to deduce that, for some appropriate positive functions γ, γ_* and γ_i similar to ξ, ξ_* and the ξ_i ($i = 0, 1, 2$) that grow to $+\infty$ as $t \to T^-$, one has:

$$\begin{cases} \displaystyle\iint_Q \left[\gamma_2^2(|\mathbf{u}_t|^2 + |\Delta \mathbf{u}|^2) + \gamma_1^2 |\nabla \mathbf{u}|^2 + \gamma_0^2 |\mathbf{u}|^2\right] + \iint_{\omega\times(0,T)} \gamma^2 |\mathbf{v}|^2 \\ \qquad \leq C\left(\displaystyle\iint_Q \gamma_*^2 |\mathbf{f}|^2 + \|\mathbf{u}_0\|_V^2\right). \end{cases}$$

This provides additional estimates for $\mathbf{u}$ and $\mathbf{v}$. □

Remark 6.17 As already mentioned, a consequence of the linearity of (6.96) is that, for this system, the null controllability property implies exact controllability to the trajectories. Namely, if $\{\overline{\mathbf{u}}, \overline{p}\}$ is a solution to (6.96) and we introduce the change of variables $\mathbf{u} = \mathbf{y} + \overline{\mathbf{u}}$, $p = q + \overline{p}$, we immediately see that $\{\mathbf{y}, q\}$ must solve the same system of PDEs with initial condition

$$\mathbf{y}(\mathbf{x}, 0) = \mathbf{u}_0(\mathbf{x}) - \overline{\mathbf{u}}(\mathbf{x}, 0), \quad \mathbf{x} \in \Omega.$$

There exist controls $\mathbf{v} \in L^2(\omega \times (0, T))^N$ such that the associated states $\mathbf{y}$ vanish at $t = T$ and this obviously yields (6.99). □

The null controllability of (6.96) and (6.113) with appropriate $\mathbf{f}$ can be used to deduce a related property for (6.95). Unfortunately, the result is local, in the sense that some smallness assumption must be imposed to the data.

Thus, the following holds:

Theorem 6.13 *Let $\overline{\mathbf{u}}$ be, together with some $\overline{p}$, a solution to the Navier-Stokes problem*

$$\begin{cases} \rho_0\left(\overline{\mathbf{u}}_t + (\overline{\mathbf{u}}\cdot\nabla)\overline{\mathbf{u}}\right) - \mu\Delta\overline{\mathbf{u}} + \nabla\overline{p} = \mathbf{0}, & (\mathbf{x}, t) \in Q, \\ \nabla\cdot\overline{\mathbf{u}} = 0, & (\mathbf{x}, t) \in Q, \\ \overline{\mathbf{u}} = \mathbf{0}, & (\mathbf{x}, t) \in \Sigma, \\ \overline{\mathbf{u}}(\mathbf{x}, 0) = \overline{\mathbf{u}}_0(\mathbf{x}), & \mathbf{x} \in \Omega, \end{cases} \tag{6.116}$$

satisfying (6.102). There exists $\varepsilon > 0$ such that, for any initial state

$$\mathbf{u}_0 \in H \cap L^{2N-2}(\Omega)^N$$

satisfying $\|\mathbf{u}_0 - \overline{\mathbf{u}}_0\|_{L^{2N-2}} \le \varepsilon$, *we can find controls* $\mathbf{v} \in L^2(\omega \times (0,T))^N$ *and associated states* $\{\mathbf{u}, p\}$ *such that one has (6.95) and*

$$\mathbf{u}(\mathbf{x},T) = \overline{\mathbf{u}}(\mathbf{x},T) \quad \text{in} \quad \Omega. \tag{6.117}$$

Sketch of the Proof The proof is given in [48] and uses some ideas from [72]. For brevity, we will consider the case $N = 3$ and will assume $\overline{\mathbf{u}}_0 \in V$. We will work with the variables $\mathbf{z} := \mathbf{u} - \overline{\mathbf{u}}$ and $q := p - \overline{p}$.

Note that $\mathbf{z}$ and q satisfy

$$\begin{cases} \rho_0\,(\mathbf{z}_t + (\overline{\mathbf{u}}\cdot\nabla)\mathbf{z} + (\mathbf{z}\cdot\nabla)\overline{\mathbf{u}} + (\mathbf{z}\cdot\nabla)\mathbf{z}) - \mu\Delta\mathbf{z} + \nabla q = \mathbf{v}1\!\!1_\omega, & (\mathbf{x},t) \in Q, \\ \nabla\cdot\mathbf{z} = 0, \quad (\mathbf{x},t) \in Q, \\ \mathbf{z} = \mathbf{0}, \quad (\mathbf{x},t) \in \Sigma, \\ \mathbf{z}(\mathbf{x},0) = \mathbf{z}_0(\mathbf{x}), \quad \mathbf{x} \in \Omega, \end{cases} \tag{6.118}$$

where $\mathbf{z}_0 := \mathbf{u}_0 - \overline{\mathbf{u}}_0$. The goal is therefore to prove that, if $\mathbf{z}_0$ is sufficiently small (in an appropriate norm), there exist controls $\mathbf{v} \in L^2(\omega \times (0,T))^N$ such that $\mathbf{z}(\cdot\,,T) = \mathbf{0}$.

Let us consider the weights ξ, ξ_* and ξ_i given by Theorem 6.11 and let us introduce the weights γ, γ_* and γ_i that appear in Remark 6.16 and introduce the spaces E and F, with

$$\begin{aligned} E := \{(\mathbf{v},\mathbf{z},p) : \;& \gamma_0\mathbf{z},\, \gamma_1\partial_i\mathbf{z},\, \gamma_2\partial_i\partial_j\mathbf{z},\, \gamma_2\partial_t\mathbf{z} \in L^2(Q)^N, \\ & \xi_*\mathbf{v} \in L^2(\omega\times(0,T))^N,\;\; \xi_*(L\mathbf{z} + \nabla p - \mathbf{v}1\!\!1_\omega) \in L^2(Q)^N\} \end{aligned} \tag{6.119}$$

and

$$F := \{(\mathbf{f},\mathbf{u}_0) : \xi_*\mathbf{f} \in L^2(Q)^N,\;\; \mathbf{u}_0 \in V\}. \tag{6.120}$$

Obviously, E and F are Hilbert spaces for their natural norms

$$\begin{cases} \|(\mathbf{v},\mathbf{z},p)\|_E^2 := \displaystyle\iint_Q \left[\gamma_2^2(|\mathbf{z}_t|^2 + |\Delta\mathbf{z}|^2) + \gamma_1^2|\nabla\mathbf{z}|^2 + \gamma_0^2|\mathbf{z}|^2\right] \\ \qquad\qquad + \displaystyle\iint_{\omega\times(0,T)} \xi_*^2|\mathbf{v}|^2 + \iint_Q \xi_*|L\mathbf{z} + \nabla p - \mathbf{v}1\!\!1_\omega|^2 \end{cases}$$

and

$$\|(\mathbf{f},\mathbf{u}_0)\|_F^2 := \iint_Q \xi_*^2|\mathbf{f}|^2 + \|\mathbf{u}_0\|_V^2\,.$$

Let us consider the mapping $\mathcal{A} : E \mapsto F$ given by

$$\mathcal{A}(\mathbf{v},\mathbf{z},p) = (L(\mathbf{z},p) + (\mathbf{z}\cdot\nabla)\mathbf{z} - \mathbf{v}1\!\!1_\omega, \mathbf{z}(\cdot\,,0)) \quad \forall (\mathbf{v},\mathbf{z},p) \in E. \tag{6.121}$$

It can be seen that $\mathcal{A}$ is well defined and C^1 in E.

The result will be ensured if we prove that, if $\|(\mathbf{f}, \mathbf{u}_0)\|_F$ is sufficiently small, the equation

$$\mathcal{A}(\mathbf{v}, \mathbf{z}, p) = (\mathbf{f}, \mathbf{u}_0), \quad (\mathbf{v}, \mathbf{z}, p) \in E$$

is solvable.

At this point, we will use the following result, usually known as *Liusternik's Inverse Function Theorem* (see [1, 33, 73]):

Theorem 6.14 *Let E and F be Banach spaces and let $\mathcal{A} : E \mapsto F$ be continuously differentiable in E. Assume that $e_0 \in E$, $\mathcal{A}(e_0) = h_0$ and $\mathcal{A}'(e_0) : E \mapsto F$ is onto. Then, there exist $\delta > 0$, $r > 0$, a mapping $\mathcal{H} : B_\delta(h_0) \subset F \mapsto E$ and a constant $K > 0$ satisfying*

$$\begin{cases} \mathcal{H}(z) \in B_r(e_0) \ \text{ and } \ \mathcal{A}(\mathcal{H}(z)) = z & \forall z \in B_\delta(h_0), \\ \|\mathcal{H}(z) - e_0\|_F \le K\|z - \mathcal{A}(e_0)\|_F & \forall z \in B_\delta(h_0). \end{cases}$$

In the particular case of the spaces given by (6.119) and (6.120) and the mapping in (6.121), we note that the identity

$$R(\mathcal{A}'(\mathbf{0}, \mathbf{0}, 0)) = F$$

is equivalent to the null controllability result stated in Remark 6.16. Therefore, we can apply Theorem 6.14 to $\mathcal{A}$ with $e_0 = (\mathbf{0}, \mathbf{0}, 0)$ and $h_0 = (\mathbf{0}, \mathbf{0})$ and deduce that, if $(\mathbf{f}, \mathbf{z}_0) \in F$ and $\|(\mathbf{f}, \mathbf{z}_0)\|_F$ is sufficiently small, there exist triplets $(\mathbf{v}, \mathbf{z}, p) \in E$ such that $\mathcal{A}(\mathbf{v}, \mathbf{z}, p) = (\mathbf{f}, \mathbf{z}_0)$.

But this means that, if $\|\mathbf{u}_0\|_V$ is small (and $\xi_* \mathbf{f}$ is small in $L^2(Q)^N$), there exist controls and associated solutions to (6.95) satisfying (6.117).

This ends the proof. □

Remark 6.18 In particular, we see from this result that (6.95) is locally null controllable. In other words, there exists $\varepsilon > 0$ such that, whenever $\mathbf{u}_0 \in H \cap L^{2N-2}(\Omega)^N$ and $\|\mathbf{u}_0\|_{L^{2N-2}} \le \varepsilon$, we can find controls $\mathbf{v} \in L^2(\omega \times (0, T))^N$ and associated states $\mathbf{u}$ satisfying

$$\mathbf{u}(\mathbf{x}, T) = \mathbf{0} \quad \text{in} \quad \Omega.$$

On the other hand, from Theorem 6.13 and the parabolic dissipativity of the solutions to (6.95), it is easy to deduce that this system is null controllable at large time. In other words, the following holds:

> *Let $\mathbf{u}_0 \in H$ be given. There exist $T > 0$, $\mathbf{v} \in L^2(\omega \times (0, T))^N$ and associated states $\mathbf{u}$ such that $\mathbf{u}(\cdot\,, T) = \mathbf{0}$.*

The proof of this assertion is very simple. It suffices to take first $\mathbf{v} = 0$ up to a first time $T_1 > 0$ with $\mathbf{u}(\cdot\,, T_1)$ sufficiently small in $L^{2N-2}(\Omega)^N$ and then apply Theorem 6.13 in the interval $(T_1, T_1 + 1)$. □

As already mentioned, at present there is no general global controllability result for (6.95); that is, it is unknown whether Theorem 6.13 holds when $\|\overline{\mathbf{u}}_0 - \mathbf{u}_0\|_{L^{2N-2}}$ is large, even in the case $\overline{\mathbf{u}} \equiv \mathbf{0}$.

This was conjectured by J.-L. Lions in 1990 but, as far as we know, global results are only known for some modified or particular systems.

Here are some of them:

- Global approximate and null controllability for (6.96), as well as exact controllability to the trajectories. As indicated, this is a consequence of the Carleman estimates (6.103).
- Global exact controllability for finite-dimensional Galerkin approximations of the form

$$\begin{cases} (\rho_0(\mathbf{u}_t + (\mathbf{u} \cdot \nabla)\mathbf{u}), \mathbf{w}_j) + \mu(\nabla \mathbf{u}, \nabla \mathbf{w}_j) = (\mathbf{v}1\!\!1_\omega, \mathbf{w}), \quad 1 \le j \le m, \\ \mathbf{u} : [0, T] \mapsto V_m, \quad \mathbf{u}(0) = \mathbf{u}_{0m} \in V_m, \end{cases} \tag{6.122}$$

where V_m is the space spanned by $\mathbf{w}_1, \ldots, \mathbf{w}_m$ and $\{\mathbf{w}_j\}$ is (for instance) an orthogonal base of V; see [84].
- Global approximate controllability for

$$\begin{cases} \rho_0(\mathbf{u}_t + \nabla \cdot (T_M(\mathbf{u})\mathbf{u})) - \mu\Delta\mathbf{u} + \nabla p = \mathbf{v}1\!\!1_\omega, \quad (\mathbf{x}, t) \in Q, \\ \nabla \cdot \mathbf{u} = 0, \quad (\mathbf{x}, t) \in Q, \\ \mathbf{u} = \mathbf{0}, \quad (\mathbf{x}, \mathbf{t}) \in \Sigma, \\ \mathbf{u}(\mathbf{x}, 0) = \mathbf{u}_0(\mathbf{x}), \quad \mathbf{x} \in \Omega, \end{cases} \tag{6.123}$$

for any $M > 0$, proved by Fabre [41]. Here, for any $\mathbf{a} \in \mathbb{R}^N$, $T_M(\mathbf{a})$ denotes the vector whose i-th component is given by

$$(T_M(\mathbf{a}))_i = \begin{cases} M & \text{if } a_i > M, \\ a_i & \text{if } -M \le a_i \le M, \\ -M & \text{if } a_i < -M. \end{cases}$$

The proof relies on a (more or less standard) fixed-point argument that can be used in combination with unique continuation.
- Global approximate controllability in a larger space for

$$\begin{cases} \rho_0(\mathbf{u}_t + (\mathbf{u} \cdot \nabla)\mathbf{u}) - \mu\Delta\mathbf{u} + \nabla p = \mathbf{v}1\!\!1_\omega, \quad (\mathbf{x}, t) \in Q, \\ \nabla \cdot \mathbf{u} = 0, \quad (\mathbf{x}, t) \in Q, \\ \nabla \times \mathbf{u} = 0, \quad \mathbf{u} \cdot \mathbf{n} = 0, \quad (\mathbf{x}, t) \in \Sigma, \\ \mathbf{u}(\mathbf{x}, 0) = \mathbf{u}_0(\mathbf{x}), \quad \mathbf{x} \in \Omega, \end{cases} \tag{6.124}$$

when $N = 2$, proved by Coron [21]. Specifically, it is established that for any $\mathbf{u}_0, \mathbf{u}_T \in H$ there exist controls $\mathbf{v}_n \in L^2(\omega \times (0, T))^2$ for $n = 1, 2, \ldots$ such that the states $\mathbf{u}_n$ associated to $\mathbf{u}_0$ and the $\mathbf{v}_n$ satisfy $\mathbf{u}_n(\cdot, T) \to \mathbf{u}_T$ in $W^{-1,\infty}(\Omega)^2$ as $n \to +\infty$.

- Global null controllability for the Navier-Stokes equations in a 2D manifold without boundary, proved by Coron and Fursikov [23]. Also, a similar result has been established in [56] when $N = 3$ and periodic boundary conditions are imposed on $\mathbf{u}$.
- Global null controllability and exact controllability to the trajectories for the systems

$$\begin{cases} \rho_0(\mathbf{u}_t + (\mathbf{u} \cdot \nabla)\mathbf{u}) - \mu\Delta\mathbf{u} + \nabla p = \mathbf{v}1\!\!1_\omega, & (\mathbf{x}, t) \in Q, \\ \nabla \cdot \mathbf{u} = 0, \quad (\mathbf{x}, t) \in Q, \\ [\mu D\mathbf{u} \cdot \mathbf{n} + M\mathbf{u}]_{\tan} = 0, \quad \mathbf{u} \cdot \mathbf{n} = 0, \quad (\mathbf{x}, t) \in \Sigma, \\ \mathbf{u}(\mathbf{x}, 0) = \mathbf{u}_0(\mathbf{x}), \quad \mathbf{x} \in \Omega \end{cases} \tag{6.125}$$

and

$$\begin{cases} \rho_0(\mathbf{u}_t + (\mathbf{u} \cdot \nabla)\mathbf{u}) - \mu\Delta\mathbf{u} + \nabla p = \mathbf{0}, \quad (\mathbf{x}, t) \in Q, \\ \nabla \cdot \mathbf{u} = 0, \quad (\mathbf{x}, t) \in Q, \\ [\mu D\mathbf{u} \cdot \mathbf{n} + M\mathbf{u}]_{\tan} = 0, \quad \mathbf{u} \cdot \mathbf{n} = 0, \quad (\mathbf{x}, t) \in \Sigma_0, \\ \mathbf{u} = \mathbf{h}, \quad (\mathbf{x}, t) \in \Sigma_1, \\ \mathbf{u}(\mathbf{x}, 0) = \mathbf{u}_0(\mathbf{x}), \quad \mathbf{x} \in \Omega \end{cases} \tag{6.126}$$

when $N = 2$ or $N = 3$, proved by Coron et al. [28]. Here, M is symmetric and definite positive, $\Sigma_i := \Gamma_i \times (0, T)$ for $i = 0, 1$, $\{\Gamma_0, \Gamma_1\}$ is a partition of $\partial\Omega$ and the control is either distributed or the Dirichlet data on a part of the boundary.
- In the particular case of a rectangular domain, some specific (additional) results can be established. Thus, let us assume that Ω is (for instance) the unit cube, let us denote by σ one face of $\partial\Omega$ and let us set $\gamma := \partial\Omega \setminus \sigma$. It has been proved in [65] that, for any $\mathbf{u}_0 \in V$, there exists a sequence $\{\mathbf{f}_\varepsilon\}$ that converges weakly to zero in a space of the form $L^q(0, T; H^{-1}(\Omega)^3)$ such that the systems

$$\begin{cases} \rho_0(\mathbf{u}_t + (\mathbf{u} \cdot \nabla)\mathbf{u}) - \mu\Delta\mathbf{u} + \nabla p = \mathbf{f}_\varepsilon, \quad (\mathbf{x}, t) \in Q, \\ \nabla \cdot \mathbf{u} = 0, \quad (\mathbf{x}, t) \in Q, \\ \mathbf{u} = \mathbf{h}1\!\!1_\gamma, \quad (\mathbf{x}, t) \in \Sigma, \\ \mathbf{u}(\mathbf{x}, 0) = \mathbf{u}_0(\mathbf{x}) \quad \mathbf{x} \in \Omega \end{cases} \tag{6.127}$$

are null-controllable. Some extensions and improvements of this result have been recently obtained in [27, 52].

6.4.2 *The Variable Density Case. A Local Result*

For the variable density problem (6.1), the controllability results are still less in number and harder to obtain. At present, only a very few partial results have been established.

Note that, for a system like (6.1) where the velocity field vanishes on Σ, it does not seem easy that a single distributed control on the motion equation be able to govern the behavior of the density ρ at time $t = T$. Indeed, the *mass distribution* of $\rho(\cdot\,, t)$ is completely determined by ρ_0 for all t. In other words, if $\{\mathbf{u}, \rho, p\}$ is a weak solution to (6.1), for any $\lambda > 0$,

$$t \mapsto \text{meas}\,\{\mathbf{x} \in \Omega : \rho(\mathbf{x}, t) \le \lambda\} \text{ is constant.} \tag{6.128}$$

This is implied by the results in Chap. 2; see Exercise 6.23.

This explains and justifies that, for variable density fluids, we only consider partial controllability problems, that is, control problems where only $\mathbf{u}(\cdot\,, T)$ is required to satisfy a desired property.

An alternative strategy is followed in [5], where the velocity field and the mass density are controlled on large parts of the lateral boundary Σ and local exact controllability to regular trajectories in all variables is established. However, it is required there that the target velocity is sufficiently large in order to force all particles to leave the spatial domain before the arrival time.

Let us give a local partial null controllability result for (6.1). In order to fix ideas, we will assume that $N = 3$ (for $N = 2$, the same result holds with a slightly simpler proof).

The result is the following:

Theorem 6.15 *Let α, β be given with $0 < \alpha < \beta$. There exists $\delta > 0$ such that, for any initial state $(\rho_0, \mathbf{u}_0)$ with*

$$\rho_0 \in C^1(\overline{\Omega}), \quad \alpha \le \rho_0 \le \beta, \quad \mathbf{u}_0 \in D(A), \quad \|\mathbf{u}_0\|_{H^2} \le \delta, \tag{6.129}$$

we can find controls $\mathbf{v} \in L^2(\omega \times (0, T))^3$ and associated solutions $\{\mathbf{u}, \rho, p\}$ to (6.1) satisfying

$$\mathbf{u}(\mathbf{x}, T) = \mathbf{0} \quad \text{in } \Omega. \tag{6.130}$$

Proof The proof relies on a fixed-point formulation of the controllability problem and arguments and estimates similar to those in Sect. 6.4.1.

For the sake of clarity, we divide the proof into four parts.

Part 1 Let us consider the Banach space $W := L^2(0, T; V) \cap C^0(\overline{Q})^3$.

In this first step, we fix $M > 0$ and, for every $\tilde{\mathbf{u}} \in W$ with

$$\|\tilde{\mathbf{u}}\|_W := \|\tilde{\mathbf{u}}\|_{L^2(0,T;V)} + \|\tilde{\mathbf{u}}\|_{C^0(\overline{Q})} \le M,$$

we consider the family of ordinary differential systems

$$\begin{cases} \dfrac{\partial \mathbf{X}}{\partial s} = \tilde{\mathbf{u}}(\mathbf{X}, s), \\ \mathbf{X}\big|_{s=t} = \mathbf{x}, \end{cases} \tag{6.131}$$

where $\mathbf{x} \in \Omega$ and $t \in [0, T]$.

From the results in [32], we know that there exists a unique solution $\tilde{\mathbf{X}} = \tilde{\mathbf{X}}(\mathbf{x}, t; s)$ with $\tilde{\mathbf{X}} \in C^1(\overline{Q} \times [0, T]; \mathbb{R}^3)$ and

$$\tilde{\mathbf{X}}(\mathbf{x}, t_1; t_2) = \tilde{\mathbf{X}}(\tilde{\mathbf{X}}(\mathbf{x}, t_1; t_3), t_3; t_2) \ \ \forall \mathbf{x} \in \overline{\Omega}, \ \ \forall t_1, t_2, t_3 \in [0, T].$$

Furthermore, $\tilde{\mathbf{X}}(\mathbf{x}, t; s) \in \overline{\Omega}$ for all $(\mathbf{x}, t, s) \in \overline{Q} \times [0, T]$ and the partial derivatives $\mathbf{Z} := \tilde{\mathbf{X}}_t$ and $\mathbf{Y}^\ell := \partial_\ell \tilde{\mathbf{X}}$ respectively satisfy

$$\begin{cases} \dfrac{\partial \mathbf{Z}}{\partial s} = \nabla \tilde{\mathbf{u}}(\mathbf{X}, s)\mathbf{Z}, \\ \mathbf{Z}\big|_{s=t} = -\tilde{\mathbf{u}}(\mathbf{x}, t) \end{cases} \quad \text{and} \quad \begin{cases} \dfrac{\partial \mathbf{Y}^\ell}{\partial s} = \nabla \tilde{\mathbf{u}}(\mathbf{X}, s)\mathbf{Y}^\ell, \\ \mathbf{Y}^\ell\big|_{s=t} = \mathbf{e}^\ell \end{cases}$$

(recall that $\mathbf{e}^\ell$ is the ℓ-th vector of the canonical basis of $\mathbb{R}^3$). Consequently, if we denote by $\mathbf{F} = \mathbf{F}(\mathbf{x}, t; s)$ the associated fundamental matrix, defined by

$$\begin{cases} \dfrac{\partial \mathbf{F}}{\partial s} = \nabla \tilde{\mathbf{u}}(\mathbf{X}, s)\mathbf{F}, \\ \mathbf{F}\big|_{s=t} = \mathbf{Id}., \end{cases}$$

one has:

$$\mathbf{Z}(\mathbf{x}, t; s) = -\mathbf{F}(\mathbf{x}, t; s)\tilde{\mathbf{u}}(\mathbf{x}, t) \ \text{ and } \ \mathbf{Y}^\ell(\mathbf{x}, t; s) = \mathbf{F}(\mathbf{x}, t; s)\mathbf{e}^\ell \tag{6.132}$$

for all $(\mathbf{x}, t, s) \in \overline{Q} \times [0, T]$.

The results in [32] also imply that $\mathbf{F} \in C^0(\overline{Q} \times [0, T]; \mathbb{R}^{3\times 3})$ and $|\mathbf{F}(\mathbf{x}, t; s)| \le C(M)$ in $\overline{Q} \times [0, T]$, whence similar estimates also hold for $\mathbf{Z}$ and the $\mathbf{Y}^\ell$:

$$|\mathbf{Z}(\mathbf{x}, t; s)| + \sum_{\ell=1}^{3} |\mathbf{Y}^\ell(\mathbf{x}, t; s)| \le C(M) \ \ \forall (\mathbf{x}, t, s) \in \overline{Q} \times [0, T]. \tag{6.133}$$

Let us set

$$\tilde{\rho}(\mathbf{x}, t) := \rho_0(\tilde{\mathbf{X}}(\mathbf{x}, t; 0)) \quad \forall (\mathbf{x}, t) \in \overline{Q}. \tag{6.134}$$

We then have the following result:

Lemma 6.3 *The function $\tilde{\rho}$ belongs to $C^1(\overline{Q})$ and is the unique solution to*

$$\begin{cases} \dfrac{\partial \rho}{\partial t} + \tilde{\mathbf{u}} \cdot \nabla \rho = 0, \\ \rho\big|_{t=0} = \rho_0 . \end{cases} \tag{6.135}$$

Furthermore, the following estimates hold:

$$\alpha \le \tilde{\rho} \le \beta \ \text{ in } Q, \ \ \|\tilde{\rho}\|_{C^1(\overline{Q})} \le C(M) \tag{6.136}$$

and the mapping $\tilde{\mathbf{u}} \mapsto \tilde{\rho}$ is continuous from W into $C^1(\overline{Q})$.

Proof From (6.134), (6.132), and (6.133), we see that $\tilde{\rho}$ is continuously differentiable in $\overline{Q}$, satisfies (6.135) and also (6.136).

Indeed, note that

$$\partial_t \tilde{\rho}(\mathbf{x}, t) = \nabla \tilde{\rho}_0(\tilde{\mathbf{X}}(\mathbf{x}, t; 0)) \cdot \mathbf{Z}(\mathbf{x}, t; 0) = -\nabla \tilde{\rho}_0(\tilde{\mathbf{X}}(\mathbf{x}, t; 0))\mathbf{F}(\mathbf{x}, t; 0)\tilde{\mathbf{u}}(\mathbf{x}, t)$$

and

$$\partial_\ell \tilde{\rho}(\mathbf{x}, t) = \nabla \tilde{\rho}_0(\tilde{\mathbf{X}}(\mathbf{x}, t; 0)) \cdot \mathbf{Y}^\ell(\mathbf{x}, t; 0) = \nabla \tilde{\rho}_0(\tilde{\mathbf{X}}(\mathbf{x}, t; 0))\mathbf{F}(\mathbf{x}, t; 0)\mathbf{e}^\ell$$

for $\ell = 1, 2, 3$.

On the other hand, the results in [32] also show that, if $\tilde{\mathbf{u}}^n \to \tilde{\mathbf{u}}$ strongly in W as $n \to +\infty$, then the corresponding $\tilde{\mathbf{X}}^n$ converge to $\tilde{\mathbf{X}}$ strongly in $C^1(\overline{Q} \times [0, T]; \mathbb{R}^3)$ and consequently, with a self-explicative notation, we also have $\tilde{\rho}^n \to \tilde{\rho}$ strongly in $C^1(\overline{Q})$. □

Once more, we note that the definition of $\tilde{\rho}$ is completely natural for a flow transported by $\tilde{\mathbf{u}}$: we are stating in (6.134) that the mass density of a particle positioned at $\mathbf{x}$ at time t is the initial density of the particle that is at $\tilde{\mathbf{X}}(\mathbf{x}, t; 0)$ at time 0; but this is the same particle.

Part 2 In a second step, for a given $\tilde{\rho}$ in $C^1(\overline{Q})$ with $\alpha \le \tilde{\rho} \le \beta$, we consider the linear system

$$\begin{cases} \dfrac{\partial \tilde{\rho}\mathbf{u}}{\partial t} + \nabla \cdot (\tilde{\rho}\mathbf{u} \otimes \tilde{\mathbf{u}}) - \mu \Delta \mathbf{u} + \nabla p = \mathbf{v}\chi_\omega, & (\mathbf{x}, t) \in Q, \\ \nabla \cdot \mathbf{u} = 0, \quad (\mathbf{x}, t) \in Q, \\ \mathbf{u} = \mathbf{0}, \quad (\mathbf{x}, \mathbf{t}) \in \Sigma, \\ \mathbf{u}(\mathbf{x}, 0) = \mathbf{u}_0(\mathbf{x}), \quad \mathbf{x} \in \Omega, \end{cases} \tag{6.137}$$

where $\chi_\omega \in \mathcal{D}(\omega)$ is a cut-off function satisfying $\chi_\omega = 1$ in a non-empty open set $\omega_0 \subset\subset \omega$ and $0 \le \chi_\omega \le 1$.

We prove the "uniform" approximate controllability with sufficiently regular controls.

More precisely, we establish the following result:

Lemma 6.4 *Let* $\tilde{\mathbf{u}} \in W$ *be given with* $\|\tilde{\mathbf{u}}\|_W \leq M$ *and let us assume that* $\mathbf{u}_0 \in D(A)$. *For every* $\kappa > 0$, *one can define a control* $\mathbf{v}_\kappa$ *and an associated solution* $\{\mathbf{u}_\kappa, p_\kappa\}$ *to* (6.137) *satisfying*

$$\begin{cases} \mathbf{v}_\kappa 1\!\!1_\omega \in L^2(0,T;H^1(\Omega)^3), \quad \partial_t(\mathbf{v}_\kappa 1\!\!1_\omega) \in L^2(0,T;H^{-1}(\Omega)^3), \\ \mathbf{u}_\kappa \in L^2(0,T;H^3(\Omega)^3) \cap C^0([0,T];D(A)), \quad \partial_t \mathbf{u}_\kappa \in L^2(0,T;V) \end{cases} \tag{6.138}$$

and estimates in these spaces only depending on M, $\|\mathbf{u}_0\|_{H^2}$ *and* κ, *such that*

$$\|\mathbf{u}_\kappa(\cdot\,,T)\| \leq \kappa. \tag{6.139}$$

Moreover, the $\mathbf{v}_\kappa$ *and* $\mathbf{u}_\kappa$ *can be found satisfying*

$$\|\mathbf{v}_\kappa\|_{L^6(\omega\times(0,T))^3} + \|\mathbf{u}_\kappa\|_{L^6(0,T;W^{2,6}(\Omega)^3)} + \|\frac{\partial \mathbf{u}_\kappa}{\partial t}\|_{L^6(Q)^3} \leq C_0(M)\|\mathbf{u}_0\|_{H^2}, \tag{6.140}$$

$$\|\mathbf{u}_\kappa\|_{L^2(0,T;V)} + \|\mathbf{u}_\kappa\|_{C^0(\overline{Q})^3} = \|\mathbf{u}_\kappa\|_W \leq C_1(M)\|\mathbf{u}_0\|_{H^2} \tag{6.141}$$

and the following property: if $\tilde{\mathbf{u}}^n \to \tilde{\mathbf{u}}$ *in* W *and the* $(\mathbf{v}_\kappa^n, \mathbf{u}_\kappa^n, p_\kappa^n)$ *are the associated triplets, then* $\mathbf{v}_\kappa^n \to \mathbf{v}_\kappa$ *strongly in* $L^2(\omega \times (0,T)^3)$ *and* $\mathbf{u}_\kappa^n \to \mathbf{u}_\kappa$ *strongly in* W, *where* $(\mathbf{v}_\kappa, \mathbf{u}_\kappa, p_\kappa)$ *is a control-state associated to* $\tilde{\mathbf{u}}$ *through* (6.135) *and* (6.137).

Proof First, note that the adjoint of (6.137) is

$$\begin{cases} -\tilde{\rho}\left(\dfrac{\partial \mathbf{w}}{\partial t} + (\tilde{\mathbf{u}}_\varepsilon \cdot \nabla)\mathbf{w}\right) - \mu\Delta\mathbf{w} + \nabla q = \mathbf{0}, \quad (\mathbf{x},t) \in Q, \\ \nabla \cdot \mathbf{w} = 0, \quad (\mathbf{x},t) \in Q, \\ \mathbf{w} = \mathbf{0}, \quad (\mathbf{x},\mathbf{t}) \in \Sigma, \\ \mathbf{w}(\mathbf{x},0) = \mathbf{w}_T(\mathbf{x}), \quad \mathbf{x} \in \Omega. \end{cases} \tag{6.142}$$

Taking into account (6.136) and arguing as in [48, 96] and Exercise 6.18, the following Carleman estimate similar to (6.103) can be deduced for the associated solutions:

$$\begin{cases} I(\mathbf{w}) := \displaystyle\iint_Q \left[\xi_2^{-2}(|\partial_t \mathbf{w}|^2 + |\Delta\mathbf{w}|^2) + \xi_1^{-2}|\nabla\mathbf{w}|^2 + \xi_0^{-2}|\mathbf{w}|^2\right] \\ \qquad \leq C(M) \displaystyle\iint_{\omega_0\times(0,T)} \xi^{-2}|\mathbf{w}|^2 \end{cases} \tag{6.143}$$

for some weights $\xi_i = \xi_i(\mathbf{x},t)$ that blow up as $t \to T^-$ exponentially.

The ξ_i can be chosen of the form $\xi_i(\mathbf{x}, t) = \xi_{0i}(t)e^{\beta(\mathbf{x},t)}$ for some bounded functions ξ_{0i} and some $\beta(\mathbf{x}, t) = \beta_0(\mathbf{x})(T - t)^{-m}$ with

$$0 < \beta_{00} \le \beta_0 \le \beta_{01}.$$

Let us introduce β_*, with

$$\beta_*(t) = C_*(T - t)^{-(m+1)} \text{ and } C_* > \beta_{01} \le \max_{\mathbf{x}\in\overline{\Omega}} |\beta_0(\mathbf{x})|$$

and let us set $\xi_*(t) = e^{\beta_*(t)}$. Then one has

$$\iint_Q \left[|\partial_t(\xi_*^{-1}\mathbf{w})|^2 + |\Delta(\xi_*^{-1}\mathbf{w})|^2 + |\xi_*^{-1}\mathbf{w}|^2\right] \le CI(\mathbf{w}) \le C(M) \iint_{\omega_0\times(0,T)} \xi^{-2}|\mathbf{w}|^2$$

for any solution to (6.142).

Recall that the Hilbert space

$$W(2, 2; D(A), H) := \{\mathbf{v} \in L^2(0, T; D(A)) : \partial_t\mathbf{v} \in L^2(0, T; H)\}$$

is continuously embedded in $C^0([0, T]; V)$ and this space is continuously embedded in $C^0([0, T]; L^6(\Omega)^3)$.

Therefore, the solutions to (6.142) satisfy

$$\begin{aligned}
\iint_{\omega\times(0,T)} |\xi^{-1}\chi_\omega\mathbf{w}|^2 &\le \int_0^T \left(\int_\omega |\xi^{-2}\xi_*\chi_\omega\mathbf{w}|^{6/5}\,d\mathbf{x}\right)^{5/6} \left(\int_\omega |\xi_*^{-1}\mathbf{w}|^6\,d\mathbf{x}\right)^{1/6} dt \\
&\le C\left(\iint_{\omega\times(0,T)} |\tilde{\xi}^{-1}\chi_\omega\mathbf{w}|^{6/5}\right)^{5/6} I(\mathbf{w})^{1/2} \\
&\le C(M)\left(\iint_{\omega\times(0,T)} |\tilde{\xi}^{-1}\chi_\omega\mathbf{w}|^{6/5}\right)^{5/6} \left(\iint_{\omega\times(0,T)} |\xi^{-1}\chi_\omega\mathbf{w}|^2\right)^{1/2},
\end{aligned}$$

where we have set $\tilde{\xi} := \xi^2\xi_*^{-1}$ (a new weight that blows exponentially as $t \to T^-$). This leads to the following inequalities for the solutions to (6.142):

$$I(\mathbf{w}) \le C(M) \iint_{\omega_0\times(0,T)} \xi^{-2}|\mathbf{w}|^2 \le C(M)\left(\iint_{\omega\times(0,T)} |\tilde{\xi}^{-1}\chi_\omega\mathbf{w}|^{6/5}\right)^{5/3}$$

and

$$\|\mathbf{w}(\cdot\,, 0)\|^2 \le C(M) \iint_{\omega_0\times(0,T)} \xi^{-2}|\mathbf{w}|^2 \le C(M)\left(\iint_{\omega\times(0,T)} |\tilde{\xi}^{-1}\chi_\omega\mathbf{w}|^{6/5}\right)^{5/3}.$$

As in Sect. 6.4.1 (and, more precisely, as in the proof of Lemma 6.2), this shows after some work that (6.137) is approximately controllable at time T and, furthermore, the control can be taken either equal to zero or equal to the restriction to $\omega \times (0, T)$ of $-(\tilde{\xi}^{-1}\chi_\omega)^{6/5}|\mathbf{w}|^{-4/5}\mathbf{w}$, where $\mathbf{w}$ is the solution to (6.142) corresponding to a well chosen final state $\mathbf{w}_T \in H$.

In both cases, we have $\mathbf{v}_\kappa \in L^6(\omega \times (0, T))^3$, with a norm in this space bounded as in (6.140). Note that this control produces a solution $\{\mathbf{u}_\kappa, p_\kappa\}$ with

$$\mathbf{u}_\kappa \in L^6(0, T; W^{2,6}(\Omega)^3) \text{ and } \partial_t \mathbf{u}_\kappa \in L^6(Q)^3$$

and norms in these spaces also bounded as in (6.140). This is a consequence of the regularity theory for (a variant of) the Stokes problem; see for instance the results in [83] (Sec. 2.2) for the details.

Let us finally observe that, for each $\kappa > 0$, $\mathbf{v}_\kappa \mathbb{1}_\omega$ belongs to $L^2(0, T; H_0^1(\Omega)^3)$ and its time derivative belongs to $L^2(0, T; H^{-1}(\Omega)^3)$, with norms bounded by a constant that only depends on M and κ times $\|\mathbf{u}_0\|_{H^2}$.

This, together with the arguments in the proof of Theorem 5.2 (and Lemma 5.1), yields (6.138).

This ends the proof. □

Part 3 Let us denote by B_M the closed ball in W centered at zero of radius M.

For every $\kappa > 0$, consider the mapping $\Xi_\kappa : B_M \mapsto W$ where, for each $\tilde{\mathbf{u}} \in W$, $\mathbf{u}_\kappa = \Xi_\kappa(\tilde{\mathbf{u}})$ is the solution to (6.137) corresponding to the solution $\tilde{\rho}$ to (6.135) and $\mathbf{v}_\kappa$ is the control indicated in the previous step.

In this part of the proof, our task is to find a fixed-point of Ξ_κ, that is, a solution $\mathbf{u}_\kappa$ to the equation in W

$$\Xi_\kappa(\mathbf{u}_\kappa) = \mathbf{u}_\kappa, \quad \mathbf{u}_\kappa \in W,$$

satisfying $\|\mathbf{u}_\kappa\|_W \le M$.

To this purpose, we will apply to Ξ_κ Schauder's Fixed-Point (Theorem 2.9).

It is clear from Lemma 6.3 and the last assertion in Lemma 6.4 that Ξ_κ is well defined and continuous. Moreover, in view of (6.140), it maps B_M into a compact set. More precisely, into a bounded set of the space

$$\{\mathbf{z} \in L^6(0, T; W^{2,6}(\Omega)^3) \cap L^2(0, T; D(A)) : \partial_t \mathbf{z} \in L^6(Q)^3\},$$

that is compactly embedded in W.

Therefore, recalling (6.141) we see that, if we fix $M > 0$ and $\mathbf{u}_0 \in D(A)$ is such that $\|\mathbf{u}_0\|_V \le M/C_1(M)$, Ξ_κ possesses at least one fixed-point in W.

Note that this means that, if $\|\mathbf{u}_0\|_{H^2}$ is sufficiently small, for any $\kappa > 0$, we can find controls $\mathbf{v}_\kappa$ uniformly bounded in $L^6(\omega \times (0, T))^3$ with respect to κ such that the corresponding solutions to (6.137) satisfy (6.139) and (6.140).

Part 4 Let $\mathbf{u}_\kappa$ be for each $\kappa > 0$ a fixed-point of the mapping Ξ_κ with

$$\|\mathbf{u}_\kappa\|_W \le M$$

and let ρ_κ and $\mathbf{v}_\kappa$ be the associated solution to (6.135) and the corresponding control.

Then $(\mathbf{v}_\kappa, \mathbf{u}_\kappa, p_\kappa)$ satisfies (6.137), (6.140), and (6.140).

From the usual estimates for the solutions of quasi-Stokes problems (6.137), we see that, as $\kappa \to 0$, one has at least for a subsequence the following

$$\begin{aligned}
&\mathbf{v}_\kappa \to \mathbf{v} \text{ weakly in } L^6(\omega \times (0,T))^3, \\
&\mathbf{u}_\kappa \to \mathbf{u} \text{ weakly in } L^6(0,T; W^{2,6}(\Omega)^3), \\
&\partial_t \mathbf{u}_\kappa \to \partial_t \mathbf{u} \text{ weakly in} Ł^6(Q)^3, \\
&\rho\kappa \to \rho \text{ weakly in } L^r(Q) \text{ for all finite } r,
\end{aligned}$$

where $\mathbf{u}$ and ρ furnish, together with some p, a solution to (6.1) associated with $\mathbf{v}$.

This is more than sufficient to take limits in (6.137) and (6.139) and conclude the proof of Theorem 6.15. □

It would be interesting to know if a similar result concerning the partial local exact controllability to the regular trajectories can be obtained. To our knowledge, this is unknown.

6.5 Some Additional Remarks and Open Questions

There are many other interesting questions concerning the optimal control and the controllability of viscous Newtonian fluids. They will not be considered here in detail for reasons of space. However, some related comments will be given. Hopefully, they will be a source of inspiration for future work.

- **Recent optimal control methods for fluids.**

  Recently, there has been an appreciable increase of interest to introduce machine learning techniques to solve optimal control problems.

  Roughly speaking, machine learning is the study of algorithms that can improve automatically from experience and data. In the framework of control, they can be of great help, for instance, to solve problems with origin in robotics, neural network modelling and others.

  However, these techniques are being investigated only since a few years. Some information on the applications to the control of fluids can be found in [37].

- **Controllability of semi-Galerkin approximations.**

  It makes sense to stay at the level of the semi-Galerkin approximations and try to control the solutions at time $t = T$.

  For instance, for appropriate (maybe large) ρ_0 and $\mathbf{u}_0$ and fixed k, it is completely natural to look for controls $\mathbf{v} \in L^2(\omega \times (0,T))^N$ such that the solution to (5.9) with $\rho^k \mathbf{f}$ replaced by $\mathbf{v}1_\omega$ satisfies

$$\mathbf{u}^k(\mathbf{x}, T) = \mathbf{0} \quad \text{in } \Omega. \tag{6.144}$$

As already mentioned, this question was considered by J.-L. Lions and Zuazua in [84] for the usual Galerkin approximation of the classical (constant density) Navier-Stokes equations. It is not too difficult to extend the results to this setting.

This is the task proposed in Exercise 6.25.

- **From optimal control to controllability through penalization.**

  Let us consider again (6.1) and the optimal control problem (6.3), where J is given by

$$J(\mathbf{v}, \mathbf{u}, \rho, p) = \frac{1}{2} \int_\Omega |\mathbf{u}(\mathbf{x}, T)|^2 + \frac{\varepsilon}{2} \iint_{\omega \times (0,T)} |\mathbf{v}|^2, \tag{6.145}$$

with $\varepsilon > 0$.

For each $\varepsilon > 0$, there exists at least one solution $(\mathbf{v}^\varepsilon, \mathbf{u}^\varepsilon, \rho^\varepsilon, p^\varepsilon)$ to this problem. If the controls $\mathbf{v}^\varepsilon$ remain bounded as $\varepsilon \to 0^+$, then, at least for a subsequence, one may expect to have the convergence of $(\mathbf{v}^\varepsilon, \mathbf{u}^\varepsilon, \rho^\varepsilon, p^\varepsilon)$ in a weak sense towards a control-state $(\mathbf{v}, \mathbf{u}, \rho, p)$ satisfying (6.130). Consequently, it is natural to attack the resolution of the null controllability problem for (6.1) by penalization, by previously solving (6.3) and (6.145) for each $\varepsilon > 0$ and then proving that $\mathbf{v}^\varepsilon$ is uniformly bounded in some space.

A similar idea can be applied to the exact controllability problem to the trajectories.

This strategy has been implemented successfully, for instance, in [46] in the framework of the heat equation. However, as far as we know, these questions still have not been investigated in depth for Navier-Stokes systems.

- **Boundary controllability.**

  From both the theoretical and practical viewpoints, it is interesting to introduce boundary controls and try to drive not only $\mathbf{u}$ but also ρ to a prescribed target at final time. Thus, consider the system

$$\begin{cases} \dfrac{\partial \rho \mathbf{u}}{\partial t} + \nabla \cdot (\rho \mathbf{u} \otimes \mathbf{u}) - \mu \Delta \mathbf{u} + \nabla p = 0, \quad (\mathbf{x}, t) \in Q, \\ \nabla \cdot \mathbf{u} = 0, \quad (\mathbf{x}, t) \in Q, \\ \dfrac{\partial \rho}{\partial t} + \nabla \cdot (\rho \mathbf{u}) = 0, \quad (\mathbf{x}, t) \in Q, \\ \mathbf{u} = \mathbf{a} 1\!\!1_\gamma, \quad (\mathbf{x}, t) \in \Sigma, \\ \rho = b 1\!\!1_\gamma, \quad (\mathbf{x}, t) \in \Sigma_{\text{in}}(\mathbf{a}), \\ \rho|_{t=0} = \rho_0, \quad (\rho \mathbf{u})|_{t=0} = \rho_0 \mathbf{u}_0, \quad \mathbf{x} \in \Omega, \end{cases} \tag{6.146}$$

where $\gamma \subset \partial\Omega$ and, by definition,

$$\Sigma_{\text{in}}(\mathbf{a}) := \{(\mathbf{x}, t) : \mathbf{x} \in \gamma, \ t \in (0, T), \ \mathbf{a}(\mathbf{x}, t) \cdot \mathbf{n}(\mathbf{x}) < 0\}.$$

The controls are **a** and b and the initial data ρ_0 and $\mathbf{u}_0$ are given. It makes sense to ask whether a local controllability result holds true for (6.146).

The situation here is technically more complex than in Sect. 6.4.2 because, for a nonlinear problem of this kind, the nonhomogeneous boundary conditions can be satisfied in a standard way only if the data are regular enough. If we want to use reasonable and not too regular controls, it is thus appropriate to work with functions that solve the problem in a very weak sense.

In the case of constant density, the local boundary controllability problem is, as indicated in Sect. 6.4, a consequence of Theorem 6.13, see [61].

For variable density fluids, some results are given in [5].

- **Controllability of other similar systems: Boussinesq, Burgers, etc.**

  If the fluid under consideration is sufficiently sensible to heat effects, it is not appropriate to consider motion equations independent of the temperature. Indeed, it is advisable to assume that a force field depending on the temperature acts on the fluid.

  In order to illustrate the relevance that the temperature distribution can have, we have depicted in Fig. 6.11 the results corresponding to numerical approximations of the solutions to the Navier-Stokes system and the so called Boussinesq system (see the equations below) in the same domain.

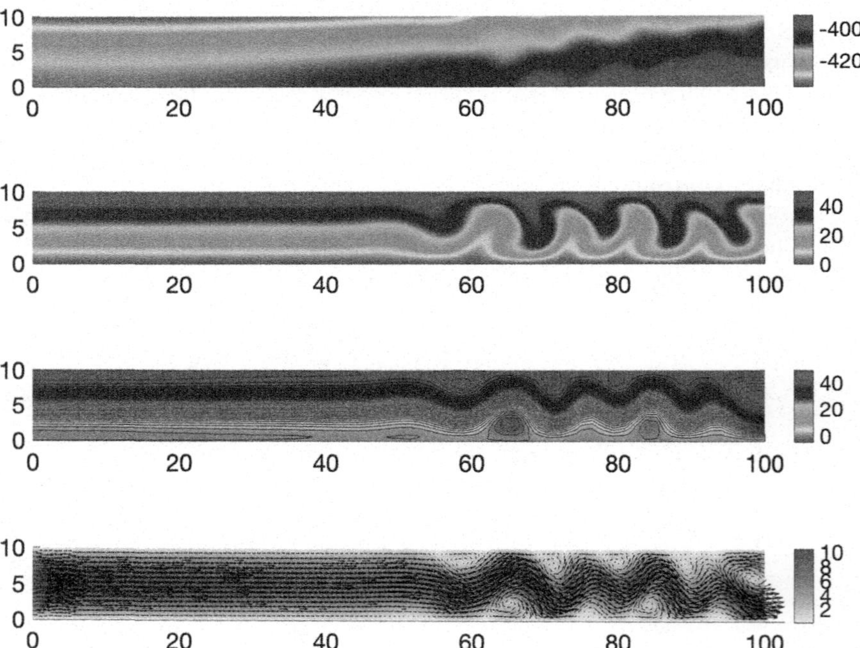

Fig. 6.11 Numerical approximation of the solution to a Boussinesq system in a 2D channel at large time. From top to bottom: density, temperature, streamlines and velocity field

Let us suppose that the extra forces depend linearly on the temperature and let us consequently consider a homogeneous viscous Newtonian fluid governed by the following equations and boundary and initial conditions:

$$\begin{cases} \rho_0(\mathbf{u}_t+(\mathbf{u}\cdot\nabla)\mathbf{u})-\mu\Delta\mathbf{u}+\nabla p=\theta\mathbf{g}_0+\mathbf{v}\mathbb{1}_\omega, & (\mathbf{x},t)\in Q, \\ \nabla\cdot\mathbf{u}=0, \quad (\mathbf{x},t)\in Q, \\ \rho_0(\theta_t+\mathbf{u}\cdot\nabla\theta)-\kappa\Delta\theta=h\mathbb{1}_\omega, \quad (\mathbf{x},t)\in Q, \\ \mathbf{u}=\mathbf{0}, \quad \theta=0, \quad (\mathbf{x},t)\in\Sigma, \\ \mathbf{u}(\mathbf{x},0)=\mathbf{u}_0(\mathbf{x}), \quad \theta(\mathbf{x},0)=\theta_0(\mathbf{x}), \quad \mathbf{x}\in\Omega, \end{cases} \tag{6.147}$$

where $\mathbf{g}_0$ is a constant vector and the right-hand sides $\mathbf{v}\mathbb{1}_\omega$ and $h\mathbb{1}_\omega$ (the controls) are obviously respectively interpreted as a field of external forces and a heat source.

In (6.147), it is natural to ask whether local exact controllability properties hold. The answer is affirmative; see [63] for detailed results.

In particular, the following holds: if $\{\overline{\mathbf{u}},\overline{p},\overline{\theta}\}$ is a solution to the Boussinesq problem (6.147) satisfying (6.102) and

$$\overline{\theta}\in L^\infty(Q), \quad \overline{\theta}_t\in L^2(0,T;L^\sigma(\Omega)) \quad \begin{pmatrix} \sigma>6/5 \text{ if } N=3 \\ \sigma>1 \quad \text{if } N=2 \end{pmatrix},$$

there exists $\varepsilon>0$ such that, for any initial state satisfying

$$(\mathbf{u}_0,\theta_0)\in(L^{2N-2}(\Omega)^N\cap H)\times L^{2N-2}(\Omega), \quad \|(\overline{\mathbf{u}}_0,\overline{\theta}_0)-(\mathbf{u}_0,\theta_0)\|_{L^{2N-2}}\le\varepsilon,$$

we can find controls $\mathbf{v}\in L^2(\omega\times(0,T))^N$ and $h\in L^2(\omega\times(0,T))$ and associated states $\{\mathbf{u},\theta,p\}$ such that

$$\mathbf{u}(\mathbf{x},T)=\overline{\mathbf{u}}(\mathbf{x},T), \quad \theta(\mathbf{x},T)=\overline{\theta}(\mathbf{x},T) \text{ in } \Omega.$$

The global exact control to the bounded trajectories can be proved for (6.147) if we modify the boundary conditions and impose (for instance)

$$\begin{cases} [\mu D\mathbf{u}\cdot\mathbf{n}+M\mathbf{u}]_{\tan}=0, \quad \mathbf{u}\cdot\mathbf{n}=0, \quad (\mathbf{x},t)\in\Sigma, \\ \kappa\nabla\theta\cdot\mathbf{n}+m\,\theta=0, \quad (\mathbf{x},t)\in\Sigma, \end{cases}$$

for some symmetric and definite positive M and some $m>0$. This has been proved in [16] by adapting the ideas in [28].

It is also natural (and meaningful) to consider partial null controllability problems for the variable density Boussinesq system

$$\begin{cases} \dfrac{\partial \rho \mathbf{u}}{\partial t} + \nabla \cdot (\rho \mathbf{u} \otimes \mathbf{u}) - \mu \Delta \mathbf{u} + \nabla p = \rho \theta \mathbf{k} + \mathbf{v} 1\!\!1_{\omega}, \quad (\mathbf{x}, t) \in Q, \\ \nabla \cdot \mathbf{u} = 0, \quad (\mathbf{x}, t) \in Q, \\ \dfrac{\partial \rho \theta}{\partial t} + \nabla \cdot (\rho \theta \mathbf{u}) - \kappa \Delta \theta = h 1\!\!1_{\omega}, \quad \nabla \cdot \mathbf{u} = 0, \quad (\mathbf{x}, t) \in Q, \\ \dfrac{\partial \rho}{\partial t} + \nabla \cdot (\rho \mathbf{u}) = 0, \quad (\mathbf{x}, t) \in Q, \\ \mathbf{u} = 0, \quad \theta = 0, \quad (\mathbf{x}, t) \in \Sigma, \\ \rho|_{t=0} = \rho_0, \quad (\rho \mathbf{u})|_{t=0} = \rho_0 \mathbf{u}_0, \quad (\rho \theta)|_{t=0} = \rho_0 \theta_0, \quad \mathbf{x} \in \Omega, \end{cases} \tag{6.148}$$

where the controls are again $\mathbf{v}$ and h. A result similar to Theorem 6.15 can be obtained by adapting the arguments in the previous proof. Specifically, the following holds:

Theorem 6.16 *Let α, β be given, with $0 < \alpha < \beta$. There exists $\delta > 0$ such that, for any initial state $(\rho_0, \mathbf{u}_0, \theta_0)$, with*

$$\begin{cases} \rho_0 \in C^1(\overline{\Omega}), \quad \alpha \le \rho_0(\mathbf{x}) \le \beta, \quad \mathbf{u}_0 \in V, \quad \theta_0 \in H_0^1(\Omega), \\ \|\mathbf{u}_0\|_V + \|\theta_0\|_V \le \delta, \end{cases}$$

we can find controls $\mathbf{v} \in L^2(\omega \times (0, T))^N$ and $h \in L^2(\omega \times (0, T))$ and associated solutions $\{\mathbf{u}, \rho, \theta, p\}$ to (6.148) satisfying

$$\mathbf{u}(\mathbf{x}, T) = \mathbf{0}, \quad \theta(\mathbf{x}, T) = 0 \quad \text{in } \Omega.$$

It also makes sense to consider the Euler system

$$\begin{cases} \rho_0 \left(\dfrac{\partial \mathbf{u}}{\partial t} + (\mathbf{u} \cdot \nabla) \mathbf{u} \right) + \nabla p = 0 & \text{in } Q, \\ \nabla \cdot \mathbf{u} = 0 & \text{in } Q \end{cases} \tag{6.149}$$

and try to prove boundary controllability results similar to those above.

Note that this PDE system is nonlinear, hyperbolic and first-order in time and space. Accordingly, the choice of boundary conditions is more delicate.

In [22] and [59], the authors established global exact controllability results respectively for $N = 2$ and $N = 3$. More precisely, they proved that, for any non-empty open set $\gamma \subset \partial\Omega$ such that $\partial\Omega \setminus \gamma$ is a connected set and any $\mathbf{u}_0$ and $\mathbf{u}_T$ in $C^\infty(\overline{\Omega})^N$ with $\nabla \cdot \mathbf{u}_0 \equiv \nabla \cdot \mathbf{u}_T \equiv 0$ and $\mathbf{u}_0 \cdot \mathbf{n} = \mathbf{u}_1 \cdot \mathbf{n} = 0$ on $\partial\Omega \setminus \gamma$, there exist regular functions $\mathbf{u}$ and p satisfying (6.149),

$$\begin{cases} \mathbf{u} \cdot \mathbf{n} = 0 \ \text{ on } (\partial\Omega \setminus \gamma) \times (0, T), \\ \mathbf{u}(\mathbf{x}, 0) = \mathbf{u}_0(\mathbf{x}) \ \text{ and } \ \mathbf{u}(\mathbf{x}, T) = \mathbf{u}_T(\mathbf{x}) \ \text{ in } \Omega. \end{cases}$$

The boundary controls can be obtained from $\mathbf{u}$. For example, the following are admissible:

$$h = (\mathbf{u} \cdot \mathbf{n})\mathbb{1}_{\gamma}, \quad \mathbf{k} = \left(\nabla \times \mathbf{u}\big|_{\tan}\right) \mathbb{1}_{\Gamma_+},$$

where $\mathbf{z} \cdot \mathbf{n}$ denotes the lateral normal trace of z on Σ, $\mathbf{z}\big|_{\tan}$ is the tangential component of $\mathbf{z}$ (i.e., $\mathbf{z}\big|_{\tan} := \mathbf{z} - (\mathbf{z} \cdot \mathbf{n})\mathbf{n}$) and

$$\Sigma_{\text{in}} := \{(\mathbf{x}, t) \in \gamma \times (0, T) : \mathbf{u}(\mathbf{x}, t) \cdot \mathbf{n}(\mathbf{x}) < 0\}.$$

Similar results have been obtained in [50] for a coupled Euler-heat system where, like in (6.147), the state is a triplet of the form $\{\mathbf{u}, p, \theta\}$.

A simplified version of (6.95) is the controlled Burgers' system

$$\begin{cases} u_t + uu_x - \nu u_{xx} = 0, & \text{in } (0, L) \times (0, T), \\ u(0, t) = v_0(t), \quad u(L, t) = v_1(t), & \text{on } (0, T), \\ u(x, 0) = u_0(x), & \text{in } (0, L), \end{cases} \tag{6.150}$$

where v_0 and v_1 are the controls and $u = u(x, t)$ is the state. This can be regarded as a model for several different phenomena in mechanics and other areas.

Thus, it can be assumed that u is the velocity of the particles of a gas constrained to move in one direction at points $x \in (0, L)$ and times $t \in (0, T)$. In a completely different context, it can also be assumed that u describes the traffic density in a road under some particular hypotheses. For these and many other applications, see for instance [76].

Several positive and negative control results can be established for (6.150) and some variants; see [44]. In particular, it was proved in [64] that (6.150) is not null-controllable unless u_0 is sufficiently small (or T is sufficiently large).

When we only have one boundary control, i.e., with $v_1(t) \equiv 0$, the null controllability of (6.150) is equivalent to the same property for a similar distributed system

$$\begin{cases} u_t + uu_x - \nu u_{xx} = v\mathbb{1}_{\omega}, & \text{in } (0, L) \times (0, T), \\ u(0, t) = 0, \quad u(L, t) = 0, & \text{on } (0, T), \\ u(x, 0) = u_0(x), & \text{in } (0, L). \end{cases} \tag{6.151}$$

This has been analyzed in [44], with explicit and sharp estimates for the minimal controllability time.

Let us finally mention that it was proved in [86] that, if we set $v_1(t) \equiv 0$ and we add an additional control $f = f(t)$ to the system in the right-hand side of the PDE, global null controllability holds.

- **Controlling with a reduced number of controls.**

  This is a very interesting question. For the moment, although some advances have been made, it is not completely solved.

  For the usual (constant density) Stokes, Navier-Stokes and Boussinesq equations, there have been some contributions (see the results in [24, 25, 49]; also, two interesting papers by Carreño and Guerrero and Coron and Lissy provide related arguments and results, see [13, 26].)

  In view of these works, it seems reasonable to expect that, at least when there exist $\mathbf{x}_0 \in \partial\Omega$ and $r > 0$ such that

$$\overline{\omega} \cap \overline{\Omega} \supset B(\mathbf{x}_0; r) \cap \overline{\Omega}, \tag{6.152}$$

  Theorem 6.15 holds with controls where at least one component is identically zero (see Exercise 6.29).

- **The control of compressible fluids.**

  In some recent works, the control of compressible Navier-Stokes fluids has been investigated under several different circumstances.

  In most cases, the fluids are assumed to be *isentropic* (this means in practice that the pressure is a function of the density) and the goal is to drive the state exactly to a nontrivial constant-in-space solution.

  In [39, 89, 90], the local exact controllability of a one-dimensional flow has been established. In the first of these references, the control is exerted on the whole boundary and appears in the motion and the continuity equations; in the second one, the control appears only on the motion equation as an external force, locally supported in space.

  The two and three dimensional compressible systems have been analyzed, from the viewpoint of controllability, in [40] and [90].

  Again, in the first case the control is exerted on the boundary (the control on the motion equation is distributed over the entire boundary, while one acts on the density only on the inward velocity part). On the other hand, [90] deals with internal controls with a small support acting only on the motion equation.

At present, it remains to extend these results to similar non-isentropic compressible fluids (involving an additional energy equation and a *state equation* for the pressure, density and temperature).

6.6 Exercises of Chapter 6

Exercise 6.1* Prove that the cost functions (6.4) and (6.5) and the admissible sets $\mathcal{U}_{\rm ad}$ given in (6.14), (6.15), (6.16), and (6.17) satisfy the hypotheses in Theorem 6.1.

Exercise 6.2* Prove a result similar to Theorem 6.2, for the following cost functional:

$$J(\mathbf{v}, \mathbf{u}, \rho, p) = \frac{a}{2} \iint_Q |\nabla \mathbf{u} - \nabla \mathbf{u}_d|^2 + \frac{b}{2} \iint_{\omega \times (0,T)} |\mathbf{v}|^2,$$

where $\mathbf{u}_d \in L^2(0, T; H^1(\Omega)^N)$.

Exercise 6.3* Prove a result similar to Theorem 6.2, for a cost functional of the form (6.4) with $a' > 0$.

Exercise 6.4* Same statement for the cost functional (6.36).

Exercise 6.5* Same statement for the cost functional (6.38).

Exercise 6.6** Prove that, under the assumptions of Theorem 6.2, (6.25) possesses exactly one solution.

HINT: Use the argument in the proof of Theorem 3.1 (resp. Theorem 5.3) to establish the existence (resp. uniqueness) of a solution.

Exercise 6.7** Prove that, under the assumptions of Theorem 6.2, one has (6.31). Also, prove that one can deduce (6.24)–(6.26) from (6.31).

HINT: Use *Lagrange Multipliers Theorem* (see for instance [31]).

Exercise 6.8** Prove that 6.26 is equivalent to (6.32).

Exercise 6.9* Prove Theorem 6.3.

HINT: Argue as in the proof of Theorem 6.1.

Exercise 6.10* Same statement and hint for Theorem 6.4.

Exercise 6.11* Same statement and hint for Theorem 6.5.

Exercise 6.12* Prove Theorem 6.6.

Exercise 6.13** Prove Theorem 6.7.

Exercise 6.14** Prove Theorem 6.8.

Exercise 6.15* Prove the local turnpike property that is, a result similar to Theorem 6.8 for the 2D classical Navier-Stokes equations.

Exercise 6.16* Prove the four assertions in Sect. 6.4.1 concerning the existence, uniqueness and regularity of the solutions to (6.95) and (6.96).

Exercise 6.17* Prove that, if $\tilde{\mathbf{u}}$ is regular enough, for any $\mathbf{w}_T \in H$ and any $\mathbf{g} \in L^2(Q)^N$, (6.101) possesses exactly one weak solution that is strong if $\mathbf{w}_T \in V$.

Exercise 6.18** Prove Theorem 6.11 for some weights of the form

$$\eta(\mathbf{x}, t) e^{\beta(\mathbf{x})/(T-t)^r}$$

with bounded positive functions η and β and $r > 1$.

HINT: Do the following:

1. Consider the backwards heat operator $\mathcal{H}$, with $\mathcal{H}w := -w_t - c\Delta w$. Find positive functions $\gamma = \gamma(\mathbf{x}, t)$ and $\beta = \beta(\mathbf{x})$ and positive constants s_0, λ_0 and K_0 such that

$$\iint_Q e^{-2s\beta/(T-t)^r} \left[(s\gamma)^{-1}(|w_t|^2 + |\Delta w|^2) + (s\gamma)\lambda^2 |\nabla w|^2 + (s\gamma)^3 \lambda^4 |w|^2 \right] \leq K_0 \left(\iint_Q e^{-2s\beta/(T-t)^r} |\mathcal{H}w|^2 + \iint_{\omega \times (0,T)} e^{-2s\beta/(T-t)^r} (s\gamma)^3 \lambda^4 |w|^2 \right)$$

for any $s \geq s_0$, any $\lambda \geq \lambda_0$ and any $w \in L^2(0, T; H^2(\Omega))$ with $w_t \in L^2(Q)$ and $w = 0$ on Σ.
2. Deduce a similar inequality for any $\mathbf{w}$ that solves, together with some q, the backwards system (6.101), with a right-hand side of the form

$$K_1 \left(\iint_Q e^{-2s\beta/(T-t)^r} (|\mathbf{g}|^2 + |\nabla q|^2) + \iint_{\omega \times (0,T)} e^{-2s\beta/(T-t)^r} (s\gamma)^3 \lambda^4 |\mathbf{w}|^2 \right).$$

3. Improve these estimates, by replacing the previous right-hand side by

$$K_2 \left(\iint_Q e^{-2s\beta/(T-t)^r} |\mathbf{g}|^2 + \iint_{\omega \times (0,T)} e^{-2s\beta/(T-t)^r} (s\gamma)^3 \lambda^4 (|\mathbf{w}|^2 + |q|^2) \right).$$

4. Finally, modify the weights appropriately, give an estimate of the local integral dealing with q and deduce (6.103).

Exercise 6.19* Prove (6.106).

Exercise 6.20* Prove the assertion in Remark 6.14.

Exercise 6.21* Prove the assertion in Remark 6.18.

Exercise 6.22** Prove the global null controllability of (6.123).
HINT: Introduce a fixed-point reformulation of the controllability problem. Using the control properties of systems like (6.96), check that Schauder's Theorem can be applied.

Exercise 6.23** Prove that, for any weak solution $\{\mathbf{u}, \rho, p\}$ to (6.1), (6.128) holds.
HINT: First, prove a differential equality for any function of the form $f(\rho)$, where $f = f(r)$ is a regular approximation of $\mathbb{1}_{\{r \leq \lambda\}}$.

Exercise 6.24** Prove a controllability result for the Galerkin approximations (6.122).
HINT: Rewrite the null controllability result as a fixed-point equation in an appropriate space and apply *Schauder's Theorem*. Also, apply this idea to other controllability results.

Exercise 6.25** Prove a partial controllability result for the semi-Galerkin approximations (5.9).

HINT: Argue as in Exercise 6.24.

Exercise 6.26** Under the assumptions of Theorem 6.15 and keeping the same notation, prove that for each $T > 0$ the linear system (6.137) is approximately controllable at time T, with a control that can be taken either equal to zero or equal to the restriction to $\omega \times (0, T)$ of $-(\tilde{\xi}^{-1}\chi_\omega)^{6/5}|\mathbf{w}|^{-4/5}\mathbf{w}$, where $\mathbf{w}$ is the solution to (6.142) corresponding to a well chosen $\mathbf{w}_T \in H$.

HINT: Introduce an appropriate (dual) function of $\mathbf{w}_T$ whose unique minimizer provides the desired final adjoint state.

Exercise 6.27* Again, let us place ourselves in the conditions of the proof of Lemma 6.4. Arguing as in the proofs of Theorem 5.2 and Lemma 5.1, check that the regularity of the control $\mathbf{v}_\kappa 1\!\!1_\omega$ leads to (6.138).

Exercise 6.28** Prove Theorem 6.16.

HINT: Adapt the argument of the proof of Theorem 6.15.

Exercise 6.29** Prove a result similar to Theorem 6.15 with controls $\mathbf{v}$ for which one component vanishes.

HINT: Assume that ω satisfies (6.152) for some $\mathbf{x}_0 \in \partial\Omega$ and some $r > 0$ and try to prove the uniform approximate controllability of (6.137) with $N - 1$ scalar controls.

Table 6.1 The optimal step gradient method with projection

Algorithm 6.1		
a.	Choose $\mathbf{v}^0 \in \mathcal{U}_{\rm ad}$;	
b.	Then, for given $n \geq 0$ and $\mathbf{v}^n \in \mathcal{U}_{\rm ad}$, do until convergence:	
	1.	Solve (6.24) with $\mathbf{v} = \mathbf{v}^n$, to obtain $(\mathbf{u}^n, \rho^n)$;
	2.	Solve (6.25) with $(\mathbf{u}, \rho) = (\mathbf{u}^n, \rho^n)$, to obtain $(\mathbf{w}^n, \eta^n)$;
	3.	Set $\mathbf{d}^n = (\mathbf{w}^n + b\mathbf{v}^n)\|_{\omega\times(0,T)}$ and find r^n such that $j^n(r^n) = \inf_{r>0} j^n(r)$. Here, $j^n(r)$ is the value of J at any $(\mathbf{v}^n - r\mathbf{d}^n, \mathbf{u}^n(r), \rho^n(r))$, where $(\mathbf{u}^n(r), \rho^n(r))$ is a state associated to $\mathbf{v}^n - r\mathbf{d}^n$;
	4.	Set $\mathbf{v}^{n+1} = \mathbb{P}_{\rm ad}(\mathbf{v}^n - r^n\mathbf{d}^n)$.

Table 6.2 A "mixed-loop" alternative to Algorithm 6.1

Algorithm 6.2		
a.	Choose $\mathbf{v}^0 \in \mathcal{U}_{\text{ad}}$ and $(\mathbf{u}^{-1}, \rho^{-1}) \in E_0$;	
b.	Then, for given $n \geq 0$ and $\mathbf{v}^n \in \mathcal{U}_{\text{ad}}$, do until convergence:	
	1.	Solve (6.24) with $\mathbf{v} = \mathbf{v}^n$ and $\mathbf{u} \cdot \nabla$ replaced by $\mathbf{u}^{n-1} \cdot \nabla$, to obtain $(\mathbf{u}^n, \rho^n)$;
	2.	Solve (6.55) with $(\mathbf{u}, \rho) = (\mathbf{u}^n, \rho^n)$, to obtain $(\mathbf{w}^n, \eta^n)$;
	3.	Do as in step 3 of ALG 6.1;
	4.	Do as in step 4 of ALG 6.1.

Table 6.3 The optimal step conjugate gradient method with projection

Algorithm 6.3		
a.	Choose $\mathbf{v}^0 \in \mathcal{U}_{\text{ad}}$;	
b.	Perform one gradient step, i.e.	
	1.	Solve (6.24) with $\mathbf{v} = \mathbf{v}^0$, to obtain $(\mathbf{u}^0, \rho^0)$;
	2.	Solve (6.25) with $(\mathbf{u}, \rho) = (\mathbf{u}^0, \rho^0)$, to obtain $(\mathbf{w}^0, \eta^0)$;
	3.	Set $\mathbf{d}^n = (\mathbf{w}^n + b\mathbf{v}^n)\|_{\omega\times(0,T)}$, etc.
c.	Then, for given $n \geq 1$ and $\mathbf{v}^n \in \mathcal{U}_{\text{ad}}$, do until convergence:	
	1.	Solve (6.24) with $\mathbf{v} = \mathbf{v}^n$, to obtain $(\mathbf{u}^n, \rho^n)$;
	2.	Solve (6.25) with $(\mathbf{u}, \rho) = (\mathbf{u}^n, \rho^n)$, to obtain $(\mathbf{w}^n, \eta^n)$;
	3.	Set $\mathbf{f}^n = (\mathbf{w}^n + b\mathbf{v}^n)\|_{\omega\times(0,T)}$, $\zeta^n = G(\mathbf{f}^n, \mathbf{f}^{n-1})$, $\mathbf{d}^n = \mathbf{f}^n + \zeta^n \mathbf{d}^{n-1}$ and compute r^n as in step 3 of ALG 6.1 with this new $\mathbf{d}^n$;
	4.	Do as in step 4 of ALG 6.1.

Table 6.4 A parallelized version of ALG 1

Algorithm 6.4		
a.	Choose $\mathbf{v}^0 \in \mathcal{U}_{\text{ad}}$ and $(\mathbf{u}^{-1}, \rho^{-1}) \in E_0(T_0)$.	
b.	Then, for given $n \geq 0$ and $(\mathbf{u}^{n-1}, \rho^{n-1})$, do until convergence:	
	1.	Solve in parallel (6.24) with $\mathbf{v} = \mathbf{v}^n$ and $\mathbf{u} \cdot \nabla$ replaced by $\mathbf{u}^{n-1} \cdot \nabla$ and (6.25) with $\mathbf{u} \cdot \nabla$, $(\nabla \mathbf{u})$ and $\nabla \rho$ resp. replaced by $\mathbf{u}^{n-1} \cdot \nabla$, $(\nabla \mathbf{u}^{n-1})$ and $\nabla \rho^{n-1}$;
	2.	Continue as in ALG 1

References

1. Alekseev, V.M., Tikhomirov, V.M., Fomin, S.V.: Optimal Control. Contemporary Soviet Mathematics. Consultants Bureau, New York (1987). Translated from the Russian by V.M. Volosov
2. Amel, B., Imad, R.: Identification of the source term in Navier-Stokes system with incomplete data. AIMS Math. **4**(3), 516–526 (2019)
3. Anh, C.T., Nguyet, T.M.: Time optimal control of the unsteady 3D Navier-Stokes-Voigt equations. Appl. Math. Optim. **79**(2), 397–426 (2019)

4. Arnautu, V., Neittaanmaki, P.: Optimal Control from Theory to Computer Programs, volume 111 of Solid Mechanics and its Applications. Kluwer Academic Publishers Group, Dordrecht (2003)
5. Badra, M., Ervedoza, S., Guerrero, S.: Local controllability to trajectories for non-homogeneous incompressible Navier-Stokes equations. Ann. Inst. H. Poincaré Anal. Non Linéaire **33**(2), 529–574 (2016)
6. Barbu, V.: The time optimal control of Navier-Stokes equations. Syst. Control Lett. **30**(2–3), 93–100 (1997)
7. Barbu, V.: Controllability and Stabilization of Parabolic Equations, volume 90 of Progress in Nonlinear Differential Equations and their Applications. Birkhäuser/Springer, Cham (2018). Subseries in Control
8. Barnes, B.A.: Majorization, range inclusion and factorization for bounded linear operators. Proc. Amer. Math. Soc. **133**(1), 155–162 (2005)
9. Bermúdez, A.: Mathematical techniques for some environmental problems related to water pollution control. In: Mathematics, Climate and Environment (Madrid, 1991), volume 27 of RMA Res. Notes Appl. Math., pp. 12–27. Masson, Paris (1993)
10. Boldrini, J.L., Caretta, B.M.C., Fernández-Cara, E.: Some optimal control problems for a two-phase field model of solidification. Rev. Mat. Comput. **23**(1), 49–75 (2010)
11. Boyer, F.: On the penalised HUM approach and its applications to the numerical approximation of null-controls for parabolic problems. In: CANUM 2012, 41e Congrès National d'Analyse Numérique, volume 41 of ESAIM Proc., pp. 15–58. EDP Sci., Les Ulis (2013)
12. Burns, J.A.: Introduction to the Calculus of Variations and Control—With Modern Applications. Chapman & Hall/CRC Applied Mathematics and Nonlinear Science Series. CRC Press, Boca Raton, FL (2014)
13. Carreño, N., Guerrero, S.: Local null controllability of the N-dimensional Navier-Stokes system with $N-1$ scalar controls in an arbitrary control domain. J. Math. Fluid Mech. **15**(1), 139–153 (2013)
14. Casas, E., Kunisch, K.: Optimal control of the two-dimensional evolutionary Navier-Stokes equations with measure valued controls. SIAM J. Control Optim. **59**(3), 2223–2246 (2021)
15. Chan, W.L., Guo, B.Z.: Optimal birth control of population dynamics. J. Math. Anal. Appl. **144**(2), 532–552 (1989)
16. Chaves-Silva, F.W., Le Balc'h, K., Machado, J.L.F., Fernández-Cara, E., Souza, D.A.: Global controllability of the Boussinesq system with Navier-slip-with-friction and Robin boundary conditions. SIAM J. Control Optim. **61**(2), 484–510 (2023)
17. Chowdhury, S., Ervedoza, S.: Open loop stabilization of incompressible Navier-Stokes equations in a 2d channel using power series expansion. J. Math. Pures Appl. (9) **130**, 301–346 (2019)
18. Ciarlet, P.G.: Introduction to Numerical Linear Algebra and Optimisation. Cambridge Texts in Applied Mathematics. Cambridge University Press, Cambridge (1989). With the assistance of Bernadette Miara and Jean-Marie Thomas, Translated from the French by A. Buttigieg
19. Ciarlet, P.G., Lions, J.L. (eds.): Handbook of Numerical Analysis. Vol. IX. Handbook of Numerical Analysis, vol. IX. North-Holland, Amsterdam (2003). Numerical methods for fluids. Part 3.
20. Constantin, P., Foias, C.: Navier-Stokes Equations. Chicago Lectures in Mathematics. University of Chicago Press, Chicago, IL (1988)
21. Coron, J.-M.: On the controllability of the 2-D incompressible Navier-Stokes equations with the Navier slip boundary conditions. ESAIM Contrôle Optim. Calc. Var. **1**, 35–75 (1995/1996) (electronic)
22. Coron, J.-M.: On the controllability of 2-D incompressible perfect fluids. J. Math. Pures Appl. (9) **75**(2), 155–188 (1996)
23. Coron, J.-M., Fursikov, A.V.: Global exact controllability of the 2D Navier-Stokes equations on a manifold without boundary. Russian J. Math. Phys. **4**(4), 429–448 (1996)

24. Coron, J.-M., Guerrero, S.: Local null controllability of the two-dimensional Navier-Stokes system in the torus with a control force having a vanishing component. J. Math. Pures Appl. (9) **92**(5), 528–545 (2009)
25. Coron, J.-M., Guerrero, S.: Null controllability of the N-dimensional Stokes system with $N - 1$ scalar controls. J. Differ. Equ. **246**(7), 2908–2921 (2009)
26. Coron, J.-M., Lissy, P.: Local null controllability of the three-dimensional Navier-Stokes system with a distributed control having two vanishing components. Invent. Math. **198**(3), 833–880 (2014)
27. Coron, J.-M., Marbach, F., Sueur, F., Zhang, P.: Controllability of the Navier-Stokes equation in a rectangle with a little help of a distributed phantom force. Ann. PDE **5**(2), 17 (2019)
28. Coron, J.-M., Marbach, F., Sueur, F.: Small-time global exact controllability of the Navier-Stokes equation with Navier slip-with-friction boundary conditions. J. Eur. Math. Soc. **22**(5), 1625–1673 (2020)
29. Criminale, W.O., Jackson, T.L., Joslin, R.D.: Theory and Computation of Hydrodynamic Stability. Cambridge Monographs on Mechanics. Cambridge University Press, Cambridge (2003)
30. Crooke, P.S., Kaynar, A.M., Hotchkiss, J.R.: A mathematical model of air-flow induced regional over-distention during mechanical ventilation: comparing pressure-controlled and volume-controlled modes. In: Advances in the Theory of Control, Signals and Systems with Physical Modeling, volume 407 of Lect. Notes Control Inf. Sci., pp. 269–282. Springer, Berlin (2010)
31. Dell'Isola, F., Di Cosmo, F.: Lagrange multipliers in infinite-dimensional systems. In: Encyclopedia of Continuum Mechanics, 9 p. Springer, Berlin (2018)
32. DiPerna, R.J., Lions, P.-L.: Ordinary differential equations, transport theory and Sobolev spaces. Invent. Math. **98**(3), 511–547 (1989)
33. Dontchev, A.L., Rockafellar, R.T.: A view from variational analysis. In: Implicit Functions and Solution Mappings. Springer Series in Operations Research and Financial Engineering, second edn. Springer, New York (2014)
34. Douglas, R.G.: On majorization, factorization and range inclusion of operators on Hilbert space. Proc. Amer. Math. Soc. **17**, 413–415 (1966)
35. Dubovickii, A.J., Miljutin, A.A.: Extremal problems with constraints. Ž. Vyčisl. Mat. Mat. Fiz. **5**, 395–453 (1965)
36. Dubovickii, A.J., Miljutin, A.A.: Necessary conditions for the extremum in some linear problems with mixed constraints. In: Probability Processes and Control (Russian), pp. 42–74. Nauka, Moscow (1978)
37. Duriez, T., Brunton, S.L., Noack, B.R.: Machine learning control–taming nonlinear dynamics and turbulence, volume 116 of Fluid Mechanics and its Applications. Springer, Cham (2017)
38. Ekeland, I., Témam, R.: Convex Analysis and Variational Problems, volume 28 of Classics in Applied Mathematics. Society for Industrial and Applied Mathematics (SIAM), Philadelphia, PA (1999)
39. Ervedoza, S., Glass, O., Guerrero, S., Puel, J.-P.: Local exact controllability for the one-dimensional compressible Navier-Stokes equation. Arch. Ration. Mech. Anal. **206**(1), 189–238 (2012)
40. Ervedoza, S., Glass, O., Guerrero, S.: Local exact controllability for the two- and three-dimensional compressible Navier-Stokes equations. Comm. Partial Differ. Equ. **41**(11), 1660–1691 (2016)
41. Fabre, C.: Résultats d'unicité pour les équations de Stokes et applications au contrôle. C. R. Acad. Sci. Paris Sér. I Math. **322**(12), 1191–1196 (1996)
42. Fabre, C., Puel, J.-P., Zuazua, E.: Approximate controllability of the semilinear heat equation. Proc. Roy. Soc. Edinb. Sect. A **125**(1), 31–61 (1995)
43. Fernández-Cara, E., Guerrero, S.: Global Carleman inequalities for parabolic systems and applications to controllability. SIAM J. Control Optim. **45**(4), 1399–1446 (2006)
44. Fernández-Cara, E., Guerrero, S.: Null controllability of the Burgers system with distributed controls. Syst. Control Lett. **56**(5), 366–372 (2007)

45. Fernández-Cara, E., Marín-Gayte, I.: Analysis and numerical solution of some minimal time control problems. Res. Appl. Math. **26**, 100582 (2025)
46. Fernández-Cara, E., Zuazua, E.: The cost of approximate controllability for heat equations: the linear case. Adv. Differ. Equ. **5**(4–6), 465–514 (2000)
47. Fernández-Cara, E., Zuazua, E.: A new proof of the existence of suitable weak solutions and other remarks for the Navier-Stokes equations. Bol. Soc. Esp. Mat. Appl. (26), 79–140 (2003)
48. Fernández-Cara, E., Guerrero, S., Imanuvilov, O.Yu., Puel, J.-P.: Local exact controllability of the Navier-Stokes system. J. Math. Pures Appl. (9) **83**(12), 1501–1542 (2004)
49. Fernández-Cara, E., Guerrero, S., Imanuvilov, O.Yu., Puel, J.-P.: Some controllability results for the N-dimensional Navier-Stokes and Boussinesq systems with $N - 1$ scalar controls. SIAM J. Control Optim. **45**(1), 146–173 (2006) (electronic)
50. Fernández-Cara, E., Santos, M.C., Souza, D.A.: Boundary controllability of incompressible Euler fluids with Boussinesq heat effects. Math. Control Signals Syst. **28**(1), Art. 7, 28 (2016)
51. Fernández-Cara, E., Santos, M.C., Souza, D.A.: On the numerical controllability of the two-dimensional heat, Stokes and Navier-Stokes equations. J. Sci. Comput. **70**(2), 819–858 (2017)
52. Fernández-Cara, E., Sousa, I.T., Viera, F.B.: Remarks concerning the approximate controllability of the 3D Navier–Şstokes and Boussinesq systems. SeMA J. **74**(3), 237–253 (2017)
53. Forough, M.: Majorization, range inclusion and factorization for unbounded operators on Banach spaces. Linear Algebra Appl. **449**, 60–67 (2014)
54. Fursikov, A.V.: Flow of a viscous incompressible fluid around a body: boundary value problems and work minimization for a fluid. Sovrem. Mat. Fundam. Napravl. **37**, 83–130 (2010)
55. Fursikov, A.V., Imanuvilov, O.Yu.: Controllability of Evolution Equations, volume 34 of Lecture Notes Series. Seoul National University, Research Institute of Mathematics, Global Analysis Research Center, Seoul (1996)
56. Fursikov, A.V., Imanuvilov, O.Yu.: Exact controllability of the Navier-Stokes and Boussinesq equations. Uspekhi Mat. Nauk. **54**(3(327)), 93–146 (1999)
57. Galdi, G.P.: Steady-state Navier-Stokes problem past a rotating body: geometric-functional properties and related questions. In: Topics in Mathematical Fluid Mechanics, volume 2073 of Lecture Notes in Math., pp. 109–197. Springer, Heidelberg (2013)
58. Geshkovski, B., Zuazua, E.: Turnpike in optimal control of PDEs, ResNets and beyond. Acta Numer. **31**, 135–263 (2022)
59. Glass, O.: Exact boundary controllability of 3-D Euler equation. ESAIM Control Optim. Calc. Var. **5**, 1–44 (2000)
60. Glowinski, R., Lions, J.-L., He, J.: A numerical approach. In: Exact and Approximate Controllability for Distributed Parameter Systems, volume 117 of Encyclopedia of Mathematics and its Applications. Cambridge University Press, Cambridge (2008)
61. González-Burgos, M., Guerrero, S., Puel, J.-P.: Local exact controllability to the trajectories of the Boussinesq system via a fictitious control on the divergence equation. Commun. Pure Appl. Anal. **8**(1), 311–333 (2009)
62. Gorshkov, A.V.: Stabilization of the solution of a two-dimensional system of Navier-Stokes equations in the exterior of a bounded domain by means of boundary control. Mat. Sb. **203**(9), 15–40 (2012)
63. Guerrero, S.: Local exact controllability to the trajectories of the Boussinesq system. Ann. Inst. H. Poincaré Anal. Non Linéaire **23**(1), 29–61 (2006)
64. Guerrero, S., Imanuvilov, O.Yu.: Remarks on global controllability for the Burgers equation with two control forces. Ann. Inst. H. Poincaré Anal. Non Linéaire **24**(6), 897–906 (2007)
65. Guerrero, S., Imanuvilov, O.Yu., Puel, J.-P.: A result concerning the global approximate controllability of the Navier-Stokes system in dimension 3. J. Math. Pures Appl. (9) **98**(6), 689–709 (2012)
66. Gunzburger, M.D.: Perspectives in Flow Control and Optimization, volume 5 of Advances in Design and Control. Society for Industrial and Applied Mathematics (SIAM), Philadelphia, PA (2003)

67. Higaki, M., Maekawa, Y., Nakahara, Y.: On stationary Navier-Stokes flows around a rotating obstacle in two-dimensions. Arch. Ration. Mech. Anal. **228**(2), 603–651 (2018)
68. Hinze, M., Kunisch, K.: Second order methods for optimal control of time-dependent fluid flow. SIAM J. Control Optim. **40**(3), 925–946 (2001)
69. Hinze, M., Kunisch, K.: Second order methods for boundary control of the instationary Navier-Stokes system. ZAMM Z. Angew. Math. Mech. **84**(3), 171–187 (2004)
70. Hishida, T.: The Navier-Stokes flow around a rotating obstacle with time-dependent body force. In: Nonlocal and Abstract Parabolic Equations and their Applications, volume 86 of Banach Center Publ., pp. 149–162. Polish Acad. Sci. Inst. Math., Warsaw (2009)
71. Hubert, F., Boyer, F., Le Rousseau, J.: Uniform controllability properties for space/time-discretized parabolic equations. Numer. Math. **118**(4), 601–661 (2011)
72. Imanuvilov, O.Yu.: Remarks on exact controllability for the Navier-Stokes equations. ESAIM Control Optim. Calc. Var. **6**, 39–72 (2001) (electronic)
73. Ioffe, A.D.: Theory and applications. In: Variational Analysis of Regular Mappings. Springer Monographs in Mathematics. Springer, Cham (2017)
74. Kasumba, H., Kunisch, K.: Vortex control in channel flows using translational invariant cost functionals. Comput. Optim. Appl. **52**(3), 691–717 (2012)
75. Kerry, I.: Boyan Slat: Pioneeing the Ocean Cleanup. Capston Press, North Mankato, MN (2021)
76. Kessels, F.: Introduction to traffic flow theory through a genealogy of models. In: Traffic Flow Modelling. EURO Advanced Tutorials on Operational Research. Springer, Cham (2019)
77. Kien, B.T.: Second-order optimality conditions and solution stability to optimal control problems governed by stationary Navier-Stokes equations. Acta Math. Vietnam. **44**(2), 431–448 (2019)
78. Kowalewski, A., Kotarski, W.: On application of Milutin-Dubovicki's theorem to an optimal control problem for systems described by partial differential equations of hyperbolic type with time delay. Syst. Sci. **7**(1), 55–74 (1981)
79. Krstic, M., Smyshlyaev A. A course on backstepping designs. In: Boundary Control of PDEs, volume 16 of Advances in Design and Control. Society for Industrial and Applied Mathematics (SIAM), Philadelphia, PA (2008)
80. Kunisch, K., Leugering, G., Sprekels, J., Tröltzsch, F. (eds.): Optimal Control of Coupled Systems of Partial Differential Equations, volume 158 of International Series of Numerical Mathematics. Birkhäuser Verlag, Basel (2009). Papers from the International Conference held in Oberwolfach, March 2–8, 2008
81. Ledzewicz, U., Schättler, H.: Pareto optimality conditions for abnormal optimization and optimal control problems. In: Optimal Control of Differential Equations (Athens, OH, 1993), volume 160 of Lecture Notes in Pure and Appl. Math., pp. 217–235. Dekker, New York (1994)
82. Lévine, J.: A flatness-based approach. In: Analysis and Control of Nonlinear Systems. Mathematical Engineering. Springer, Berlin (2009)
83. Lions, P.-L.: Mathematical Topics in Fluid Mechanics. Vol I: Incompressible Models, volume 3 of Oxford Lecture Series in Mathematics and its Applications. The Clarendon Press/Oxford University Press, New York (1996)
84. Lions, J.-L., Zuazua, E.: Exact boundary controllability of Galerkin's approximations of Navier-Stokes equations. Ann. Scuola Norm. Sup. Pisa Cl. Sci. (4) **26**(4), 605–621 (1998)
85. Maitelli, A.L., Yoneyama, T.: Adaptive control techniques for arterial blood pressure by means of vasoactive drugs. In: Computational Methods in Biophysics, Biomaterials, Biotechnology and Medical Systems: Algorithm Development, Mathematical Analysis and Diagnostics, vol. 3, pp. 151–175. Kluwer Acad. Publ., Boston, MA (2003)
86. Marbach, F.: Small time global null controllability for a viscous Burgers' equation despite the presence of a boundary layer. J. Math. Pures Appl. (9) **102**(2), 364–384 (2014)
87. Marinoschi, G.: Exact controllability in minimal time of the Navier-Stokes periodic flow in a 2D-channel. SIAM J. Control Optim. **58**(6), 3658–3683 (2020)

88. Mathew, T.P.A.: Domain Decomposition Methods for the Numerical Solution of Partial Differential Equations, volume 61 of Lecture Notes in Computational Science and Engineering. Springer, Berlin (2008)
89. Mitra, D., Renardy, M.: Interior local null controllability for multi-dimensional compressible flow near a constant state. Nonlinear Anal. Real World Appl. **37**, 94–136 (2017)
90. Mitra, D., Ramaswamy, M., Renardy, M.: Interior local null controllability of one-dimensional compressible flow near a constant steady state. Math. Methods Appl. Sci. **40**(10), 3445–3478 (2017)
91. Munteanu, I.: Boundary Stabilization of Parabolic Equations, volume 93 of Progress in Nonlinear Differential Equations and their Applications. Birkhäuser/Springer, Cham (2019). Subseries in Control
92. Nečasová, S., Kračmar, S.: Mathematical analysis of its asymptotic behavior. In: Navier-Stokes Flow around a Rotating Obstacle, volume 3 of Atlantis Briefs in Differential Equations. Atlantis Press, Paris (2016)
93. Porretta, A., Zuazua, E.: Long time versus steady state optimal control. SIAM J. Control Optim. **51**(6), 4242–4273 (2013)
94. Porretta, A., Zuazua, E.: Remarks on long time versus steady state optimal control. In: Mathematical Paradigms of Climate Science, volume 15 of Springer INdAM Ser., pp. 67–89. Springer, Cham (2016)
95. Puel, J.-P.: Controllability of Navier-Stokes equations. In: Lectures on the Analysis of Nonlinear Partial Differential Equations. Part 5, volume 5 of Morningside Lect. Math., pp. 263–282. Int. Press, Somerville, MA (2018)
96. Puel, J.-P., Imanuvilov, O.Yu., Yamamoto, M.: Carleman estimates for parabolic equations with nonhomogeneous boundary conditions. Chin. Ann. Math. Ser. B **30**(4), 333–378 (2009)
97. Raković, S.V., Levine, W.S. (eds.): Handbook of Model Predictive Control. Control Engineering. Birkhäuser/Springer, Cham (2019)
98. Russell, D.L.: A unified boundary controllability theory for hyperbolic and parabolic partial differential equations. Stud. Appl. Math. **52**, 189–211 (1973)
99. Russell, D.L.: Controllability and stabilizability theory for linear partial differential equations: recent progress and open questions. SIAM Rev. **20**(4), 639–739 (1978)
100. Salsa, S.: Partial Differential Equations in Action, volume 86 of Unitext, second edn. Springer, Cham (2015). From modelling to theory, La Matematica per il 3+2
101. Saut, J.-C., Scheurer, B.: Unique continuation for some evolution equations. J. Differ. Equ. **66**(1), 118–139 (1987)
102. Slat, B.: https://theoceancleanup.com
103. Sritharan, S.S.: An optimal control problem in exterior hydrodynamics. In: Distributed Parameter Control Systems (Minneapolis, MN, 1989), volume 128 of Lecture Notes in Pure and Appl. Math., pp. 385–417. Dekker, New York (1991)
104. Stavre, R.: Optimization of the blood pressure with the control in coefficients. Evol. Equ. Control Theory **9**(1), 131–151 (2020)
105. Témam, R.: Theory and numerical analysis. In: Navier-Stokes Equations. AMS Chelsea Publishing, Providence, RI (2001). Reprint of the 1984 edition
106. Trélat, E., Zuazua, E.: The turnpike property in finite-dimensional nonlinear optimal control. J. Differ. Equ. **258**(1), 81–114 (2015)
107. Vázquez, R., Trélat, E., Coron, J.-M.: Control for fast and stable laminar-to-high-Reynolds-numbers transfer in a 2D Navier-Stokes channel flow. Discrete Contin. Dyn. Syst. Ser. B **10**(4), 925–956 (2008)
108. Velin, J.: Discriminating distributed sentinel involving a Navier-Stokes problem and parameter identification. In: CSVAA 2004—Control Set-valued Analysis and Applications, volume 17 of ESAIM Proc., pp. 143–166. EDP Sci., Les Ulis (2007)
109. Williams, N.D., Mehlsen, J., Tran, H.T., Olufsen, M.S.: An optimal control approach for blood pressure regulation during head-up tilt. Biol. Cybernet. **113**(1–2), 149–159 (2019)

110. Xu, Y., Wang, G., Wang, L., Zhang, Y.: Time Optimal Control of Evolution Equations, volume 92 of Progress in Nonlinear Differential Equations and their Applications. Birkhäuser/Springer, Cham (2018). Subseries in Control
111. Yousefi, K., Saleh, R.: Three-dimensional suction flow control and suction jet length optimization of NACA 0012 wing. Meccanica **50**(6), 1481–1494 (2015)
112. Zamorano, S.: Turnpike property for two-dimensional Navier-Stokes equations. J. Math. Fluid Mech. **20**(3), 869–888 (2018)

Index

P. Braz e Silva et al., *Analysis and Control of the Variable Density Incompressible Navier-Stokes Equations*, MS&A 22, https://doi.org/10.1007/978-3-032-14510-9

The manufacturer's authorised representative in the EU is Springer Nature Customer Service Centre GmbH, Europaplatz 3, 69115 Heidelberg, Germany. If you have any concerns regarding our products, please contact ProductSafety@springernature.com

Printed and bound by CPI Group (UK) Ltd, Croydon, CR0 4YY
07/07/2026
02160908-0001